GAMES, THEORY AND APPLICATIONS

Mathematics and its Applications

Series Editor: G. M. BELL, Professor of Mathematics, King's College (KQC), University of London

Statistics and Operational Research

Editor: B. W. CONOLLY, Professor of Operational Research, Queen Mary College, University of London

Mathematics and its applications are now awe-inspiring in their scope, variety and depth. Not only is there rapid growth in pure mathematics and its applications to the traditional fields of the physical sciences, engineering and statistics, but new fields of application are emerging in biology, ecology and social organisation. The user of mathematics must assimilate subtle new techniques and also learn to handle the great power of the computer efficiently and economically.

The need of clear, concise and authoritative texts is thus greater than ever and our series will endeavour to supply this need. It aims to be comprehensive and yet flexible. Works surveying recent research will introduce new areas and up-to-date mathematical methods. Undergraduate texts on established topics will stimulate student interest by including applications relevant at the present day. The series will also include selected volumes of lecture notes which will enable certain important topics to be presented earlier than would otherwise be possible.

In all these ways it is hoped to render a valuable service to those who learn, teach, develop and use mathematics.

Statistics and Operational Research

Editor: B. W. CONOLLY, Professor of Operational Research, Queen Mary College, University of London

Beaumont, G.P.	**Introductory Applied Probability**
Beaumont, G.P.	**Probability and Random Variables***
Conolly, B.W.	**Techniques in Operational Research: Vol. 1, Queueing Systems***
Conolly, B.W.	**Techniques in Operational Research: Vol. 2, Models, Search, Randomization**
Conolly, B.W.	**Lecture Notes in Queueing Systems**
French, S.	**Sequencing and Scheduling: Mathematics of the Job Shop**
French, S.	**Decision Theory: An Introduction to the Mathematics of Rationality**
Griffiths, P. & Hill, I.D.	**Applied Statistics Algorithms**
Hartley, R.	**Linear and Non-linear Programming**
Jolliffe, F.R.	**Survey Design and Analysis**
Jones, A.J.	**Game Theory**
Kemp, K.W.	**Dice, Data and Decisions: Introductory Statistics**
Oliveira-Pinto, F.	**Simulation Concepts in Mathematical Modelling***
Oliveira-Pinto, F. & Conolly, B.W.	**Applicable Mathematics of Non-physical Phenomena**
Schendel, U.	**Introduction to Numerical Methods for Parallel Computers**
Stoodley, K.D.C.	**Applied and Computational Statistics: A First Course**
Stoodley, K.D.C., Lewis, T. & Stainton, C.L.S.	**Applied Statistical Techniques**
Thomas, L.C.	**Games, Theory and Applications**
Whitehead, J.R.	**The Design and Analysis of Sequential Clinical Trials**

**In preparation*

Mathematics and its Applications

Series Editor: G. M. BELL, Professor of Mathematics, King's College (KQC), University of London

Artmann, B.	**The Concept of Number***
Balcerzyk, S. & Joszefiak, T.	**Commutative Rings***
Balcerzyk, S. & Joszefiak, T.	**Noetherian and Krull Rings***
Baldock, G.R. & Bridgeman, T.	**Mathematical Theory of Wave Motion**
Ball, M.A.	**Mathematics in the Social and Life Sciences: Theories, Models and Methods**
de Barra, G.	**Measure Theory and Integration**
Bell, G.M. and Lavis, D.A.	**Co-operative Phenomena in Lattice Models Vols. I & II***
Berkshire, F.H.	**Mountain and Lee Waves**
Berry, J.S., Burghes, D.N., Huntley, I.D., James, D.J.G. & Moscardini, A.O.	**Teaching and Applying Mathematical Modelling**
Burghes, D.N. & Borrie, M.	**Modelling with Differential Equations**
Burghes, D.N. & Downs, A.M.	**Modern Introduction to Classical Mechanics and Control**
Burghes, D.N. & Graham, A.	**Introduction to Control Theory, including Optimal Control**
Burghes, D.N., Huntley, I. & McDonald, J.	**Applying Mathematics**
Burghes, D.N. & Wood, A.D.	**Mathematical Models in the Social, Management and Life Sciences**

Continued at back of book

GAMES, THEORY AND APPLICATIONS

L. C. THOMAS, B.A., M.A., D.Phil.
Senior Lecturer in Decision Theory
University of Manchester

ELLIS HORWOOD LIMITED
Publishers · Chichester

Halsted Press: a division of
JOHN WILEY & SONS
New York · Chichester · Brisbane · Toronto

First published in 1984 and
Reprinted and issued in paperback for the first time in 1986 by
ELLIS HORWOOD LIMITED
Market Cross House, Cooper Street,
Chichester, West Sussex, PO19 1EB, England
The publisher's colophon is reproduced from James Gillison's drawing of the ancient Market Cross, Chichester.

Distributors:
Australia and New Zealand:
JACARANDA WILEY LIMITED
GPO Box 859, Brisbane, Queensland 4001, Australia
Canada:
JOHN WILEY & SONS CANADA LIMITED
22 Worcester Road, Rexdale, Ontario, Canada
Europe and Africa:
JOHN WILEY & SONS LIMITED
Baffins Lane, Chichester, West Sussex, England
North and South America and the rest of the world:
Halsted Press: a division of
JOHN WILEY & SONS
605 Third Avenue, New York, NY 10158, USA

British Library Cataloguing in Publication Data
Thomas, L.C.
Games, theory and applications. —
(Ellis Horwood series in mathematics and its applications)
1. Game theory
I. Title
519.3 QA269

Library of Congress Card No. 83–26673

ISBN 0–85312–515–5 (Ellis Horwood Limited — Library Edn.)
ISBN 0–7458–0142–0 (Ellis Horwood — Student Edn.)
ISBN 0–470–27507–3 (Halsted Press — Library Edn.)
ISBN 0–470–20741–8 (Halsted Press — Paperback Edn.)

Printed in Great Britain by Unwin Brothers of Woking

To Carey, Iris, Fred and Glenys,
and in memory of Eunice

TABLE OF CONTENTS

Author's Preface

The preface of a book is positioned first, written last, and read little. It should also be the author's guide to and *raison d'être* for the following manuscript. So here goes.

The theory of games, as a way of modelling problems with two or more decision-makers involved, originated at the end of the Second World War, though the playing of games themselves has been one of the enduring strands in the history of mankind. At the same time as game theory was being developed, new mathematical techniques were being invented in other types of decision-making. This whole area of 'applicable mathematics' has continued to be one of the most active branches of research and development up to the present day. Game theory, despite an uneven reputation during this period, has continued to develop, in depth of understanding and breadth of applications.

Its fascination, for me, is that it is the one area where quite difficult logical ideas can be expressed rigorously, with the minimum of mathematical sophistication. These ideas can be quickly comprehended because there are manifestations of them in real life, with which people are familiar. Thus, of itself, game theory has much to recommend it, apart from its applications. However, it is the applications which are of paramount importance. Games can be used as models of situations as diverse as international conflict, advertising budgets, pricing policy, airport loading fees, and the evolution of animal behaviour. In some cases, the analysis of the game will lead to a specific strategy for each participant being considered as his best way of playing the game: in other games and types of analysis, all that can be done is an exploration of what were the players' understanding of each other in order for a certain outcome to occur. So there is something for everyone.

The book evolved from a lecture course I have given for several years to undergraduates at Manchester. The students came from many different departments—mathematics, science, economics, accountancy, liberal studies, and psychology. Thus, the book assumes a modest mathematical background—an ability to differentiate and integrate simple functions, the idea of maxima and minima of functions, familiarity with matrix notation (though not the properties of matrices) and summation signs, and a basic knowledge of probability. Although my upbringing means I like to use theorems and lemmas to structure and point out the important results, proofs are only given where they can be understood with this background, and employ ideas and techniques which are used again in the text.

The book's aim is to give an introduction to game theory to as wide a numerate audience as possible. Thus, I do not concentrate on linear programming solution techniques, which would weight it more towards the mathematician; and I treat the economic applications of game theory, which several authors have concentrated on, as only one of several such applications.

Chapter 1 introduces the terminology of game theory and some examples of games which will recur throughout the book. Chapters 2, 3, and 4 give the theory of two-person and *n*-person games. The remaining chapters are independent of one another, but do use the theory developed in the first few chapters. However, I do point out connections between work in different chapters when they arise. All chapters end with a section of further reading which indicates where one can pursue the ideas of that chapter further, and some problems. I have included solutions to the problems to encourage the reader to attempt them. Doing the problems is the way of proving you understand the ideas in the text.

I am grateful to my colleagues Doug White, Roger Hartley, and Chris Birchenall for many useful conversations. I am especially grateful to Simon French, who taught a course from the draft of this book for a year, and whose comments and corrections of my errors proved very valuable. All remaining errors are 'deliberate mistakes' by me! My thanks are also due to Kate Baker and Bob Lande, who typed the manuscripts with great care, on both sides of the Atlantic in between babies and racquetball championships, respectively—anything to get away from my handwriting! Finally, my thanks go to my family who taught me the best cooperative game of all while I was writing this book.

L. C. Thomas

Glossary of Symbols

$x \in X$ means x is a member of the set X
$x \notin X$ means x is not a member of the set X
$\{x|....\}$ is the set of x with the property ...
$S \cup T = \{x | x \in S \text{ or } x \in T\}$
$S \cap T = \{x | x \in S \text{ and } x \in T\}$
$S \subseteq T =$ set S is included in set T
$S - T = \{x | x \in S \text{ and } x \notin T\}$
$\emptyset$ is the empty set
$\#S$ the number of elements in the set S
$|x| = \max\{x, -x\}$
$x > y$ x greater than y
$x \geqslant y$ x greater than or equal to y
$x \not> y$ x not greater than y
$n!$ $1. 2. 3. 4. .... (n-1). n$
$\sum_{i=1}^{n} x_i = x_1 + x_2 + ... + x_n$
If $\underline{x} = (x_1, ..., x_n)$
$\|x\| = \max_{\underline{k}} |x_k|$

$\|x\|_2 = \left(\sum_{i=1}^{n} x_i^2\right)^{1/2}$
$\frac{df(x)}{dx} = f'(x)$ is the derivative of function f

Two-person zero-sum games
I, II are the players
I_i (II_i) ith pure strategy of player I (II)
e_{ij} payoff to player I of I_i versus II_j
v_{L}, v_{U} lower and upper values of the game
$\mathbf{x} = (x_1, ..., x_n)$ mixed strategy for player I
$\mathbf{y} = (y_1, y_2, ..., y_m)$ mixed strategy for player II
v value of game
$\mathbf{x}^*(\mathbf{y}^*)$ optimal strategy for player I (II)
$X(Y)$ set of all mixed strategies for player I (II)

Two-person non-zero-sum game

$e_i(\mathbf{x},\mathbf{y})$ payoff to player i if I plays mixed strategy x and II plays mixed strategy y
v_{I} (v_{II}) maximin values for player I (II)
B – bargaining set

n-person games

$N = \{1, 2, \ldots n\}$ is the set of players
$S \subseteq \text{N}$ coalition of players
$v\,(\cdot)$ characteristic function
X_S set of coordinated strategies for players in S
$x = (x_1, x_2, \ldots, x_n)$ imputation
$E(v)$ set of imputations in game with characteristic function v
$\text{C}(v)$ core of game with characteristic function of v
$N(v)$ nucleolus of game with characteristic function of v
$S(v)$ stable set of game with characteristic function v
$\phi_i(v)$ Shapley value of player i in game with characteristic function v
$x \underset{\bar{S}}{\geqslant} y$ imputation x dominates y on coalition S
$x > y$ imputation x dominates imputation y

metagames

$k_1k_2 \ldots k_nG$ the metagame $k_1k_2 \ldots k_nG$ based on game G
$R_i(k_1k_2 \ldots k_nG)$ rational outcomes for player i in $k_1k_2 \ldots k_nG$
$\hat{R}_i(k_1k_2 \ldots k_nG)$ metarational outcomes for player i in $k_1k_2 \ldots k_nG$
$\hat{E}(G)$ metaequilibria in game G

multi-stage games

Γ_i ith subgame of multi-stage game Γ
$v = (v_1 \ldots v_n)$ values of game Γ
val $\Gamma\,(w)$ value of game $\Gamma(w)$
h_k history of kth stage of the game
H_k history of game up to kth stage
$E_i\,(\cdot,\cdot)$ payoff to player i in supergame

evolutionary games

$e(x,y)$ fitness payoff of playing strategy x against a population playing strategy y
$e_i(r,(x,y))$ fitness payoff to player of type i of playing strategy r against a population where type 1's play x and type 2's play y

CHAPTER 1

'The game's afoot'

(Shakespeare)

1.1 WHAT IS A GAME?

Have you read the newspaper today? If so, the front page probably contains a report of some political controversy, or a strike, or perhaps describes armed conflict and violent actions by groups of people or countries. The inside pages will report actions by various pressure groups to change social policy, or will describe government decisions about such policies—improvements in housing, allocation of finance to the various branches of health and social services. The financial pages will be full of take-overs, firms' attempts to improve their market share, changes in prices of goods, or government attempts to control the financial markets. Finally, there will be the sports page or the chess or bridge column to read.

What do all these reports have in common? They all describe conflicts of interests between people or groups of people such as political parties, governments, and businesses. We call the theoretical models of such conflicts of interests **games**. Recalling the famous definition of a model as a small imitation of the real thing (as in 'model husband') indicates that in a game one is trying to extract the essential problems in the conflict of interest. **Game theory** consists of ways of analysing these problems. Obviously, most conflicts are too complicated to include all the facets involved in the corresponding model, but a game could still be useful in describing the main types of decisions open to the participants and the sort of results that could occur. For some games, game theory will suggest a 'solution' to the game, that is a best way of playing the game for each person involved; but for most games describing real problems all it can do is to rule out some types of decisions and perhaps suggest, which players will work together.

These theoretical models of conflict are called games because we can identify easily the conflicts of interest in recreational games like Poker, Noughts and Crosses, or Monopoly; and some of the board games, like

Chess, did develop historically as models of warfare. It is in a way an unfortunate choice of name, because it has the connotations of amusement, light-heartedness, and a recreational contest. I hope the reader will occasionally be amused, but games cover a much wider area than just board games or 'bored' games. We shall consider economic and business problems, the tactics and logistics of warfare, international and national politics, and social policy all as candidates to be modelled as games.

1.2 EXAMPLES OF GAMES

To get a feeling for what is involved in a game, let's look at some simple examples. Try to think what are their important features and what if anything they have in common.

Example 1.1 — Noughts and Crosses. Draw an $n \times n$ grid. The first player puts a nought in any square, and then the second player puts a cross in some unused square. Continue in sequence until one player has a line (column, row, or diagonal) of n noughts or n crosses. That player then wins.

You probably haven't played this game for a few years, so now is your chance. You will find that if $n = 1$ or 2, the first player always wins, and if $n = 3$, you can always draw unless you make a mistake. What happens for $n = 4$ and higher n? What about the three-dimensional version, i.e. $n \times n \times n$?

Example 1.2—Simplified Poker. There are only two cards involved—an 'Ace' and a 'Two'—and only two players—whom we label I and II. Each puts £1 in the pot and I deals II one card, which II then looks at. If it is the Ace, II must say 'Ace', but if it is the 'Two', II can say 'Ace' or 'Two'. If he says 'Two' he loses the game and his £1 in the pot. If, however, II says 'Ace' no matter what the card is, II must put another £1 in the 'pot'. In this case player I can either believe him, i.e. 'fold' and so lose his £1 in the 'pot', or else he can demand to 'see' the card. In this case I has to put another £1 into the pot, and then the card is shown. If II had the Ace he wins the 'pot' and so takes £2 from I, but if II has the Two, I wins the 'pot' and so takes £2 from II. This game involves the elements of 'bluffing' and 'calling' involved in real poker, but not surprisingly has not yet taken Las Vegas by storm.

Example 1.3 — Nim. A number of matches are set out in two piles, and two players take turns at taking matches from the piles. At each turn a player must take at least one match, but he can take more provided they are all from the same pile. The loser is the player who picks up the last match.

Example 1.4—Prisoners' Dilemma. Two people, arrested with stolen property in their possession, are being interviewed separately by the police. They both know that if they keep quiet there is not enough evidence for them to be convicted of theft, and so they will only get a one-year gaol sentence for being in possession of stolen property. If both confess to the theft, they will both get nine years in prison. However, if one confesses and the other keeps quiet, the one who turned Queen's Evidence will go free, while the other will have a ten-year gaol sentence (the extra year is for not assisting the police). What should they do? Would it make any difference if they could talk to each other between being arrested and interviewed? You can argue that this game also embodies the dilemma over nuclear disarmament or whether unions should pursue high or low wage claims for their members. Can you see why?

Example 1.5—Pick a number. Each person in a group of people chooses a number. The one with the highest number wins. What number would you choose? Infinity is not allowed. Could you guarantee to win the game somehow?

Example 1.6—Duellists. Two duellists stand $2N$ paces apart with loaded pistols and start to walk toward each other. At each pace they can decide whether or not to fire their one bullet, and the chance of killing their opponent increases as they get nearer. If they fire and miss, honour demands that they still keep walking nearer. When should each man fire? Is this affected by whether their aim [*sic*] is to kill their opponent or to stay alive themselves? In a duel, you would know when your opponent had fired and missed, but suppose instead you were in planes, far apart, armed with a missile, so you didn't know if your opponent had fired and missed. Would this make any difference to when you should fire?

1.3 TERMINOLOGY OF GAME THEORY

Let us emphasise again that game theory is not a prescriptive way of how to play a game. Rather it is a set of ideas and techniques for analysing these mathematical models of conflict of interest. It doesn't tell you how to play the game, but describes properties that certain ways of playing the game have, and which you might think desirable. Even when the analysis suggests a best way of playing the game, it only does it assuming that everyone is playing in the 'best way' they can. It never allows for ways of punishing your opponent if he makes a mistake, which is the way most games, whether board ones or real life conflict situations, end.

Before we start on this analysis let us look again at the six examples in the previous section, and see what features they have in common. This will help us in defining the terminology of game theory.

Firstly there are always at least two participants (as many as you like in Example 1.5) called the **players** hereafter labelled I, II, III, etc. or, in later chapters, 1, 2, 3, etc. Each game consists of a sequence of **moves**, some simultaneous, which are either decisions by the players or outcomes of chance events. Thus, in the simplified poker of Example 1.2 the first move is the chance event of which card is picked. If it is a Two, this is followed by the decision by II whether to say 'Ace' or 'Two'. If II says 'Ace', the last move is whether I believes him or not.

At the end of the game, each player receives a **payoff**. We will always assume that the payoff is given by a real number. In many games you could associate this number with the amount won, or say the payoff is $+1$ if you won the game, 0 if it is a draw, and -1 if you lose it. However, in other games the result is more complicated or intangible.

When von Neumann and Morgenstern (1947) introduced the basics of game theory, they also developed the idea of the **utility** of an outcome of the game, so that these numbers reflect your preferences. Thus, suppose the outcomes of the game were 'going to a football match' or 'going to the cinema' and you preferred the football match. You would choose two numbers $u(\mathrm{FM})$—the utility of going to the football match—and $u(\mathrm{C})$—the utility of going to the cinema—so that $u(\mathrm{FM}) > u(\mathrm{C})$. Say $u(\mathrm{FM}) = 4$, $u(\mathrm{C}) = 2$. Obviously you can choose almost any pair of numbers here, but if you start asking more of the utility function this cuts down the number of possible numerical representations. Thus, you may prefer seeing the football match in the dry, to going to the cinema; but would rather go to the cinema than get soaked watching a football match. This requires the utilities to satisfy

$$u(\mathrm{FM\ Dry}) > u(\mathrm{C}) > u(\mathrm{FM\ Wet}). \tag{1.1}$$

If you think the chance of rain tonight is ½, and you can't decide whether to go to the football match or the cinema (you are indifferent) this would be represented by the equation

$$\tfrac{1}{2}u(\mathrm{FM\ Dry}) + \tfrac{1}{2}u(\mathrm{FM\ Wet}) = u(\mathrm{C}). \tag{1.2}$$

Thus, if you choose $u(\mathrm{FM\ Dry}) = 4$, $u(\mathrm{C}) = 2$, this means $u(\mathrm{FM\ Wet})$ must equal 0. This is the situation in most versions of utility theory, where you can choose the utility of two outcomes as you like, provided they reflect your preference, but the utilities of all the other outcomes are then fixed completely by your preferences. Readers who want to learn more about utility theory and the axioms underlying it should turn to Chapter 2 of Luce and Raiffa (1957) or to Raiffa (1968).

.We will always assume that the players' preferences over the outcomes of a game satisfy the rules underlying utility theory, and so each outcome can be represented numerically by its utility value. Remember that this utility reflects all the aspects of the outcome, including your regret or joy about what happened to your opponent. So if you get a payoff of utility value 4 and your opponent gets one of 2, you are as happy with that as if you get 4 and he gets 200 or −200.

Returning to the common features in the various games, notice that each player has to make decisions at some moves of the game. A **strategy** for a player is a description of the decisions he will make at all the possible situations that can arise in the game. Thus, having chosen his strategy it will tell him what to do at every situation that can arise no matter what his opponent does or what are the outcomes of the chance events. In 2×2 Noughts and Crosses, a strategy tells the first player which of the four squares to put his first nought in; and for each of the three replies by his opponent it must say which of the remaining two squares to put the second nought in. Think of it as a set of instructions which enables a computer to play the game for you. It is obvious that for many games, chess for instance, although in principle you can conceive of a strategy, in practice it is too long to write down. (If you were 'Black' in chess you would have to write down your response to all 20 possible opening moves by White, and at your second move reply to all the 400 different situations White can be in after two moves, and so on.)

If the sum of the players' payoffs is zero no matter what strategy they use, the game is called **zero-sum**. In these games, like Nim, or Poker (Examples 1.2 and 1.3), the players are completely opposed to one another in that what one wins the other loses. Games which don't have this property are, not surprisingly, called **non-zero-sum** games. In Prisoners' Dilemma the total payoff is two years in prison if they both keep silent and 18 years if they confess.

Finally, if at every move in the game all the players know all the moves that have already occurred, the game is said to be one of **perfect information**. Thus, Noughts and Crosses and Nim have perfect information, whereas in Poker player I doesn't know which card player II has picked up. Later we shall show that this difference leads to a difference in the type of strategy you might think is best in each game. Is it a good idea always to play the same strategy in Poker? What about Noughts and Crosses?

1.4 HISTORY OF GAME THEORY

Game theory started with two papers by von Neumann (Von Neumann, 1928, 1937), though Borel in the 1920s had also looked at similar problems. (See Borel, 1953, for a translation.) It really sprang to life, however, with the publication of von Neumann and Morgenstern's book *Theory of Games and Economic Behaviour* in 1944, and especially the second edition in 1947. Rarely has the first book in a subject contained so many of the ideas that are still the main areas of interest in the subject or made so great an impact as this one. The reason was mainly the Second World War, because during this there had been considerable activity in modelling decision situations which involved one or more decision-makers. Hence, the rise of Operations Research. Most of the military problems that can be modelled as games are of the two-player zero-sum type, and these are the very ones for which game theory can suggest a specific 'solution'. Thus, at the end of the war, people were thinking of how to model decision situations, and there was a view that game theory had had a successful, if secret, track record in the military area.

In the next few years there was a great deal of work in game theory as people sought to show it was the mathematical panacea in all areas of human conflict. They tried to expand the mathematics to incorporate other problems, such as bargaining and arbitration, into the framework. They tried to solve the open theoretical question, and to apply the theory to areas like politics and economic competition. The four volumes: *Contributions to the Theory of Games* (Kuhn and Tucker, 1950, 1953; Drescher, Tucker and Wolfe, 1957; and Tucker and Luce, 1959), give a good idea of the problems that were being examined at that time.

However, as people concentrated more on games with more than two people, it was realised that there was no 'nice' solution concept for these games; nor did the game models capture all the features of real-life conflict situations. Thus, when in 1957 Luce and Raiffa wrote the other classic book in the area, *Games and Decisions*, they were careful to point out the limitations of game theory. This fall from grace continued through the 1960s, and the nadir was reached when Lucas (1967) found a ten-person game that did not have a 'solution' in the sense suggested by von Neuman and Morgenstern.

Since then game theory has recovered some of its popularity. It is no longer considered the 'panacea' to solve all human conflict, but it is certainly the best way of thinking about conflict situations. It is a useful test-bench for looking at what new concepts about conflict actually imply, and it highlights the important decisions that have to be made in real-life situations. There are still large groups of researchers, especially in the USA and Israel, working on 'classical' game theory and bargaining. This involves developing and exploring new properties that they feel resolutions of conflict situations should possess, and widening the areas of application of game theory even to include religion (Brams, 1980).

There has also been tremendous development in using game theory to record how people actually react in conflict situations. Psychologists and game theorists devise particular games to test what people's decisions are in these situations and how they compare with game theory predictions. Prisoners' Dilemma is a popular candidate for such a game, and the number of published papers concerning experiments with it is over two hundred. This has led to the idea of devising games as a teaching and learning tool, and in Chapter 10 we shall look at **gaming**, as this aspect of game theory is called. Another development from the experiments of actually playing games is to try to explain how people arrive at outcomes which are not in agreement with the 'good' solutions of classical game theory. The most obvious of these theoretical extensions is the idea of metagames introduced by Howard (1971).

In the last ten years completely new applications of game theory have been developed in unexpected areas. Thus, Maynard-Smith (1974) described how a game-theoretic framework is useful in describing the evolution of genes, which affect breeding patterns.

Thus, games and game theory can look forward to an exciting future, not as a way of solving all conflict problems, but as the most useful collection of techniques for analysing these problems.

PROBLEMS FOR CHAPTER 1

1.1. Consider the following types of games.

(a) Games with perfect information, where some of the moves are chance events.

(b) Games with perfect information, with no chance moves.

(c) Games which do not have perfect information, but have chance moves.

(d) Games which do not have perfect information and do not have chance moves.

Which type is each of the following games Chess, Bridge, Monopoly, Draughts (Checkers), Scissors–Stone–Paper, Ludo, Poker, Noughts and Crosses? Write down one more game of each type.

1.2 Give examples of a situation (not necessarily a board game) which can be modelled as a game

(a) with two players and is zero-sum;

(b) with two players and is non-zero-sum;

(c) with more than two players and is zero-sum;
(d) with more than two players and is non-zero-sum.

1.3. Explain why there are really two strategies for each player in 2×2 Noughts and Crosses (Example 1.1). What are they?

1.4. In the duellists' game (Example 1.6), write down all the strategies for each player in a 'quiet' duel when the players don't know if their opponent has fired. What is the difference between these strategies and the ones in a 'noisy' duel where they know if their opponent has fired?

1.5. Let $u(x)$ be a person's utility function of wining £x: and it is standardised by calling $u(0) = 0$ and $u(100) = 100$. If he is indifferent between winning £40 for certain or taking part in a gamble where he has a 50% chance of winning £100 and a 50% of winning £0, what is $u(40)$? If he is willing to pay £10 to take part in a gamble where you have a ¼ chance of winning £50, otherwise you get no prize, what can you say about his utility of losing £10?

CHAPTER 2

TWO-PERSON ZERO-SUM GAMES

'The game is never lost till won' *George Crabb*

2.1 EXTENSIVE FORM

We concentrate in this chapter on games with only two players, I and II, where what I wins, II loses, i.e. a two-person zero-sum game. One way of describing such a game (in fact it will work for any game) is by recording all possible sequences of moves that can occur in it, and the payoff at the end of each sequence. This can be represented diagrammatically by a **tree graph**. In such a graph each point represents a point in the game where a move must be made. Remember, a move is either a decision by one of the players or a chance event. The possible moves that can be made at that point (either the outcomes of the chance event or the possible decisions by the player) are represented by lines drawn from that point (Fig.2.1).

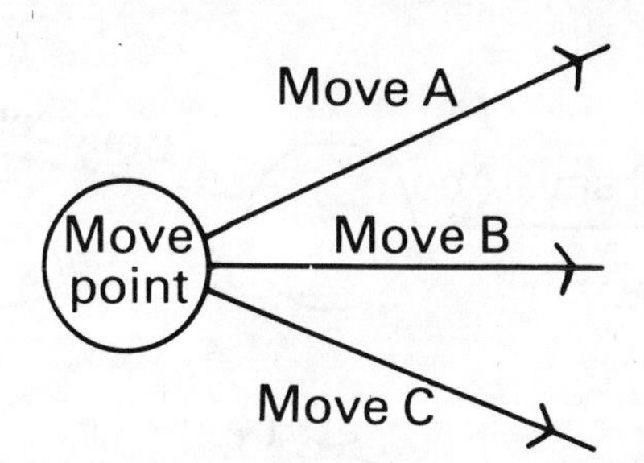

Fig. 2.1—Typical move on a tree graph.

If we can arrive at the same point in the game by two different sequences of moves, we will represent it by two different points, since this allows the decisions made to depend on what has already happened in the game, and so be different for the two different sequences. Thus, we cannot have the situation shown in Fig. 2.2.

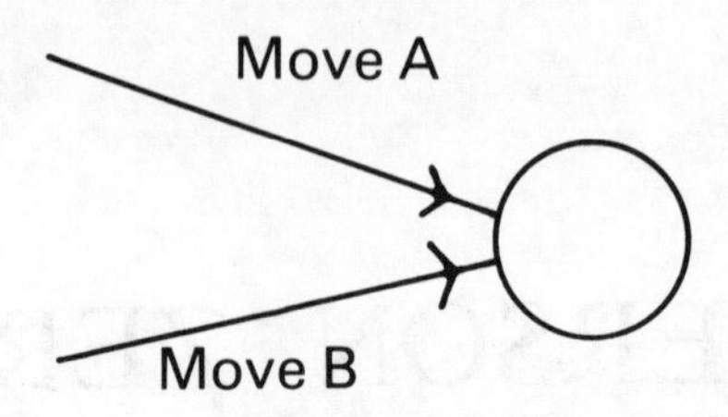

Fig. 2.2—This cannot happen.

Let's look at Examples 1.2 and 1.3 in this way, which is called the **extensive** form of the game.

Example 2.1 — Simplified Poker (as in Example 1.2). In the tree graph for this game we label the chance move points M_0; those that involve a decision by player I, $M_{\rm I}$; and those that involve a decision by II, $M_{\rm II}$ (see Fig. 2.3). The payoffs are written (a,b) where a is I's payoff and b is II's. Looking at the tree, one can see that I has two strategies, namely:

$\rm I_1$—believe II when he says Ace,

$\rm I_2$—don't believe II when he says Ace;

while II also has two strategies:

$\rm II_1$ — say 'Two' when he has a 'Two',

$\rm II_2$ — say 'Ace' when he has a 'Two'.

The real crux of the game is that player I doesn't know which of the $M_{\rm I}$ points he is at when he has to make his decisions. Given the information I has available he cannot distinguish between these points. Points with this property are said to be in the same **information set**, and all a player knows is that he is in that set somewhere.

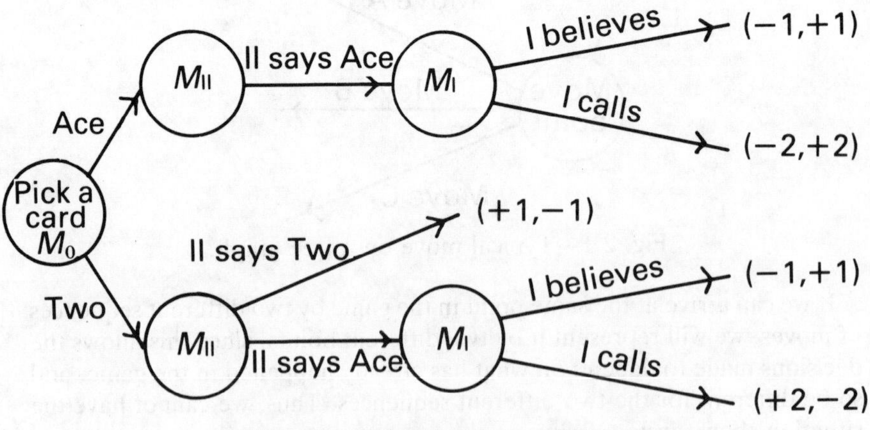

Fig. 2.3—Extensive form of Poker.

Example 2.2 — 2-2 Nim (as in Example 1.3). Take the very simple example where there are two matches in each pile. This example was discussed in Jones (1980) and leads to the tree graph shown in Fig. 2.4, where (=, =) is the situation with two matches in each pile. From the tree, we can again identify the different strategies for each player—three for player I, and six for player II—as follows:

I_1 — take 1 match in the (=, =) case and 1 in the (=,) case,
I_2 — take 1 match in the (=, =) case and 2 in the (=,) case,
I_3 — take 2 matches in the (=, =) case.

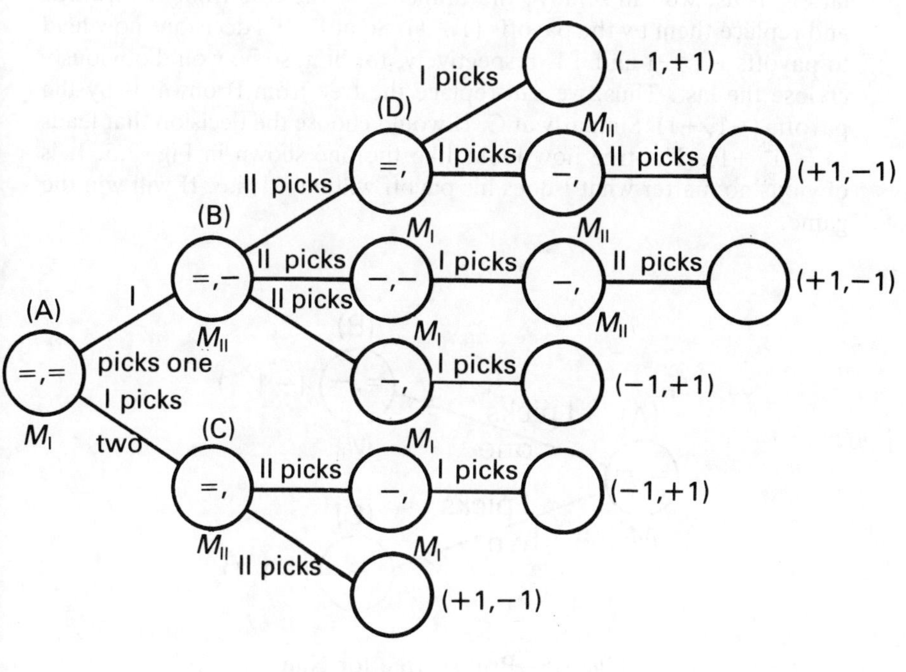

Fig. 2.4.—

Notice that in I_3 you do not have to state what to do in the (=,) case since it will not arise for I. II's strategies are:

II_1 — if (=,) take 2 matches, and if (=, −) take 1 from smaller pile,
II_2 — if (=,) take 2 matches, and if (=, −) take 1 from larger pile,
II_3 — if (=,) take 2 matches, and if (=, −) take 2 from larger pile,

II_4 — if (=,) take 1 match, and if (=, −) take 1 from smaller pile,
II_5 — if (=,) take 1 match, and if (=, −) take 1 from larger pile,
II_6 — if (=,) take 1 match, and if (=, −) take 2 from larger pile.

Also in this problem we can prune the tree somewhat by working back from the end of each branch and working out what the best decision at each move is. Thus, at point D, one of I's strategies leads to a payoff− 1 and the other to a payoff +1 for him, so he would obviously choose the latter. Thus, we can remove the branches of the tree from D onwards and replace them by the payoffs (1,−1). So at B, II's decisions now lead to payoffs −1, −1 and +1, respectively, for him, so he would obviously choose the last. Thus, we can replace the tree from B onwards by the payoffs (−1, +1). Similarly at C, II would choose the decision that leads to (−1, +1). The tree now looks like the one shown in Fig. 2.5. It is obvious no matter what I does his payoff will be −1, i.e. II will win the game.

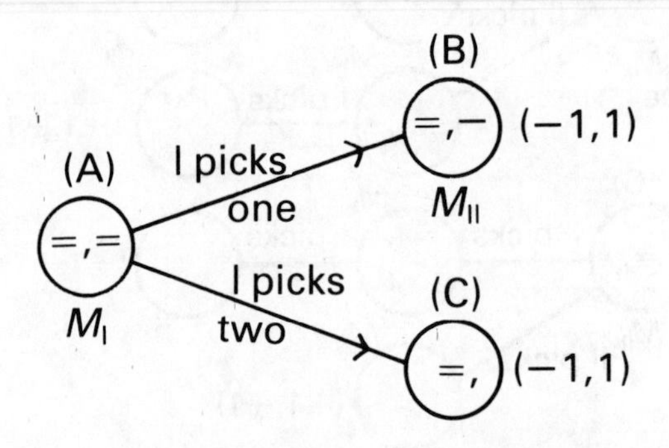

Fig. 2.5—Pruned tree for Nim.

Thus, in Nim we can use the extensive form to work out who will win the game. Notice that all the information sets consists only of one point. In Poker, we can't say who would win because we can't work back along each branch separately. What I chooses between 'believes' and 'calls' must hold for both branches. This difficulty arises because Poker is not a game of perfect information like Nim and so has information sets consisting of more than one point.

These examples highlight the advantages and disadvantages of the extensive form of games. The disadvantages are firstly, even for Nim, it was quite complicated to draw the tree and then work back through it. For most games it is just too long to draw the tree graph and work back through it to find the answer to the game. Secondly, for games like Poker, which don't have perfect information, you cannot work back through the tree because you have to make the same decision on different branches of the tree. The advantage of the extensive form is that it enables you to get a 'picture' of the game so that you can systematically work out what are the possible strategies. Also, it seems obvious that for games of perfect information we could in principle always draw the tree graph and then work back through the tree to find the solution to the game, i.e. a best strategy for each player and the expected payoff if these play each other. Chess, Draughts and Noughts and Crosses all have perfect information. What is their solution?

2.2 NORMAL FORM

Another way of analysing games starts by listing all possible strategies for each player, i.e. $I_1, I_2, \ldots, I_n$ for player I and $II_1, II_2, \ldots, II_m$ for player II. We could use the tree graph to find all these strategies. Although, as we mentioned above, this could be a very difficult if not impossible task, we can always make this list in principle, and so this practical problem will not invalidate any of the theoretical results obtained. Having made the list, if we imagine I playing strategy I_i and II playing II_j, we can work through which moves each player will play at each decision point and get to the payoffs, which we write as $(e_{ij}, -e_{ij})$. Thus, in Nim (Example 2.2) if I plays I_1 and II plays II_1, firstly I takes 1 match leaving (=, −) then II takes 1 from the smaller pile to get (=,). I_1 then tells I to take one more match leaving (−, .) and II must pick up the last match and so lose the game. So the payoff is $(1,-1)$.

If there are chance moves in the game, then when I plays I_i against II's II_j, the payoff depends on the outcome of the chance move. It seems reasonable in such a case to take the expected payoff when I_i plays II_j. So we multiply each payoff by the probabilty of the chance event that gave rise to it, and add all these products together. This gives the average or expected payoff. In Poker (Example 2.1) suppose I_1 plays II_2. Then if the outcome of the chance event is that II gets the Ace, he says 'Ace' and I believes him. The payoff then is $(-1,+1)$. If the outcome of the chance event is that II gets the Two, however, he says 'Two' and loses the game giving payoff $(+1,-1)$. We assume the chances of the Ace and Two being chosen are both ½, so the expected payoff is $\frac{1}{2}(-1,+1) + \frac{1}{2}(+1,-1) = (0,0)$. Notice that once we start using expected payoffs, we are implicitly thinking of the game being played over and over again, and looking at the long-run average. We will return to this point below.

Thus, we can think of the game as one in which each player simultaneously (or in secret from the other) chooses one of his strategies, I_i and II_j say, and the expected payoff is then $(e_{ij}, -e_{ij})$. This is the **normal form** of the game. From now on because Payoff II = −Payoff I in this chapter, we shall only record I's payoffs. We record them as an $n \times m$ matrix with the i,jth entry, e_{ij}, the **payoff matrix**. In the normal form of the game you can make all the moves you could make in the original game, and get the same payoff although the sequential nature of the moves has been lost, as has the idea of perfect information. However, it is the same game we are looking at, whichever form we take. Returning to the two examples of Section 2.1 again and writing them in normal form, we get the following:

Example 2.1 — Simplified Poker. Recall:

I_1 — believe II when he says 'Ace',

I_2 — don't believe II when he says 'Ace',

are I's strategies, and II's are:

II_1 — say 'Two' when you have a Two,

II_2 — say 'Ace' when you have a Two (bluff).

The payoff matrix is:

$$\begin{array}{c} \\ I_1 \\ I_2 \end{array} \begin{array}{c} \begin{array}{cc} II_1 & II_2 \end{array} \\ \begin{pmatrix} 0 & -1 \\ -\tfrac{1}{2} & 0 \end{pmatrix} \end{array}. \tag{2.1}$$

This is obtained as follows. For I_1 versus II_1, if II gets an Ace he says 'Ace' and I believes him, so I has payoff −1; if II gets the Two he says 'Two' and so I gets payoff +1 straightaway. The chance of an Ace is ½, the chance of a Two, is ½, so expected payoff is $1.\tfrac{1}{2} + -1.\tfrac{1}{2} = 0$. For I_1 versus II_2, no matter what II gets, he always says 'Ace' and I always believes him and gets payoff −1. In I_2 against II_1, if II has an Ace he says 'Ace', but I does not believe him and so when Ace is shown I loses −2; whereas if II gets the Two he says so and I gains +1. The expected payoff is $\tfrac{1}{2} \times -2 + \tfrac{1}{2} \times 1 = -\tfrac{1}{2}$. Lastly if I_2 plays II_2, then II always says 'Ace', and I never believes him. When the card is shown, if it is an Ace, I gets −2, but if it is a Two he gets +2. Thus, the expected payoff is $-2 \times \tfrac{1}{2} + 2 \times \tfrac{1}{2} = 0$.

Example 2.2 Nim. Recall that I has three strategies.

I_1 — take 1 match in the (=,=) case, and 1 in the (=,) case,

I_2 — take 1 match in the (=,=) case, and 2 in the (=,) case,

I_3 — take 2 matches in the (=,=) case.

Whereas II has six strategies:

II_1 — if (=,) take 2 matches, and if (=,−) take 1 from smaller pile,

II_2 — if (=,) take 2 matches, and if (=,−) take 1 from larger pile,
II_3 — if (=,) take 2 matches, and if (=,−) take 2 from larger pile,
II_4 — if (=,) take 1 match, and if (=,−) take 1 from smaller pile,
II_5 — if (=,) take 1 match, and if (=,−) take 1 from larger pile,
II_6 — if (=,) take 1 match, and if (=,−) take 2 from larger pile.

When we combine the strategies we get the following 3 × 6 payoff matrix:

$$\begin{array}{c} \\ I_1 \\ I_2 \\ I_3 \end{array} \begin{array}{c} \begin{array}{cccccc} II_1 & II_2 & II_3 & II_4 & II_5 & II_6 \end{array} \\ \begin{pmatrix} 1 & 1 & -1 & 1 & 1 & -1 \\ -1 & 1 & -1 & -1 & 1 & -1 \\ 1 & 1 & 1 & -1 & -1 & -1 \end{pmatrix} \end{array} \tag{2.2}$$

As an example, if we play I_1 against II_2 we have the sequence:

$$(=,=) \xrightarrow{I} (=,-) \xrightarrow{II} (-,-) \xrightarrow{I} (-,\ ) \xrightarrow{II} (\ ,\ ), \tag{2.3}$$

and so I wins and gets payoff +1.

2.3 MAXIMIN CRITERION

Having obtained the payoff matrix, how do we analyse the game? Which strategy is player I or player II likely to choose, and which should we advise him to choose? The important assumption made is that all game-players should be naturally pessimistic, certainly for zero-sum games. Since your opponent is trying to maximise his payoff, then in a zero sum game this means he is also trying to minimise your own payoff. So for each of I's strategies I looks at the minimum payoff he gets using that strategy. For I_i this is $\min_j e_{ij}$. He might well chose the strategy which has the largest of these minimum payoffs. I is **maxi**mising his **min**imum payoff and we call this the **maximin criterion**. Using this strategy he can guarantee he will get a payoff of at least v_L, where

$$v_L = \max_i \min_j e_{ij}. \tag{2.4}$$

This is the **lower value** of the game.

II does exactly the same, but since we are recording only I's payoff, which is the negative of II's payoff, for each of II's strategy II looks at the maximum I gets under this (which is equivalent to the minimum that II gets using that strategy). For II_j this will be $\max_i e_{ij}$. II then chooses the strategy that **mini**mises this **max**imum payoff for I, i.e. he is **minimaxing**. Using this strategy, II can guarantee that I will not get a payoff greater than v_U, the **upper value** of the game, where

$$v_U = \min_j \max_i e_{ij}. \tag{2.5}$$

Going back to our two examples we have:

Poker

$$\begin{array}{cc} & \begin{array}{cc} II_1 & II_2 \end{array} & \text{Minimum using } I_i \\ \begin{array}{c} I_1 \\ I_2 \end{array} & \begin{pmatrix} 0 & -1 \\ -\tfrac{1}{2} & 0 \end{pmatrix} & \begin{array}{c} -1 \\ -\tfrac{1}{2} \end{array} \\ \begin{array}{c}\text{Maximum} \\ \text{using } II_j\end{array} & \begin{array}{cc} 0 & 0 \end{array} & \end{array} \tag{2.6}$$

$$v_L = \max\{-1,-\tfrac{1}{2}\} = -\tfrac{1}{2}, \qquad v_U = \min\{0,0\} = 0. \tag{2.7}$$

Nim

$$\begin{array}{ccc} & \begin{array}{cccccc} II_1 & II_2 & II_3 & II_4 & II_5 & II_6 \end{array} & \text{Minimum using } I_i \\ \begin{array}{c} I_1 \\ I_2 \\ I_3 \end{array} & \begin{pmatrix} 1 & 1 & -1 & 1 & 1 & -1 \\ -1 & 1 & -1 & -1 & 1 & -1 \\ 1 & 1 & 1 & -1 & -1 & -1 \end{pmatrix} & \begin{array}{c} -1 \\ -1 \\ -1 \end{array} \\ \begin{array}{c}\text{Maximum} \\ \text{using } II_j\end{array} & \begin{array}{cccccc} 1 & 1 & 1 & 1 & 1 & -1 \end{array} & \end{array} \tag{2.8}$$

$$v_L = \max\{-1,-1,-1\} = -1, \qquad v_U = \min\{1,1,1,1,1,-1\} = -1. \tag{2.9}$$

In Nim, $v_L = v_U = -1$, and so the best payoff I can be sure of is the same as the payoff that II can restrict him to. Surely this is what is meant by the solution of the game. II should play II_6, I can play anything and the outcome must be -1 to I. This is the same result as we obtained using the extensive form.

However, in Poker $v_L = -\tfrac{1}{2} < 0 = v_U$. So I can be sure of getting $-\tfrac{1}{2}$, but II can only be sure of stopping I getting more than 0. It is not obvious which outcome will occur, because I will choose I_2 apparently, but II can choose either II_1 or II_2 under the maximin criterion. In the one case the payoff is 0, in the other it is $-\tfrac{1}{2}$, but there is no reason under the maximin criterion why II should choose II_1 rather than II_2.

2.4 MIXED STRATEGIES

If you actually think about how you would play Poker—even the simplified game of Example 2.1—you realise that you would never bluff at every turn (II_2) nor would you tell the truth all the time (II_1). Similarly, you don't always 'call' your opponent (I_2) nor never call him (I_1). You would play a mixture of these strategies—sometimes bluff, sometimes tell the truth. Von Neumann suggested that one way of overcoming the difficulty that $v_L \neq v_U$ is to enlarge the set of possible strategies to allow for these **mixed strategies**. A mixed strategy consists of performing a random experiment each time the game is played in order to choose which strategy to use that time. Thus, in Poker I might decide to toss a coin and if it comes down heads use I_1 and if it comes down tails use I_2. His expected payoffs would be $0 \times \frac{1}{2} + -\frac{1}{2} \times \frac{1}{2} = -\frac{1}{4}$ against II_1 and $-1 \times \frac{1}{2} + 0 \times \frac{1}{2} = -\frac{1}{2}$ against II_2.

A strategy that does not involve this random experiment is called a **pure** strategy. For a game where I has n pure strategies, the set of mixed strategies, X, can be represented by the n-tiples $\mathbf{x} = (x_1,x_2,\ldots,x_n)$, where $x_i \geqslant 0$, $i = 1,2,\ldots,n$, and $\sum_{i=1}^{n} x_i = 1$. This corresponds to the mixed strategy where the random experiment has a probability x_i of choosing I_i. A similar definition holds for Y, the set of mixed strategies for II.

If I plays strategy $\mathbf{x} = (x_1,x_2,\ldots,x_n)$ and II plays strategy $\mathbf{y} = (y_1,y_2,\ldots y_m)$, the expected payoff to I is:

$$e(\mathbf{x},\mathbf{y}) = \sum_{j=1}^{m}\sum_{i=1}^{n} x_i e_{ij} y_j. \tag{2.10}$$

In the game of simplified Poker, suppose I plays $\mathbf{x} = (x,1 - x)$ and II plays $\mathbf{y} = (y,1 - y)$, then with the payoff matrix

$$\begin{array}{cc} & \begin{array}{cc} y & 1-y \end{array} \\ \begin{array}{c} x \\ 1-x \end{array} & \begin{pmatrix} 0 & -1 \\ -\frac{1}{2} & 0 \end{pmatrix}, \end{array} \tag{2.11}$$

$$\begin{aligned} e(\mathbf{x},\mathbf{y}) &= 0.xy - 1\,x(y - 1) - \tfrac{1}{2}(1 - x).y + 0(1 - x).(1 - y) \\ &= -x - \tfrac{1}{2}y + \tfrac{3}{2}xy. \end{aligned} \tag{2.12}$$

Now define

$$v_{\mathrm{L}}^{M} = \max_{\mathbf{x}\in X} \min_{\mathbf{y}\in Y} e(x,y)$$

and

$$v_{\mathrm{U}}^{M} = \min_{\mathbf{y}\in Y} \max_{\mathbf{x}\in X} e(x,y)$$

to be the lower and upper values of these enlarged strategy sets, just as we defined v_{L} and v_{U}. If I plays $\mathbf{x}^* = (\frac{1}{3},\frac{2}{3})$, than substituting $x = \frac{1}{3}$ in (2.12) gives:

$$e(\mathbf{x}^*,\mathbf{y}) = -\tfrac{1}{3}, \quad \text{for all } y \in Y, \tag{2.13}$$

which implies that I is sure of at least $-\frac{1}{3}$ payoff if he plays $\mathbf{x}^*$. Alternatively, if II plays $\mathbf{y}^* = (\frac{2}{3},\frac{1}{3})$, by substituting $y = \frac{2}{3}$ in (2.12) gives:

$$e(\mathbf{x},\mathbf{y}^*) = -\tfrac{1}{3}, \quad \text{for all } x \in X. \tag{2.14}$$

So if II plays $\mathbf{y}^*$ he ensures I cannot get more than $-\frac{1}{3}$ no matter what strategy he plays.

It is intuitively obvious that $v_{\mathrm{L}}^{\mathrm{M}} \leqslant v_{\mathrm{U}}^{\mathrm{M}}$ (what I can guarantee himself is less than or equal to what II can restrict him from getting more than).

Playing $\mathbf{x}^*$ guarantees I at least $-1/3$ so $v_L^M \geqslant -1/3$, whereas if II plays $\mathbf{y}^*$ he restricts I to no more than $-1/3$, so $v_U^M \leqslant -1/3$. Hence, we must have $v_L^M = v_U^M = -1/3$, and the solution consists of II playing the mixed strategy $(1/3, 2/3)$ and I playing the mixed strategy $(2/3, 1/3)$ with expected payoff $-1/3$ to I. (So don't be the dealer in this game.)

The idea of playing a mixed strategy implies we are thinking of playing the game repeatedly. This idea has already appeared when we calculate the expected payoff in games with chance events, e.g. we never actually get $-1/2$ payoff in Poker even though this is the matrix entry of playing I_2 against II_1. So we must be very careful when applying these concepts to 'one-off' games—those which will only be played once—or even those we will play just a few times. In one-off games it is impossible to differentiate between a mixed strategy and the pure strategy which the mixed strategy actually picks in that game. Thus, it is valid to ask whether mixed strategies really do exist for such games and also whether one can take expectations over chance events for them. What one must beware of is that using mixed strategies leads to a false sense of security in one-off games. Although we say that if I plays $x^* = (1/3, 2/3)$ in the Poker game it guarantees him an expected payoff $-1/3$, we must remember the word 'expected'. In reality his payoff could be -2 (or $+2$) under this strategy. Notice that this criticism applies even to the lower value of the game using pure strategies, v_L, because this is still an expectation and the actual payoff can be worse than $v_L = -1/2$.

To point up the difference between mixed strategies in one-off games and repeatedly played games, consider the following military pseudo example. In World War II the normal tactic of fighter planes when attacking opposing bomber planes was to swoop down on their target from the direction of the sun—known in boys' comics as the 'Hun in the Sun' strategy. However, if every plane employs this strategy, the bomber pilots put on their sunglasses and just keep staring into the sun looking for the fighters. Thus, a second method of attack was suggested, which was to attack straight up from below. This is very successful if the fighter is not spotted, but since planes go much more slowly when climbing than diving, is likely to have fatal consequences for the fighter pilot if he is spotted. With hindsight, we can describe this as the Ezak-Imak strategy (reverse Kami-kaze). Taking the payoffs as the chance of survival of the fighter plane when it attacks, one matrix which might describe the situation is:

		Bomber crew	
		Look up	Look down
Fighter pilot	Hun in the Sun	0.95	1
	Ezak-imak	1	0

(2.15)

The maximin strategy for player I is $\mathbf{x}^* = (20/21, 1/21)$ which guarantees a payoff of 0.9524 (if you don't believe me read on until Section 2.8 and then try Problem 2.4). Thus, for the squadron leader in charge of the squadron of aircraft, it would be a good idea to put 20 white balls and 1 black ball in his cap, and ask each of his pilots to take out one. White means Hun in the Sun for that pilot, black means Ezak-Imak. This way he maximises the expected number of aircraft which will return. However, the pilot who picked the black ball would be very unhappy and might not find it at all a sensible procedure. The squadron leader is thinking of it as a game played many times—once by each pilot; whereas for the pilot it is a 'one-off' game, in the most fatal sense of the word. More worryingly, suppose you thought of just one pilot flying a large number of missions. Then his safest strategy is the mixed one $\mathbf{x}^* = (20/21, 1/21)$. So, before each flight he would try the squadron leader's experiment and pick one of the 21 balls to tell him what to do. If it were you, and you picked the black ball, would you really perform Ezak-imak, or would you always come in from the sun? The choice is between a certain safety level of 0.95 and an expected safety level of 0.952, but with the chance of certain death.

Again I don't think people will think very highly of a general who tosses a coin to decide whether to advance or stay where he is, but it might be a very good strategy for him to tell each platoon commander to toss a coin to decide whether to attack or not. In the one-off game of the general, playing a mixed strategy seems an abdication of responsibility, whereas in the repeatedly played 'platoon commanders'' game, the mixed strategy adds an uncertainty which could confuse the enemy.

Despite these criticisms, mixed strategies do describe how people play many games. One of the reasons may be that even if the game is only played once, if the players think that they may play it again they must think of it as a repeatable game. One of the advantages of allowing mixed strategies is that it leads to the satisfactory result that all two-person zero-sum games have a solution.

2.5 MINIMAX THEOREM

This is the most important result in game theory and was proved by von Neumann (1937). Essentially it says that what happened in Nim and Poker was no fluke. If we allow mixed strategies, then for every game we can find a best strategy for I under the maximin criterion, which guarantees him v_L^M, and a best strategy for II, which ensures I does not get more than v_U^M and $v_L^M = v_U^M$. Formally, we have:

Theorem 2.1 In a two-person zero-sum game where I has n strategies and II has m strategies (where n and m are finite), then

$$v_L^M = \max_{\mathbf{x} \in X} \min_{\mathbf{y} \in Y} e(\mathbf{x},\mathbf{y}) = \min_{\mathbf{y} \in Y} \max_{\mathbf{x} \in X} e(\mathbf{x},\mathbf{y}) = v_U^M. \quad (2.16)$$

This leads to the definition of a optimal solution to a game.

Definition 2.1 If $\mathbf{x}^* \in X$ is the strategy for I under the maximin criterion, it maximises the second term in (2.16), and if $\mathbf{y}^* \in Y$ is the optimal strategy for II under this criterion it minimises the third term in (2.16). $e(\mathbf{x}^*,\mathbf{y}^*)$ must be at least v_L^M, because of $\mathbf{x}^*$, and no more than v_U^M, because of $\mathbf{y}^*$. Since $v_L^M = v_U^M$, then $e(\mathbf{x}^*,\mathbf{y}^*) = v_L^M = v_U^M = v$, where v is called the **value** of the game. The value v, and the optimal strategies $\mathbf{x}^*$ and $\mathbf{y}^*$ are called a **solution** to the game.

Notice that the theorem requires the number of pure strategies for each player to be finite, and so it is only these finite games that are guaranteed solutions. Recall Example 1.2, and think of the game where two players choose a number and the higher number wins. Each player has an infinite number of strategies—the possible numbers he can choose—and there is no solution to this game, because there is no strategy that guarantees anything better than a loss for either player. Hereafter a **finite** game to us is one where both players have a finite number of pure strategies.

We will not give a proof of the theorem here, but will give a sketch of the proof in another context in Chapter 3, Section 3.4. For a detailed proof see Theorem 2.6 of Jones (1980). The proof of the theorem is closely related to that of two other important theorems—the duality theorem of linear programming, and the separating hyperplane theorem. Although it is an easy matter to prove any two of these as a corollary of the third, the proof of the original theorem will always involve a 'deep' (that's a euphemism for hard) result in mathematical analysis.

2.6 DOMINATION

Now we know there is a solution to all finite two-person zero-sum games in terms of mixed strategies if we want to find it. A useful idea in simplifying these games is the **domination** or **duplication** of one strategy by another. Look at the following 3 × 5 game:

$$\begin{array}{c} \\ I_1 \\ I_2 \\ I_3 \end{array} \begin{array}{c} \begin{array}{ccccc} II_1 & II_2 & II_3 & II_4 & II_5 \end{array} \\ \begin{pmatrix} 4 & 5 & 6 & 4 & 4 \\ 4 & 2 & 3 & 4 & 4 \\ 2 & 4 & 5 & 5 & 5 \end{pmatrix} \end{array} . \qquad (2.17)$$

The payoffs under II_4 and II_5 are the same for all of I's strategies, and so it doesn't matter which strategy II uses. These are **duplicated strategies**, and we can obviously remove one of them, say II_5 and it won't affect v the value of the game. Now look at II_1 and II_4. No matter what I does, the payoff to I is always the same or less under II_1 than II_4. Thus, II will always do as well if not better by using II_1 rather than II_4. We say II_1 **dominates** II_4

and so we can remove II_4 from the strategy set, because II would never want to play that. Similarly, II_2 dominates II_3, since $5 < 6$, $2 < 3$ and $4 < 5$. Thus, II's strategy set can be thought of as just II_1, II_2 and mixtures of these, i.e. we have the matrix:

$$\begin{array}{c} \\ I_1 \\ I_2 \\ I_3 \end{array} \begin{array}{c} \begin{array}{cc} II_1 & II_2 \end{array} \\ \begin{pmatrix} 4 & 5 \\ 4 & 2 \\ 2 & 4 \end{pmatrix} \end{array} . \tag{2.18}$$

Now turning to I, it is obvious he would rather play I_1 in this reduced game than I_2 since $4 \geqslant 4$ and $5 > 2$. So I_1 dominates I_2 in this reduced game. Similarly, I_1 dominates I_3. Notice I_1 didn't dominate I_3 in the original game when II had five pure strategies, but the argument goes that I has realised that II would never play II_3, II_4 or II_5 and so can concentrate on the reduced game when thinking about his own strategies. So the game is now reduced to:

$$\begin{array}{cc} & \begin{array}{cc} II_1 & II_2 \end{array} \\ I_1 & (\begin{array}{cc} 4 & 5 \end{array}) \end{array} \tag{2.19}$$

and in this game it is obvious that II_1 dominates II_2. So we are left with the strategies I_1 and II_1 and a value 4 as the solution to this reduced 1×1 game. This is also a solution to the original game.

Thus, by using domination on one player's strategies we can get a reduced game. Then we can use domination on the other player's strategies to reduce the reduced game more, and so on, going back and forward between the players until we reach a stage where we cannot remove any more strategies by domination. We won't always end up in a 1×1 game like (2.19) but we will often have cut down the size of the game considerably.

We need to prove that any solution we get for a reduced game is also a solution for the original game.

Lemma 2.1 If a dominated strategy is removed from a game, the solution of the reduced game is a solution of the original game.

Proof Suppose that I_2 dominates I_1 in the original $n \times m$ game. We can always renumber the players or strategies so this is the case. This implies that $e_{2j} \geqslant e_{1j}$, for all j, $1 \leqslant j \leqslant m$, in the payoff matrix. Now suppose $\mathbf{x}^* = (x^*_2, x^*_3, \ldots, x^*_n)$, $\mathbf{y}^* = (y^*_1, y^*_2, \ldots, y^*_m)$ and v is a solution to the reduced game when I_1 is ignored, we then have:

$$e(\mathbf{x}^*, \mathbf{y}) \geqslant v, \quad \text{for all } \mathbf{y} \in Y\text{— II's strategy set,} \tag{2.20}$$

and

$$e(\mathbf{x},\mathbf{y}^*) \leqslant v, \quad \text{for all } \mathbf{x} \in X^{\tau} \text{— I's strategy set in the reduced game.} \tag{2.21}$$

We want to show that $\mathbf{x}^* = (0,x^*_2,x^*_3,\ldots,x^*_n)$, $\mathbf{y}^* = (y^*_1,y^*_2,\ldots,y^*_m)$ and v are a solution of the original game. It is enough to show

$$e(\mathbf{x}^*,\mathbf{y}) \geqslant v, \quad \text{for all } \mathbf{y} \in Y, \tag{2.22}$$

$$e(\mathbf{x},\mathbf{y}^*) \leqslant v, \quad \text{for all } \mathbf{x} \in X \text{— I's strategy set in the original game.} \tag{2.23}$$

Equation (2.22) follows immediately from (2.20), so we have to prove (2.23). Take any $\mathbf{x} = (x_1,x_2,\ldots,x_n) \in X$, then

$$\begin{aligned} e(\mathbf{x},\mathbf{y}^*) &= \sum_{i=1}^{n}\sum_{j=1}^{m} x_i e_{ij} y^*_j = x_1 \sum_{j=1}^{m} e_{1j} y^*_j + \sum_{i=2}^{n}\sum_{j=1}^{m} x_i e_{ij} y^*_j \\ &\leqslant x_1 \sum_{j=1}^{m} e_{2j} y^*_j + \sum_{i=2}^{n}\sum_{j=1}^{m} x_i e_{ij} y^*_j = e(\mathbf{x}',\mathbf{y}), \end{aligned} \tag{2.24}$$

where $\mathbf{x}' = (0,x_1 + x_2,x_3,\ldots,x_n)$ can be thought of as a strategy in the reduced game, i.e. in X^{τ}, Thus, by (2.21):

$$e(\mathbf{x},\mathbf{y}^*) \leqslant e(\mathbf{x}',\mathbf{y}^*) \leqslant v \tag{2.25}$$

and (2.22) holds.

However, we do have to pay a price for simplifying the game by removing dominated strategies. All Lemma 2.1 says is that any solution of the reduced game is a solution of the original game, and not that any solution of the original game is also a solution of the reduced game. Thus, we may lose solutions on the way. Look at the game

$$\begin{array}{c c} & \begin{array}{cc} \mathrm{II}_1 & \mathrm{II}_2 \end{array} \\ \begin{array}{c} \mathrm{I}_1 \\ \mathrm{I}_2 \end{array} & \begin{pmatrix} 2 & 2 \\ 0 & 1 \end{pmatrix} \end{array}. \tag{2.26}$$

If we use domination we could say that II_1 dominates II_2, and then that I_1 dominates I_2 and end up with the solution $(\mathrm{I}_1, \mathrm{II}_1)$ and value 2. However, in fact there are a whole series of solution because II can really choose any strategy. So anything of the form $\mathbf{x}^* = (1,0)$, $\mathbf{y}^* = (y,1-y)$, $v = 2$ is a solution for any $y, 0 \leqslant y \leqslant 1$, but all except $y = 1$ are lost when we reduce the game.

If you only allow **strict domination** where a strategy I_1 strictly dominates I_2 if $e_{1j} > e_{2j}$, for all j, and II_1 strictly dominates II_2 if $e_{i1} < e_{i2}$ for all i, you preserve all the solutions. Since we are usually only interested in finding a solution to the game, we will concentrate on ordinary domination, which will in general reduce the game more.

We have only looked at domination of a pure strategy by a pure strategy, but you can also rule out a strategy if it is dominated by a mixed one. Finding a suitable mixed strategy is often quite difficult, and the general view is it is not worth the effort.

2.7 WORTHWHILE STRATEGIES

The second useful idea in solving small two-person zero-sum games is that of **worthwhile strategies.**

Definition 2.2 A worthwhile strategy is a pure strategy which appears with positive probability in an optimal strategy. The important property about such strategies is

Lemma 2.2 When a worthwhile strategy plays an optimal strategy the payoff is the value of the game.

Proof Suppose that, after reordering the pure strategies if necessary, I has an optimal strategy $\mathbf{x}^* = (x_1,x_2,\ldots,x_k,0,0,\ldots,0)$, where $x_i > 0$, $i = 1,\ldots,k$, $\sum_{i=1}^k x_i = 1$, II has an optimal strategy $\mathbf{y}^*$ and the value of the game is v. So $v = e(\mathbf{x}^*,\mathbf{y}^*)$. Moreover, for any pure strategy I_i, $1 \leqslant i \leqslant k$, $e(\mathrm{I}_i,\mathbf{y}^*) = v_i \leqslant v$, since II is playing an optimal strategy $\mathbf{y}^*$. Then

$$v = e(\mathbf{x}^*,\mathbf{y}^*) = \sum_{i=1}^{k} x_i e(\mathrm{I}_i,\mathbf{y}^*) = \sum_{i=1}^{k} x_i v_i \leqslant \sum_{i=1}^{k} x_i v = v. \tag{2.27}$$

Thus, we must have equality everywhere and so $v_i = v$, $i = 1,\ldots,k$, and playing the worthwhile strategy I_i against the optimal $\mathbf{y}^*$ gives a payoff that is the value of the game. The proof for I's optimal against II's worthwhile is similar.

These two ideas are enough to solve lots of games.

2.8 SOLUTION OF 2 × *m* GAMES

We are now in a position to solve all games where player I (or player II) has at most two pure strategies after domination. We proceed in three steps.

A: First remove all dominated and duplicated strategies.
B: Use a graphical method to find which are the worthwhile strategies. You can in fact read off the value of the game from the graph, but to get an accurate answer we
C: use 'worthwhile versus optimal equals value' to solve for the optimal strategies and the value of the game.

To see how this works let us solve a few games.

Example 2.3. I has two aircraft, II has four missile batteries to cover four approaches to a target. Each battery can shoot down with certainty one aircraft, if it attacks along that approach, but only one aircraft since the reloading time is long. The payoff to I is 1 if an aircraft gets through and so destroys the target, 0 otherwise.

The strategies describe the distribution of the batteries and the aircraft rather than which specific approach each is put on. Thus, the strategies for I are:

I_1 — send aircraft on different approaches.
I_2 — send aircraft on the same approach.

Whereas those for II are:

II_1 — put one battery on each approach.
II_2 — put two batteries on each of two approaches.
II_3 — put two batteries on one approach and one each on two others.
II_4 — put three batteries on one approach, one on another.
II_5 — put the four batteries on one approach.

The payoff matrix is:

$$\begin{array}{c} \\ I_1 \\ I_2 \end{array} \begin{array}{c} \begin{array}{ccccc} II_1 & II_2 & II_3 & II_4 & II_5 \end{array} \\ \begin{pmatrix} 0 & 5/6 & 1/2 & 5/6 & 1 \\ 1 & 1/2 & 3/4 & 3/4 & 3/4 \end{pmatrix} \end{array} . \tag{2.28}$$

Against I_1, II_1 is bound to shoot both aircraft down, whereas II_5 can never do so. For II_2 and II_4, the only time both aircraft are shot down is if the guns are covering the particular pair of approaches which they choose. Since there are six ways of making a pair of approaches from four different approaches, the chance of a plane getting through is 5/6. For II_3, a plane will get through if it flies along the undefended approach and of the six possible pairs of approaches three of them include a specific approach, and so the chance of success is 3 out of 6.

Against I_2, II_1 cannot shoot the second plane down, whereas II_2 can defend successfully two of the four approaches and so the chance a plane gets through is 2/4. II_3, II_4 and II_5 only defend one approach against I_2 and so the chance a plane gets through is 3/4.

Looking at the strategies it is easy to see II_3 dominates II_4 and II_5, which is what you would expect since these last two strategies have an element of overkill about them. So we have the reduced game:

$$\begin{array}{c} \\ I_1 \\ I_2 \end{array} \begin{array}{c} \begin{array}{ccc} II_1 & II_2 & II_3 \end{array} \\ \begin{pmatrix} 0 & 5/6 & 1/2 \\ 1 & 1/2 & 3/4 \end{pmatrix} \end{array} . \tag{2.29}$$

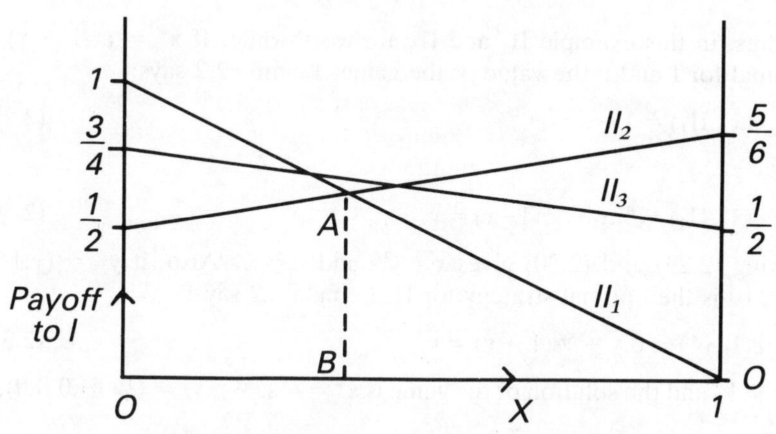

Fig. 2.6—Graphical solution of Example 2.3.

If I's optimal strategy is $(x,1 - x)$, then against II_1 the payoff is $0.x + 1(1 - x) = 1 - x$; against II_2 it is $\frac{5}{6}x + \frac{1}{2}(1 - x) = \frac{1}{2} + \frac{1}{3}x$; and against II_3 it is $\frac{1}{2}x + \frac{3}{4}(1 - x) = \frac{3}{4} - \frac{1}{4}x$. We can show this graphically, plotting each payoff as a function of x as in Fig. 2.6. For each value of x, the height of the lines at that point denotes the payoff of each of II's strategies against $(x,1 - x)$ for I. I is worried about his least payoff when he plays a particular strategy, which is the lowest of the three lines at that point, and wants to choose x so as to maximise this minimum payoff, i.e. $\max_x \min(1 - x, \frac{1}{2} + \frac{1}{3}x, \frac{3}{4} - \frac{1}{2}x)$. This is at A, where the lower envelope (lowest of the three lines at each point) is highest. The distance AB is the value of the game, v, and $x = OB$ give the optimal strategy for I.

However, if we want a more precise answer than can be read off the graph, we use the fact that all II's strategies going through this solution point A are *worthwhile*. It is obvious that any line, like II_3, not going through A is not worthwhile, because playing it against the optimal strategy gives a higher payoff than v. Moreover, if we have applied domination, at least one of the lines through A will have positive slope and at least one will have negative slope. If II takes a mixture of the pure strategies, which go through A, say II_1 with probability y_1 and II_2 with probability y_2, the payoffs of I's strategies against this mixture is also a straight line through A with gradient y_1 times the gradient of the line denoting II_1 plus y_2 times the gradient of the line denoting II_2. If one of these has positive gradient and the other negative gradient we can always choose y_1 and y_2 so that the line representing the mixture has zero gradient and so I has a payoff v against this mixture for all its strategies. This mixture is then optimal for II and hence any line through A represents a pure strategy which can be made part of an optimal strategy.

Thus, in this example II_1 and II_2 are worthwhile. If $\mathbf{x}^* = (x, 1 - x)$ is optimal for I and v the value of the game, Lemma 2.2 says:

$$e(\mathbf{x}^*, II_1) = 1 - x = v, \tag{2.29}$$

and

$$e(\mathbf{x}^*, II_2) = \tfrac{5}{6}x + \tfrac{1}{2}(1-x) = v. \tag{2.30}$$

Solving (2.29) and (2.30) gives $x = \frac{3}{8}$ and $v = \frac{5}{8}$ Also, if $\mathbf{y}^* = (y, 1 - y, 0, 0, 0)$ is the optimal strategy for II, Lemma 2.2 says:

$$e(I_1, \mathbf{y}^*) = 0.y + \tfrac{5}{6}(1 - y) = v = \tfrac{5}{8}, \tag{2.31}$$

so $y = \frac{1}{4}$ and the solution of the game is $\mathbf{x}^* = (\frac{3}{8}, \frac{5}{8})$, $\mathbf{y}^* = (\frac{1}{4}, \frac{3}{4}, 0, 0, 0)$, $v = \frac{5}{8}$.

Example 2.4 This puts some numbers into the duellist game (Example 1.6). Suppose I and II start 4 paces apart and I's chance of killing II is 0.2 at 4 paces, 0.8 at 2 paces and 1 at 0 paces, while II's chance of killing I is 0.5 at 4 paces, 0.75 at 2 paces and 1 at 0 paces. The payoff is +1 if I is still alive and II dead, −1 if II alive and I dead and 0 otherwise. In fact, there are two different games here depending on whether a duellist knows if his opponent has fired (**noisy duel**) or doesn't know if he has (**silent duel**).

(a) **Noisy duel**. The pure strategies are with abuse of the numbering:

$I_4(II_4)$ — fire when 4 steps apart,
$I_2(II_2)$ — fire when 2 steps apart if opponent has not yet fired, otherwise fire when 0 steps apart,
$I_0(II_0)$ — fire when 0 steps apart.

The payoff matrix is:

$$\begin{array}{c} \\ I_4 \\ I_2 \\ I_0 \end{array} \begin{array}{c} \begin{array}{ccc} II_4 & II_2 & II_0 \end{array} \\ \begin{pmatrix} -0.3 & -0.6 & -0.6 \\ 0 & 0.05 & +0.6 \\ 0 & -0.5 & 0 \end{pmatrix} \end{array}. \tag{2.32}$$

Checking on domination, we see I_2 dominates I_4 and I_0, and in the reduced game, II_4 (payoff 0) dominates II_2 (payoff 0.5) and II_0 (payoff 0.6), so that the solution is (I_2, II_4) with value $v = 0$. It is true (see Karlin, 1959) that all realistic noisy duels have pure optimal strategies.

(b) **Silent duel.** Here, of course, the strategy $Ii(II_i)$ is to fire when i paces apart no matter what, since you can't tell if your opponent has fired.

The payoff matrix is

$$\begin{array}{c} \\ I_4 \\ I_2 \\ I_3 \end{array} \begin{array}{c} \begin{array}{ccc} II_4 & II_2 & II_0 \end{array} \\ \begin{pmatrix} -0.3 & -0.4 & -0.6 \\ -0.1 & 0.05 & 0.6 \\ 0 & -0.5 & 0 \end{pmatrix} \end{array} . \tag{2.33}$$

I_4 is dominated by I_2, and in the reduced game without I_4, II_2 dominates II_4. Thus, we have the reduced game:

$$\begin{array}{c} \\ I_2 \\ I_0 \end{array} \begin{array}{c} \begin{array}{cc} II_4 & II_2 \end{array} \\ \begin{pmatrix} -0.1 & 0.05 \\ 0 & -0.5 \end{pmatrix} \end{array} . \tag{2.34}$$

There is no further domination and drawing the graph of the payoffs of $x^* = (0,x,1-x)$ for I against each of II's pure strategy gives Fig. 2.7, where A

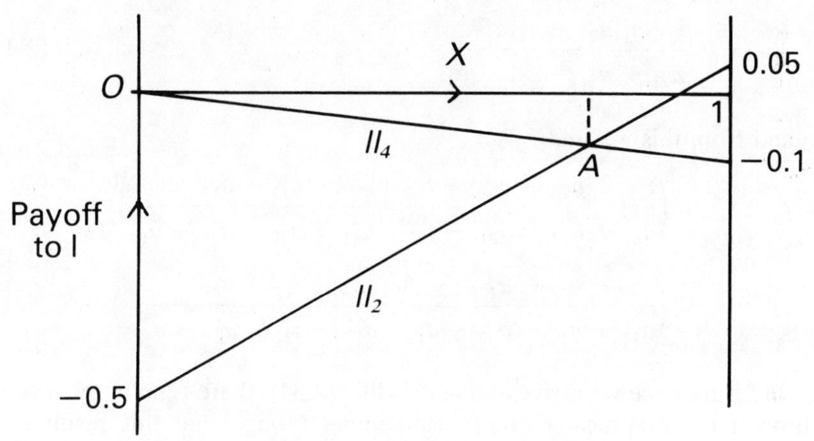

Fig. 2.7—Graphical solution of silent duel.

is the solution point so both II_4 and II_2 are worthwhile. Hence, if $\mathbf{x}^* = (0,x,1-x)$ is the optimal solution for I and v the value of the game, Lemma 2.2 says:

$$e(\mathbf{x}^*,II_4) = -0.1x = v \tag{2.35}$$

and

$$e(\mathbf{x}^*,II_2) = 0.05x - 0.5(1-x) = v. \tag{2.36}$$

Solving (2.35) and (2.36) gives $x = 10/13$, $v = -1/13$, while if $\mathbf{y}^* = (y,1-y,0)$ is the optimal solution for II, Lemma 2.2. gives:

$$e(I_0,y^*) = -0.5(1-y) = -1/13 = v. \tag{2.37}$$

So the solution is I plays $(0,10/13,3/13)$, II plays $(11/13,2/13,0)$ with $v = -1/13$.

Example 2.5 — 2×2 game. Consider a typical 2×2 game with payoff matrix

$$A = \begin{array}{c} \\ \mathrm{I}_1 \\ \mathrm{I}_2 \end{array} \begin{array}{c} \begin{array}{cc} \mathrm{II}_1 & \mathrm{II}_2 \end{array} \\ \begin{pmatrix} a_{11} & a_{12} \\ a_{21} & a_{22} \end{pmatrix} \end{array}, \tag{2.38}$$

where there is no domination. Then trivially the optimal strategy for II must be a mixture of II_1 and II_2 — otherwise one would dominate the other. If $\mathbf{x}^* = (x, 1 - x)$ is optimal for I, Lemma 2.2 gives:

$$e(\mathbf{x}^*, \mathrm{II}_1) = xa_{11} + (1 - x)a_{21} = v, \tag{2.39}$$

$$e(\mathbf{x}^*, \mathrm{II}_2) = xa_{12} + (1 - x)a_{22} = v, \tag{2.40}$$

and solving gives:

$$v = \frac{a_{11}a_{22} - a_{21}a_{12}}{a_{21} + a_{22} - a_{21} - a_{12}}, \tag{2.41}$$

and from this we find:

$$\begin{aligned} \mathbf{x}^* &= \left(\frac{a_{22} - a_{21}}{a_{22} + a_{22} - a_{21} - a_{12}}, \frac{a_{11} - a_{12}}{a_{11} + a_{22} - a_{21} - a_{12}} \right), \\ \mathbf{y}^* &= \left(\frac{a_{22} - a_{12}}{a_{11} + a_{22} - a_{21} - a_{12}}, \frac{a_{11} - a_{21}}{a_{11} + a_{22} - a_{21} - a_{12}} \right). \end{aligned} \tag{2.42}$$

So in future once we arrive at a matrix like (2.34), there is no need to work through the graphical method. Remember though that this result only holds for 2×2 games if there is no domination.

2.9 EQUILIBRIUM PAIRS

One of the interesting properties of the optimal strategies in these two-person, zero-sum games is that if both players are playing their optimal strategies, then it does not pay either to change his strategy if his opponent sticks to his optimal one. So if I plays his optimal strategy $\mathbf{x}^*$ and II plays his optimal strategy $\mathbf{y}^*$, if II keeps playing $\mathbf{y}^*$, then if I changes to some other strategy $\mathbf{x}$ he will do no better than if he stayed with $\mathbf{x}^*$ and vice versa. The strategies are in some sort of equilibrium. It also means that I can announce that he is going to play $\mathbf{x}^*$, his optimal strategy, and II can do nothing better with this information than play his optimal strategy $\mathbf{y}^*$ in response. The strategies are 'spy-proof' because it doesn't matter if your opponent knows what you are going to do. Only the optimal strategies have this property, but before proving that let us give the formal definition of a pair of strategies in equilibrium.

Definition 2.3 A pair of strategies $\mathbf{x}^* \in X$, $\mathbf{y}^* \in Y$ is an **equilibrium pair** if for any $\mathbf{x} \in X$, $\mathbf{y} \in Y$

$$e(\mathbf{x},\mathbf{y}^*) \leqslant e(\mathbf{x}^*,\mathbf{y}^*) \leqslant e(\mathbf{x}^*,\mathbf{y}). \tag{2.43}$$

You could write this symmetrically by using $e_1(\ ,\ )$ as I's payoff and $e_2(\ ,\ )$ as II's payoff. Then (2.43) becomes:

$$e_1(\mathbf{x},\mathbf{y}^*) \leqslant e_1(\mathbf{x}^*,\mathbf{y}^*); \qquad e_2(\mathbf{x}^*,\mathbf{y}) \leqslant e_2(\mathbf{x}^*,\mathbf{y}^*). \tag{2.44}$$

Thus, in the noisy duel (2.32), if II plays II_4, there is no point in I changing from I_2 as he cannot do better, whereas if I plays I_2, II should not change from II_4 as everything else is worse for him.

It is quite possible for there to be more than one equilibrium pair in a game. For example, in

$$\begin{array}{cc} & \begin{array}{ccc} \mathrm{II}_1 & \mathrm{II}_2 & \mathrm{II}_3 \end{array} \\ \begin{array}{c} \mathrm{I}_1 \\ \mathrm{I}_2 \end{array} & \begin{pmatrix} 1 & 4 & 1 \\ 0 & 2 & -1 \end{pmatrix} \end{array} \tag{2.45}$$

both $(\mathrm{I}_1,\mathrm{II}_1)$ and $(\mathrm{I}_1,\mathrm{II}_3)$ are equilibrium pairs, since both are minimum payoffs along the row $(e(\mathbf{x}^*,\mathbf{y}^*) \leqslant e(\mathbf{x}^*,\mathbf{y}))$ and maximums along the columns $(e(\mathbf{x},\mathbf{y}^*) \leqslant e(\mathbf{x}^*,\mathbf{y}^*))$. In fact, careful examination shows that $\mathbf{x}^* = (1,0)$, $\mathbf{y}^* = (y,0,1-y)$ for any y, $0 \leqslant \mathbf{y} \leqslant 1$, is an equilibrium pair. However, for all these pairs $e(\mathbf{x}^*,\mathbf{y}^*) = 1$. Is this just a fluke or does it always happen that the payoffs of all equilibrium pairs in a game have the same value? Well, we are not so short of material we need to waste three lines on flukes so:

Lemma 2.3 If $(\mathbf{x}_1,\mathbf{y}_1)$, $(\mathbf{x}_2,\mathbf{y}_2)$ are equilibrium pairs, then $e(\mathbf{x}_1,\mathbf{y}_1) = e(\mathbf{x}_2,\mathbf{y}_2)$.

Proof Since $(\mathbf{x}_1,\mathbf{y}_1)$ is an equilibrium pair we have:

$$e(\mathbf{x}_2,\mathbf{y}_1) \leqslant e(\mathbf{x}_1,\mathbf{y}_1) \leqslant e(\mathbf{x}_1,\mathbf{y}_2) \tag{2.46}$$

and since $(\mathbf{x}_2,\mathbf{y}_2)$ is an equilibrium pair:

$$e(\mathbf{x}_1,\mathbf{y}_2) \leqslant e(\mathbf{x}_2,\mathbf{y}_2) \leqslant e(\mathbf{x}_2,\mathbf{y}_1). \tag{2.47}$$

Hence trivially from (2.47) and (2.46) we must have $e(\mathbf{x}_1,\mathbf{y}_1) = e(\mathbf{x}_2,\mathbf{y}_2)$.

Now we will prove formally our statement that optimal strategies form equilibrium pairs and these are the only equilibrium pairs.

Theorem 2.2 If $(\mathbf{x}^*,\mathbf{y}^*)$ is a pair of strategies in a two-person $n \times m$

zero-sum game, then $(\mathbf{x}^*,\mathbf{y}^*)$ is an equilibrium pair if and only if $(\mathbf{x}^*,\mathbf{y}^*,e(\mathbf{x}^*,\mathbf{y}^*))$ is a solution to the game.

Proof If $(\mathbf{x}^*,\mathbf{y}^*)$ is an equilibrium pair, then it follows from (2.43) that:

$$\max_{\mathbf{x}\in X} e(\mathbf{x},\mathbf{y}^*) \leqslant e(\mathbf{x}^*,\mathbf{y}^*) \leqslant \min_{\mathbf{y}\in Y} e(\mathbf{x}^*,\mathbf{y}). \tag{2.48}$$

Then we get from (2.48):

$$\begin{aligned} v_{\mathrm{U}}^{M} &= \min_{\mathbf{y}\in Y} \max_{\mathbf{x}\in X} e(\mathbf{x},\mathbf{y}) \leqslant \min_{\mathbf{y}=\mathbf{y}^*} \max_{\mathbf{x}\in X} e(\mathbf{x},\mathbf{y}) = \max_{\mathbf{x}\in X} e(\mathbf{x},\mathbf{y}^*) \\ &\leqslant e(\mathbf{x}^*,\mathbf{y}^*) \leqslant \min_{\mathbf{y}\in Y} e(\mathbf{x}^*,\mathbf{y}) = \max_{\mathbf{x}=\mathbf{x}^*} \min_{\mathbf{y}\in Y} e(\mathbf{x}^*,\mathbf{y}) \leqslant \max_{\mathbf{x}\in X} \min_{\mathbf{y}\in Y} e(\mathbf{x},\mathbf{y}) \\ &= v_{\mathrm{L}}^{M}. \end{aligned}$$

So $v_{\mathrm{U}}^{M} \leqslant v_{\mathrm{L}}^{M}$, but it is intuitively true that $v_{\mathrm{L}}^{M} \leqslant v_{\mathrm{U}}^{M}$. (If your intuition isn't 'intuiting' today, notice that

$$\min_{\mathbf{y}\in Y} e(\mathbf{x}',\mathbf{y}) \leqslant e(\mathbf{x}',\mathbf{y}') \leqslant \max_{\mathbf{x}\in X} e(\mathbf{x},\mathbf{y}') \tag{2.50}$$

for any $\mathbf{x}' \in X$, $\mathbf{y}' \in Y$. In particular:

$$\begin{aligned} v_{\mathrm{L}}^{M} &= \max_{\mathbf{x}\in X} \min_{\mathbf{y}\in Y} e(\mathbf{x},\mathbf{y}) = \min_{\mathbf{y}\in Y} e(\mathbf{x}',\mathbf{y}) \leqslant e(\mathbf{x}',\mathbf{y}') \leqslant \max_{\mathbf{x}\in X} e(\mathbf{x},\mathbf{y}') \\ &< \min_{\mathbf{y}\in Y} \max_{\mathbf{x}\in X} e(\mathbf{x},\mathbf{y}) = v_{\mathrm{U}}^{M}, \end{aligned} \tag{2.51}$$

where $\mathbf{x}'$ is the maximiser of the first term and $\mathbf{y}'$ the minimiser of the last term in (2.51).) Thus, $v_{\mathrm{L}}^{M} = v_{\mathrm{U}}^{M} = e(\mathbf{x}^*,\mathbf{y}^*)$ and so $(\mathbf{x}^*,\mathbf{y}^*,e(\mathbf{x}^*,\mathbf{y}^*))$ is a solution of the game.

Conversely, if $(\mathbf{x}^*,\mathbf{y}^*,\ v = e(\mathbf{x}^*,\mathbf{y}^*))$ is a solution to the game, then:

$$\begin{aligned} e(\mathbf{x},\mathbf{y}^*) &\leqslant \max_{\mathbf{x}\in X} e(\mathbf{x},\mathbf{y}^*) = \min_{\mathbf{y}\in Y} \max_{\mathbf{x}\in X} e(\mathbf{x},\mathbf{y}) = e(\mathbf{x}^*,\mathbf{y}^*) \\ &= \max_{\mathbf{x}\in X} \min_{\mathbf{y}\in Y} e(\mathbf{x},\mathbf{y}) = \min_{\mathbf{y}\in Y} e(\mathbf{x}^*,\mathbf{y}) \leqslant e(\mathbf{x}^*,\mathbf{y}). \end{aligned} \tag{2.52}$$

where all the equalities in the middle follow becaue $\mathbf{x}^*,\mathbf{y}^*,\ e(\mathbf{x}^*,\mathbf{y}^*)$ is a solution. Hence, $(\mathbf{x}^*,\mathbf{y}^*)$ is an equilibrium pair.

This establishes the one-to-one correspondence between solutions of two-person zero-sum games and equilibrium, pairs. As we shall see below it is the idea of an equilibrium pair that is easier to generalise to more complicated games. Also, the result of Lemma 2.3 can now be reinterpreted in terms of solutions of a game as

Corollary 2.1 If a two-person finite zero-sum game has more than one solution, they all have the same value.

2.10 GAMES WITH PERFECT INFORMATION

At the end of Section 2.1 on extensive forms of games, we said we could solve a game with perfect information by using the decision tree. However, we had neither defined what we meant by a solution then nor introduced the idea of mixed strategies. We can now give the result correctly.

Theorem 2.3 A finite two person zero sum game with perfect information has a solution in which both players have pure optimal strategies.

This was essentially proved by Zermelo (1913) before the advent of game theory in its present form. The first proof in that context was given by von Neumann and Morgenstern (1947) and depends on working back through the extensive form of the game. It is not difficult, but is long so we shall omit it (see Jones, 1980, for a more up-to-date version). It works by induction on the number of move points still to go in the game, and says that as we work back from the end of each branch in the tree graph there is always a decision that is best for a player at each decision point. The pure strategy that chooses these best decisions at each decision point is then optimal.

Noughts and Crosses (Example 1.1), Draughts, Chess, Go and Backgammon are all games with perfect information, so the theorem applies to them. In the case of $n \times n$ Noughts and Crosses the solution to the game is well known; and, the value is a win for I if $n = 2$ and a draw if $n \geq 3$. Most people who play Draughts wouldn't like to hazard a guess what the outcome is there. However, the solution is a draw. Expert Draughts players know this and so the competition rules have been changed so that the first two moves are made by a chance experiment, in order to try and force people away from the optimal strategies. The situation in Chess is more obscure. O.K., so the theorem says there is a best way of playing it, but even Karpov or Kasparov are not sure what the outcome of these optimal strategies is, let alone the strategies. If both players play their optimal strategies, will White win, will Black win, or will it be a draw? Since in Chess you can essentially waste moves by doing in two moves something you could have done in one—like moving a pawn forward two—the argument is that Black cannot win, since if he could, White would waste a move and essentially become Black and win himself. So either White wins or it is a draw. Although the essential conservatism of mankind means most players favour a draw as the likely value of the game, no one knows.

2.11 SOLVING $n \times m$ GAMES

In Section 2.8 we learnt how to solve $2 \times m$ or $m \times 2$ games and games that reduced to these by domination. In this section we describe two methods for solving all two-person zero-sum games, no matter how many pure strategies the players have.

2.11.1 Linear Programming

Soon after he had discovered the simplex method algorithm, Dantzig (1951), as well as Gale, Kuhn and Tucker (1951), pointed out that solving a two-person zero-sum game is equivalent to solving a linear programming. This is the fastest and most useful method of solving large games.

To solve the game we want a $\mathbf{x}^* = (x_1, x_2, \ldots, x_n)$, a $\mathbf{y}^* = (y_1, y_2, \ldots, y_m)$ and a v so that

$$e(\mathbf{x},\mathbf{y}^*) \leqslant e(\mathbf{x}^*,\mathbf{y}^*) = v \leqslant e(\mathbf{x}^*,\mathbf{y}), \tag{2.53}$$

for any $\mathbf{x} \in X$, $\mathbf{y} \in Y$. The first inequality of (2.53) implies

$$e(\mathrm{I}_i,\mathbf{y}^*) = \sum_{j=1}^{m} e_{ij} y_j^* \leqslant v, \quad \text{for } i = 1, \ldots, n. \tag{2.54}$$

II chooses y^* so as to make the v that satisfies this as small as possible—i.e. he makes $v_U^M = v$ as low as he can. So the choice of v and y^* is given by:

$$\begin{aligned} &\min v \\ &\text{subject to } \sum_{j=1}^{m} e_{ij} y_j \leqslant v, \qquad i = 1, \ldots, n, \\ &\qquad\qquad y_j \geqslant 0, \qquad j = 1, \ldots, m \end{aligned} \tag{2.55}$$

Let $Y_j = y_j/v$ for $1 \leqslant j \leqslant m$, so $Y_1 + Y_2 + \ldots + Y_n = \sum_{j=1}^{m} y_j/v = 1/v$. Minimising v is equivalent to maximising $1/v$, so assuming $v > 0$, we can write (2.55) as:

$$\begin{aligned} &\max Y_1 + Y_2 + \ldots + Y_m \\ &\text{subject to } \sum_{j=1}^{m} e_{ij} Y_j \leqslant 1, \qquad i = 1, \ldots, n, \\ &\qquad\qquad Y_j \geqslant 0, \qquad j = 1, \ldots, m. \end{aligned} \tag{2.56}$$

This is a standard L.P. problem and can be solved in the usual way (for details see Jones, 1980).

It does not matter if we are not certain that $v > 0$. If we add the same amount to every payoff in a game so that $\tilde{e}_{ij} = e_{ij} + c$, the value of the game goes up by c but the optimal strategies stay the same. It is just as if II gives I a present of c to play the game. This should not affect what they do thereafter. Thus, we add a constant c to the payoffs to make them all positive. This ensures the value of this new game is positive and so we solve it by linear programming. Then take c off the value of the solution to get the value of the original game.

One of the important results about linear programming is that each time you solve a problem you get the solution to another one—its dual—'for free', because the optimal value of the objective function is the same in both. The dual of (2.56) is:

$$\min X_1 + X_2 + \ldots + X_n$$

$$\text{subject to} \sum_{i=1}^{n} X_i e_{ij} \geqslant 1, \quad j = 1,2,\ldots,m, \tag{2.57}$$

$$X_i \geqslant 0, \quad i = 1,2,\ldots,n,$$

and in exactly the same way as in (2.56) we can interpret $X_i = x_i/v$, $i = 1,\ldots n$ and rewrite this as

$$\max v$$

$$\text{subject to} \sum_{i=1}^{n} x_i e_{ij} \geqslant v, \quad j = 1,\ldots,m, \tag{2.58}$$

$$x_i \geqslant 0, \quad i = 1,\ldots,n.$$

This is like I making v_{L}^{M} as large as possible. Thus, the solution to the linear programme (2.56) gives the value of the game and II's optimal strategy $\mathbf{y}^*$, while the solution of the dual problem (2.58) gives I's optimal strategy $\mathbf{x}^*$. If we solve (2.56) by the standard simplex method, the optimal values of these dual variables also appear as part of the solution, and so we get v, $\mathbf{y}^*$ and $\mathbf{x}^*$ by solving one L.P.

2.11.2 Brown's Method

This is no way as useful as linear programming for actually calculating the solution of the game, but it is an interesting theoretical result and it is good fun to do. What Brown (1951) suggested was the following. Pretend to play the game many times, and at each play record the possible payoffs you could have got for each of your pure strategies against that which your opponent actually played. Add these up for each of your strategies over all the plays, so far made, and choose as your strategy next time the one that gave you the best total over all the plays. If both players do this, then the average payoff per game tends to the value v of the game, and the average number of times per game a strategy is played gives the probability you assign to that strategy in the random experiment which determines the optimal mixed strategy.

To see how it works recall simplified Poker (Example 2.1) where

$$\begin{array}{cc} & \begin{array}{cc} II_1 & II_2 \end{array} \\ \begin{array}{c} I_1 \\ I_2 \end{array} & \begin{pmatrix} 0 & -1 \\ -\tfrac{1}{2} & 0 \end{pmatrix} . \end{array} \tag{2.59}$$

At the first play we can choose any strategies we like, so suppose I chooses I_2, II chooses II_1. Then against II_1 I's payoffs would be 0 or $-\frac{1}{2}$, written $(0,-\frac{1}{2})$, so next time I will choose I_1 as 0 is greater than $-\frac{1}{2}$. Similarly, II looks at the payoffs against I_2 which are $-\frac{1}{2}$, or 0, written $(-\frac{1}{2},0)$. II will choose II_1 next time as he prefers I to have a payoff of $-\frac{1}{2}$ rather than 0. So the second game will consist of I_1 versus II_1 with payoff 0. This time I looks at his total payoffs against the strategies II has played (II_1 twice) which is 0 for I_1 and $-\frac{1}{2} + -\frac{1}{2} = -1$ for I_2, so he will choose I_1 again next time. II looks at the payoffs against the total of I's strategies (I_1 once and I_2 once) which is $-\frac{1}{2}$ for II_1 and -1 for II_2. So he would choose II_2 next time as he prefers I to have -1 rather than $-\frac{1}{2}$. Continuing in this way we build up Table 2.1. So at $n = 12$, our estimate for v is $v_{12} = -4\frac{1}{2}/12 = -0.375$, for $\mathbf{x}^*$, it is $\mathbf{x}^*_{12} = (\frac{2}{12},\frac{10}{12})$, and for $\mathbf{y}^*$ it is $\mathbf{y}^*_{12} = (\frac{8}{12},\frac{4}{12})$, whereas the true values (see Section 2.4) is $v = -0.333$ and $\mathbf{x}^* = (\frac{1}{3},\frac{2}{3})$, $\mathbf{y}^* = (\frac{2}{3},\frac{1}{3})$. If you have a spare evening or so, you might like to carry on for another fifty iterations, by which time the approximation will be quite good.

2.12 FURTHER READING

All the results in this chapter, except for the last section, are to be found in von Neumann and Morgenstern's (1947) *Theory of Games and Economic Behaviour*, the first book on the subject. Rarely has the theory of a subject been so completely described in its first publication. However, let me whisper that I find von Neumann's book notationally difficult to follow, and an easier discussion for non-mathematicians is Luce and Raiffa's (1957) comprehensive view of the subject or Williams' (1966) very funny book. More modern mathematical treatments are given by Owen (1968), Vorob'ev (1977), Jones (1980) and Shubik (1982).

We have not considered games with an infinite number of pure strategies, and although Theorem 2.1 does not hold in this case, it does not mean that certain types of infinite games cannot have solution. There has been a lot of work done on finding conditions under which the minimax theorem holds for infinite games (see Parthasarathy and Ragharan, 1971, or Karlin, 1959), as well as application of such games both in games of timing, like the continuous time version of the dual games, and in mathematical economics (see Aubin, 1979).

Table 2.1

Brown's method of solving a Poker game

Game No.	I's Strategy	II's Strategy	Payoff	Total payoff against I's strategies	Total payoff against II's strategies	Total payoff	Number of times played I_1	I_2	II_1	II_2
1	I_2	II_1	−½	(−½,0)	(0,−½)	−½	0	1	1	0
2	I_1	II_1	0	(−½,−1)	(0,−1)	−½	1	1	2	0
3	I_1	II_2	−1	(−½,−2)	(−1,−1)	−1½	2	1	2	1
4	I_2†	II_2	0	(−1,−2)	(−2,−1)	−1½	2	2	2	2
5	I_2	II_2	0	(−1½,−2)	(−3,−1)	−1½	2	3	2	3
6	I_2	II_2	0	(−2,−2)	(−4,−1)	−1½	2	4	2	4
7	I_2	II_1†	−½	(−2½,−2)	(−4,−1½)	−2	2	5	3	4
8	I_2	II_1	−½	(−3,−2)	(−4,−2)	−2½	2	6	4	4
9	I_2	II_1	−½	(−3½,−2)	(−4,−2½)	−3	2	7	5	4
10	I_2	II_1	−½	(−4,−2)	(−4,−3)	−3½	2	8	6	4
11	I_2	II_1	−½	(−4½,−2)	(−4,−3½	−4	2	9	7	4
12	I_2	II_1	−½	(−5,−2)	(−4,−4)	−4½	2	10	8	4

†Indicates where there was a choice of which strategy to take.

Another class of infinite games of considerable interest are differential games introduced by Isaacs (1965). An example of such games is the Princess and the Lion, where both can move within a bounded region and the object is for the Lion to try and catch the Princess, who in turn is trying to avoid capture. The strategies are the possible paths which the Lion and the Princess can follow and which are assumed to be differentiable. The problem of finding conditions under which differential games have solutions is very hard, and involves deep results in functional analysis.

Finally, there has been some work on two-person zero-sum games with partial information. In this, the players do not know all the payoffs of the game, but learn as they play the game. The main work in this area was done by Harsanyi (1967, 1968a, 1968b).

PROBLEMS FOR CHAPTER 2

2.1. In Russian roulette, I and II put 1,000 roubles each in the 'pot'. I then must either put in another 1,000 roubles or spin the cylinder of a six-shot gun which has one bullet in it and fire it at his own head. If he is able, he then passes the gun to II who also must either add another 1,000 roubles to the 'pot' or fire at his own head. If both are still alive after this, they share the 'pot' equally, otherwise the winner takes all. Set this game up in extensive form, i.e. draw the tree graph and solve it.

2.2. A gladiatorial team consisting of a Woman, a Lion and a Cat has to fight one consisting of a Man, a Dog and a Mouse. Each side can put forward one champion to fight and the chance of victory for team I (Woman, Lion, Cat) for each possible combination is

	Man	Dog	Mouse
Woman	0.5	0.6	0.1
Lion	0.6	0.7	0.8
Cat	0.2	0.5	0.9

Write down the game in normal form. Find the upper and lower values of the game and thence its solution.

2.3. Solve the following games which have payoff matrices:

$$\text{(a)} \begin{pmatrix} 3 & 1 \\ 2 & 0 \end{pmatrix}; \quad \text{(b)} \begin{pmatrix} 4 & 0 \\ 2 & 6 \end{pmatrix}; \quad \text{(c)} \begin{pmatrix} 4 & 1 & 3 \\ 0 & 2 & 6 \\ 1 & 7 & 6 \end{pmatrix}.$$

2.4. Fighter Plane versus Bomber Plane. In the pseudo-example of Section 2.4, the payoff matrix is:

		Bomber crew	
		Look up	Look down
Fighter Pilot	Man in the Sun	0.95	1
	Ezak-imak	1	0

Find the solution to the game.

2.5. Three divisions of an army defend a town, which has two roads approaching it. The town is attacked by two divisions of army I, either one along each road or two along the same road. The defenders, army II, can either put all three divisions to guard one road or two to guard one road and one the other. Whichever army has more divisions on a given road will command the road. If there are equal numbers of defenders and attackers on the same road then half the time the defenders will hold the road and half the time the attackers will enter the city. Calculate the payoff matrix of the attackers' chances of entering the town. Hence, find the value of the game and the optimal strategies.

2.6. Two firms I and II make colour and black and white television sets. I can make either 200 colour sets a week or 200 black and white sets and makes a profit of £20 on each colour set and £10 on each black and white. II can make either 400 colour sets a week, or 200 black and white and 200 colour sets, or 400 black and white sets a week. There is a market each week for 200 colour and 400 black and white sets, and the manufacturers share the appropriate market in the proportion that they manufacture that particular set (i.e. if I makes 200 colour and II makes 400 colour, then I sells $\approx 200/600 \times 200 \approx 66\frac{2}{3}$, while II sells $\approx 400/600 \times 200$. Set up the payoff matrix of I's profit per week. Calculate I and II's optimal strategy and the value of the game. Why is this an unrealistic game for II? (For a more realistic game see Problem 3.10.)

2.7. An enterprise has two companies, Fly-By-Night and Shady Dealings, which on average have tax bills of \$4,000,000 and \$12,000,000, respectively, each year. For each company the enterprise can either admit (and pay) their true tax liability or falsify their accounts to show a tax liability of zero. The internal revenue service only has the resources to investigate one of the companies each year. If they investigate a company with false returns, they will discover the fraud and that company will have to pay the required tax plus a penalty of half as much again. Set this up as a 2×4 game where the payoffs are the money that the internal revenue service receives. Show that the enterprise will pay an average of \$14,000,000 to the internal revenue service and find the optimal strategies for both players. Suppose the penalty for fraud is increased to paying the required tax plus twice as much again. Solve this game and describe the optimal strategies.

2.8. I deals II one of three cards—Ace, King or Queen—at random and face down. II looks at the card. If it is an Ace, II must say 'Ace', if a King he can say 'King' or 'Ace', and if a Queen he can say 'Queen' or 'Ace'. If II says 'Ace', I can either believe him and give him £1 or ask him to show his card. If it is an Ace, I must pay II £2, but if it is not II pays I £2. If II says 'King' neither side loses anything, but if he says 'Queen', II must pay I £1. Set this up as a 2×4 game in normal form, and solve it. If when II says 'King', instead of losing nothing he loses £a, what value must a become before it is worth II lying when he has a King?

2.9. Solve the game with payoff matrix

$$\begin{pmatrix} 1 & 4 & 3 & 2 \\ 5 & 2 & 6 & 4 \\ 0 & 2 & 4 & 2 \end{pmatrix}$$

and find all the optimal strategies.

2.10. Show that if $(\mathbf{x}_1,\mathbf{y}_1)$, $(\mathbf{x}_2,\mathbf{y}_2)$ are equilibrium pairs in the same game, then so is $(\mathbf{x}_1,\mathbf{y}_2)$ and $(\mathbf{x}_2,\mathbf{y}_1)$. Show that $e(\mathbf{x}_1,\mathbf{y}_1) = e(\mathbf{x}_1,\mathbf{y}_2)$

2.11. I must decide between 0 and 1 and announce his choice. II also decides on either 0 or 1 but stays quiet. I then chooses again 0 or 1. If the sum of the three numbers is even I gets that amount; if odd, II gets it. Write down the normal form of the game (a 4×4 payoff matrix) and then construct the equivalent linear programme. Use Brown's method to estimate the value and the optimal strategies. (Do about ten iterations starting with everyone choosing 0.)

Chapter 3

Two-person non-zero-sum games

'No enemy can match a friend'—Jonathan Swift

3.1 DIFFERENCES BETWEEN ZERO AND NON-ZERO-SUM GAMES

This chapter concentrates on non-zero-sum games, where it is no longer true that the payoff to I = − payoff to II for all outcomes. We will write the outcomes now as a pair (3,4), where the first component, 3, is I's payoff and the second, 4, is II's payoff. In these games the players are not completely antagonistic to one another, and so might both be happier with one outcome than with another one. It is also true that some of the results for zero-sum games no longer hold, namely:

(a) a maximin pair is not necessarily an equilibrium pair or vice versa,
(b) all equilibrium pairs don't have the same payoffs, and
(c) there is no obvious solution concept for the game.

We look at some famous examples of non-zero-sum games which illustrate these points.

3.2 EXAMPLES

Example 3.1—In the non-zero-sum game, the generalisation of the maximin–minimax pair of strategies for each player is where I plays his maximin strategy when considering his own payoff, and II also plays his maximin strategy when considering his (II's) own payoff. The argument is that, in the non-zero-sum game, II is interested more in maximising what he gets than in minimising what I gets since these are now different. Look at

$$\begin{array}{ccc} & II_1 & II_2 \\ \begin{array}{c} I_1 \\ I_2 \end{array} & \multicolumn{2}{c}{\begin{pmatrix} (2,2) & (3,3) \\ (1,1) & (4,4) \end{pmatrix}} \end{array} . \qquad (3.1)$$

I, in looking for his maximin strategy, thinks only of the game with his payoffs:

$$\begin{array}{cc} & \begin{array}{cc} II_1 & II_2 \end{array} \\ \begin{array}{c} I_1 \\ I_2 \end{array} & \begin{pmatrix} 2 & 3 \\ 1 & 4 \end{pmatrix} \end{array} . \tag{133.2}$$

In this game since II_1 dominates II_2, I would choose I_1 to ensure himself 2. I_1 is his **maximin strategy** and 2 is his **maximin value**. Conversely, II looks at the game with his payoffs:

$$\begin{array}{cc} & \begin{array}{cc} I_1 & I_2 \end{array} \\ \begin{array}{c} II_1 \\ II_2 \end{array} & \begin{pmatrix} 2 & 1 \\ 3 & 4 \end{pmatrix} \end{array} .$$

He would obviously choose II_2, which ensures himself 3 — his *maximin* value. Notice that playing the two *maximin strategies* I_1 and II_2 against one another gives payoffs (3,3), which are different from the maximin values, (2,3). The latter are what the various players can ensure themselves no matter what their opponents do. It does not take into account that the opponent is also trying to maximise his security level.

Turning to equilibrium pairs the only pair of strategies in which it does not pay either person to change his strategy if his opponent keeps to the same strategy is I_2 versus II_2 with payoff (4,4). So the equilibrium pair and the maximin pair are different.

Example 3.2—Battle of the Sexes. A married couple are trying to decide where to go for a night out. She would like to go to the theatre, and he would like to go to a football match — they have been married a few months! However, they are still very much in love and so they only enjoy the entertainment if their partner is with them. If the first strategy for each is to go to the theatre and the second to go to the football match, the payoff is:

$$\begin{array}{ccc} & & \text{Woman} \\ & & \begin{array}{cc} II_1 & II_2 \end{array} \\ \text{Man} & \begin{array}{c} I_1 \\ I_2 \end{array} & \begin{pmatrix} (1,4) & (0,0) \\ (0,0) & (4,1) \end{pmatrix} \end{array} . \tag{3.3}$$

In this case there are three equilibrium pairs: (I_1,II_1) giving a payoff (1,4), (I_2,II_2) giving a payoff (4,1), and a third one consisting of the mixed strategies (1/5, 4/5) versus (4/5, 1/5) which gives a payoff of (4/5, 4/5). It is obvious that the problem with the game is to persuade your opponent to do the same as you. So the man tries to arrive at the equilibrium pair

(I_2,II_2) and the woman at (I_1,II_1). These do not have the same payoff. It is not even obvious that we shall end up in an equilbrium pair, for I may play I_2 as a way of forcing II to play II_2, whereas she might play II_1 from a similar motive, and so they end up with payoffs (0,0). This course of events is reminiscent of the 0. Henry story where the husband sells his watch to buy hair brushes for his wife, who cuts off her hair and sells it to buy a watch chain for her husband.

The maximin–maximin strategies are (4/5, 1/5) for I and (1/5, 4/5) for II and the maximin values are (4/5, 4/5). Again these are not equilibrium strategies. The obvious solution would be to toss a coin and if it came down heads both go to the theatre, and if tails go to the football match, i.e. play (I_1,II_1) with probability ½ and (I_2,II_2) with probability ½. This requires the players to cooperate and we will return to this idea of cooperation in Section 3.7.

Example 3.3—Prisoners' Dilemma. The basic problem was described in example 1.4. I and II are being questioned separately after being caught with stolen goods, and can either confess to theft (I_1,II_1) or keep quiet (I_2,II_2). If both confess they get nine years in prison $(-9,-9)$, whereas if both keep quiet they get one year each $(-1,-1)$. If one confesses and the other does not, the confessor goes free while the other gets ten years in prison. Writing the payoffs, as $-n$ for n years in prison, we get:

$$\begin{array}{c} \\ I_1 \\ I_2 \end{array} \begin{array}{c} \begin{array}{cc} II_1 & \quad II_2 \end{array} \\ \begin{pmatrix} (-9,-9) & (0,-10) \\ (-10,0) & (-1,-1) \end{pmatrix} \end{array} . \tag{3.4}$$

In this example the maximin–maximin pair of strategies and the only equilibrium pair are the same — (I_1,II_1), where both confess. However, this is not a very satisfactory solution to the game since it leads to payoffs $(-9,-9)$, which is worse for both players than $(-1,-1)$, where both players keep quiet. I_1 dominates I_2, II_1 dominates II_2, but most people would regard (I_2,II_2) as the 'best' solution.

This game, originally attributed to A. W. Tucker, has exercised an overwhelming fascination with game theorists and psychologists for years —and rightly so. It encapsulates two of the major dilemmas in conflict situations and also models problems as diverse as nuclear disarmament, wage negotiation, and the controversy of whooping cough vaccinations (see Problem 3.3). The first dilemma is what should be the player's objective—to do what is best for him as an individual or him as part of a group? This conflict is between **individual rationality** which would lead one to confess in Prisoners' Dilemma or **group rationality** which would suggest keeping quiet. Which one is used depends very much on the individual involved, and his previous experiences with other people, including his opponent. This obviously explains psychologists' interest in the game.

The second problem is whether to think of Prisoners' Dilemma as a one-off game or as one that will be played repeatedly. In a one-off game it seems best to confess, because there is no reason to build up your opponent's trust in you; but what should you do when playing the game many times? It can be shown that, if we play the game a fixed number of times, any equilibrium pair of strategies will result in $(\mathrm{I}_1,\mathrm{II}_1)$ being played all the time. The argument goes: think of the last game. Since there will be no more games, both players will choose the confess strategy as in the one-off case. Having decided on what happens in the last game, look at the last but one. Since the last game's strategies are now fixed we can think of this as really the last game, and so on. If the number of games to be played is not known by the players there will be equilibrium pairs that result in the 'keep silent' strategy being played all the time. There have been many experiments to see what happens in practice when these games are played. For further details of the discussion and the experimental results, see the book by Rapoport and Chammah (1965) or the papers in the bibliography of Guyer and Perkel (1972). Players do try to encourage their opponents by 'keeping silent' in the hope they will follow suit, and once such a pattern is established they do sometimes 'punish' opponents who confessed last time by confessing themselves, but the amount this happens varies considerably depending on sex, nationality, temperament, inducements, and the way the experiment was conducted.

3.3 EQUILIBRIUM PAIRS AND MAXIMIN–MAXIMIN PAIRS

Having seen all the difficulties that arise in non-zero-sum games, let us sort out what still carries over from the zero-sum case to such games.

It is trivial that a maximin–maximin pair will always exist for games with finite numbers of pure strategies since we can split the non-zero-sum games into two zero-sum games: one with I's payoffs, the other with II's payoffs. By the von Neumann theorem such games always have a solution, and the optimal strategy for I in the game involving I's payoff is his maximin strategy, whereas its solution is v_{I}, his maximin value. The solution of the game, with II's payoffs, gives II's maximin strategies and his maximin value v_{II}. Remember $(v_{\mathrm{I}},v_{\mathrm{II}})$ need not be the payoff that results from I's maximin strategy playing II's maximin strategy, and rarely is.

Turning to equilibrium pairs, it is the symmetric version (2.45) of Definition 2.3 that extends to non-zero-sum games.

Definition 3.1. A pair of strategies $\mathbf{x}^* \in X$, $\mathbf{y}^* \in Y$ is an **equilibrium pair** for a non-zero-sum game if for any $\mathbf{x} \in X$, $\mathbf{y} \in Y$:

$$e_1(\mathbf{x},\mathbf{y}^*) \leq e_1(\mathbf{x}^*\mathbf{y}^*); \qquad e_2(\mathbf{x}^*,\mathbf{y}) \leq e_2(\mathbf{x}^*,\mathbf{y}^*), \tag{3.5}$$

where $e_1(\ ,\ )$ is I's payoff and $e_2(\ ,\ )$ is II's payoff.

Nash (1951) then proved this generalisation of the von Neumann minimax theorem (Theorem 2.1) appropriate for non-zero-sum games.

Theorem 3.1. Any two-person game (zero-sum or non-zero-sum) with a finite number of pure strategies has at least one equilibrium pair.

By Theorem 2.2, which proved the equivalence of equilibrium pairs and solutions of the game for zero-sum games, this is equivalent to the minimax theorem for these games. In non-zero-sum games, it tells us that all such games have equilibrium pairs as well as maximin–maximin pairs (possibly more than one of each type) but they need not be the same. We shirked proving the minimax theorem, but we will give a sketch of how to prove this generalisation, for those who are interested.

3.4 SKETCH PROOF OF NASH'S THEOREM

As might be expected, the easiest way of proving Theorem 3.1 is to make use of a deep result in mathematical analysis. The one we use is called the Brouwer fixed point theorem or, irreverently, the 'hairy ball' theorem.

Theorem 3.2 (Brouwer fixed point theorem). If a function f maps the points of a closed bounded convex set S onto S, and if f is continuous, then at least one point of S is mapped onto itself (the fixed point).

A set S is **closed** if it contains its boundary line. It is **bounded** if it does not go off to infinity in any direction, and is **convex** if any line joining two points in the set is itself wholly within the set. The function f is **continuous** if it maps points close together into points which are still close together.

Suppose $S = \{x | 0 \leq x \leq 1\}$, then this is closed, bounded and convex, and if we take $f(x) = 1 - x$ this maps all the points of S onto S and is continuous. Since $f(\frac{1}{2}) = \frac{1}{2}$, the point $\frac{1}{2}$ is the fixed point. Another example is to take S as a sphere or ball—your head for example. One way of describing the function f is to draw lines showing where f takes each point (see Fig. 3.1). These can be thought of as hairs on the ball and what the theorem says is that if f is continuous (so points close together are mapped to points close together)), then there must be one point that is mapped onto itself. If you think of your head as the sphere and your hair as describing the function f, which is continuous (provided you don't have a parting), then the fixed point is the crown of your hair, because at that point a hair would stick straight up. Thus, the result got its nickname of the 'hairy ball' theorem.

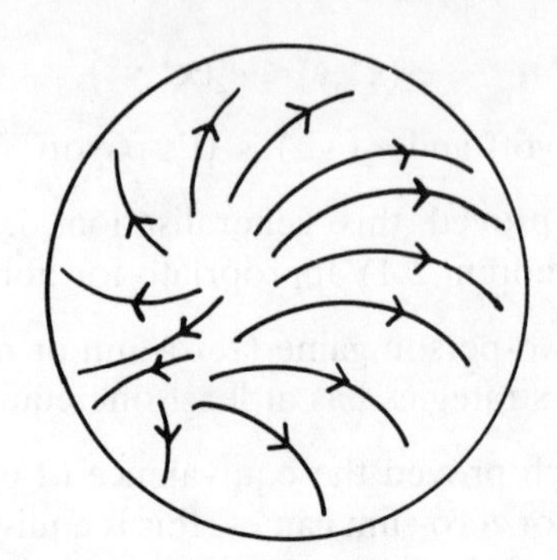

Fig. 3.1—Function f on a sphere.

Returning to the proof of Nash's theorem, let S be the set of all possible pairs of strategies:

$$S = \{(\mathbf{x},\mathbf{y}) \mid \mathbf{x} \in X, \mathbf{y} \in Y\}. \tag{3.6}$$

It is fairly easy to show that S is closed, bounded and convex. Then for any $\mathbf{x} \in X$, $\mathbf{y} \in Y$ define:

$$c_i(\mathbf{x},\mathbf{y}) = \max\{0, e_1(\mathrm{I}_i,\mathbf{y}) - e_1(\mathbf{x},\mathbf{y})\}, \qquad 1 \leqslant i \leqslant n, \tag{3.7}$$

which is the amount I gets extra by playing I_i rather than $\mathbf{x}$ against $\mathbf{y}$. Similarly, define:

$$d_j(\mathbf{x},\mathbf{y}) = \max\{0, e_2(\mathbf{x},\mathrm{II}_j) - e_2(\mathbf{x},\mathbf{y})\}, \qquad 1 \leqslant j \leqslant m, \tag{3.8}$$

which is the amount II gets extra by playing II_j rather than $\mathbf{y}$. Define the function f from S to S by $f(\mathbf{x},\mathbf{y}) = (\mathbf{x}',\mathbf{y}')$, where

$$x_i' = (x_i + c_i(x,y)) \Big/ \Big(1 + \sum_{i=1}^{n} c_i(x.y)\Big), \qquad 1 \leqslant i \leqslant n. \tag{3.9}$$

and

$$y_j' = (y_j + d_j(x,y)) \Big/ \Big(1 + \sum_{j=1}^{m} d_j(x,y)\Big), \qquad 1 \leqslant j \leqslant m. \tag{3.10}$$

f is continuous, since small changes in $\mathbf{x}$ and $\mathbf{y}$ lead to small changes in $\mathbf{x}'$ and $\mathbf{y}'$, and so by Brouwer's fixed point theorem, there is a fixed point $(\mathbf{x}^*,\mathbf{y}^*)$ where

$$f(\mathbf{x}^*,\mathbf{y}^*) = (\mathbf{x}^*,\mathbf{y}^*). \tag{3.11}$$

We cannot have $e_1(\mathrm{I}_i,\mathbf{y}^*) > e_1(\mathbf{x}^*,\mathbf{y}^*)$, for all i, $1 \leqslant i \leqslant$ n, otherwise we have:

$$e_1(\mathbf{x}^*,\mathbf{y}^*) = \sum_{i=1}^{n} x_i^* e_1(\mathrm{I}_i,\mathbf{y}^*) > \sum_{i=1}^{n} x_i^* e_1(\mathbf{x}^*,\mathbf{y}^*) = e_1(\mathbf{x}^*,\mathbf{y}^*). \tag{3.12}$$

Thus, for some i:

$$c_i(\mathbf{x}^*,\mathbf{y}^*) = 0. \tag{3.13}$$

From (3.11) we get:

$$x_i^* = (x_i^* + c_i(\mathbf{x}^*,\mathbf{y}^*))\Big/\Big(1 + \sum_{i=1}^{n} c_i(\mathbf{x}^*,\mathbf{y}^*)\Big), \tag{3.14}$$

and for the i where $c_i\ (\mathbf{x}^*,\mathbf{y}^*) = 0$, (3.14) implies:

$$\sum_{i=1}^{n} c_i\,(\mathbf{x}^*,\mathbf{y}^*) = 0. \tag{3.15}$$

Thus, for all i, $1 \leqslant i \leqslant n$, $c_i(\mathbf{x}^*,\mathbf{y}^*) = 0$, and this gives:

$$e_1(\mathbf{x}^*,\mathbf{y}^*) \geqslant e_1(\mathrm{I}_i,\mathbf{y}^*), \quad \text{for all } 1 \leqslant i \leqslant n. \tag{3.16}$$

and hence:

$$e_1(\mathbf{x}^*,\mathbf{y}^*) \geqslant e_1(\mathbf{x},\mathbf{y}^*), \quad \text{for all } \mathbf{x} \in X. \tag{3.17}$$

Similarly, we can show $e_2(\mathbf{x}^*,\mathbf{y}^*) \geqslant e_2(\mathbf{x}^*,\mathbf{y})$ for all $\mathbf{y} \in Y$ and so $(\mathbf{x}^*,\mathbf{y}^*)$ is an equilibrium pair.

3.5 HOW TO FIND THE EQUILIBRIUM PAIRS (SWASTIKA METHOD)

Although Nash's theorem tells us there is always at least one equilibrium pair in the game, it doesn't tell us how to find it, since solving (3.14) involves solving a set of quadratic equations. In fact it is quite hard to construct algorithms for finding all the equilibrium pairs in a two-person non-zero-sum game, and research is still continuing in this area (see Winkels, 1979). However for a 2×2 game we can use a graphical method for determining all the equilibrium pairs as follows.

Example 3.4

$$\begin{array}{c} \\ \mathrm{I}_1 \\ \mathrm{I}_2 \end{array} \begin{array}{c} \begin{array}{cc} \mathrm{II}_1 & \quad \mathrm{II}_2 \end{array} \\ \begin{pmatrix} (3,2) & \quad (2,1) \\ (0,3) & \quad (4,4) \end{pmatrix} \end{array}. \tag{3.18}$$

Suppose I plays $\mathbf{x}^*$ and II plays $\mathbf{y}^*$, then in an equilibrium pair, we have $e_1(\mathbf{x}^*,\mathbf{y}^*) \geqslant e_1(\mathbf{x},\mathbf{y}^*)$. So, for a particular $\mathbf{y}$ we find the $\mathbf{x}$ that maximises $e_1(\mathbf{x},\mathbf{y})$ and if that $\mathbf{y}$ is part of an equilibrium pair, then we have found the $\mathbf{x}$ that must be its partner in it. In this example if $\mathbf{x} = (x,1 - x)$, and $\mathbf{y} = (y,1 - y)$, then

$$\begin{aligned} e_1(\mathbf{x},\mathbf{y}) &= 3xy + 2x(1 - y) + 0(1 - x)y + 4(1 - x)(1 - y) \\ &= x(5y - 2) + 4 - 4y. \end{aligned} \tag{3.19}$$

If $y < 2/5$, then $e_1(\mathbf{x},\mathbf{y})$ is maximised by $x = 0$. If $y = 2/5$, then $e_1(x,y)$ is maximised by any x, $0 \leqslant x \leqslant 1$. If $y > 2/5$, then $e_1(\mathbf{x},\mathbf{y})$ is maximised for $x = 1$.

We describe this solution in Fig. 3.2.

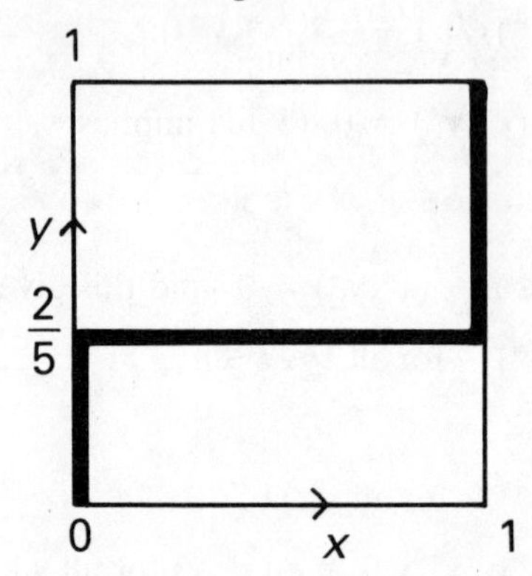

Fig. 3.2—x that maximises $e_1(x,y)$.

In an equilibrium pair $e_2(\mathbf{x}^*,\mathbf{y}^*) \geqslant e_2(\mathbf{x}^*,\mathbf{y})$. So, if for any fixed $\mathbf{x}$ we find the $\mathbf{y}$ that maximises $e_2(\mathbf{x},\mathbf{y})$, then if $\mathbf{x}$ is part of an equilibrium pair, we have found the $\mathbf{y}$ that must be its partner in it. Doing this gives:

$$\begin{aligned} e_2(\mathbf{x},\mathbf{y}) &= 2xy + 1x(1-y) + 3.(1-x)y + 4(1-x)(1-y) \\ &= y(2x-1) + (4-3x). \end{aligned} \tag{3.20}$$

If $x < 1/2$. then $e_2(\mathbf{x},\mathbf{y})$ is maximised by $y = 0$.
If $x = 1/2$, then $e_2(\mathbf{x},\mathbf{y})$ is maximised by any y, $0 \leqslant y \leqslant 1$.
If $x > 1/2$, then $e_2(\mathbf{x},\mathbf{y})$ is maximised by $y = 1$.

If we superimpose this result on that of Fig. 3.2 we get Fig. 3.3. The pairs

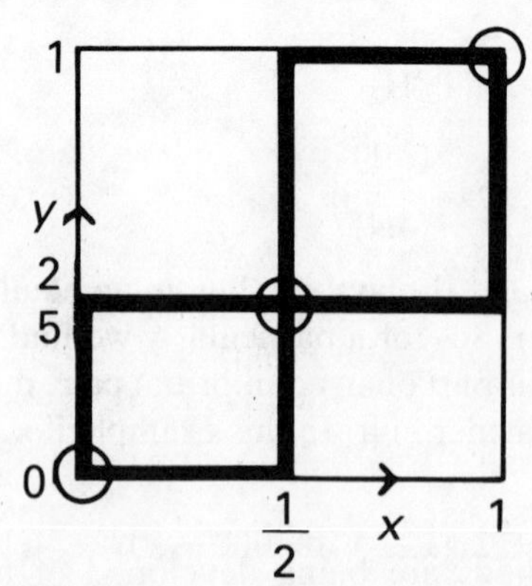

Fig. 3.3—Swastika method for equilibrium pairs.

($\mathbf{x}^* = (x^*,1 - x^*)$, $\mathbf{y}^* = (y^*,1 - y^*)$) on both curves are the equilibrium pairs, since they satisfy $e_1(\mathbf{x}^*,\mathbf{y}^*) \geq e_1(\mathbf{x},\mathbf{y}^*)$ and $e_2(\mathbf{x}^*,\mathbf{y}^*) \geq e_2(\mathbf{x}^*,\mathbf{y})$ These are

(a) $x = 0$, $y = 0$ corresponding to $(\mathrm{I}_2,\mathrm{II}_2)$ with payoff (4,4),
(b) $x = 1$, $y = 1$ corresponding to $(\mathrm{I}_1,\mathrm{II}_1)$ with payoff (3,2), and
(c) $x = \frac{1}{2}$, $y = \frac{2}{5}$ corresponding to $((\frac{1}{2},\frac{1}{2}),(\frac{2}{5},\frac{3}{5}))$ with payoff (2.4,2.5).

Sometimes the final diagram looks like a swastika and on other occasions, as here, one or other of the arms is reversed.

While we are considering this example, let us find the maximin–maximin pairs. Split it into the two games corresponding to I's and II's payoffs. For I's payoff we have:

$$\begin{array}{c c} & \begin{array}{cc} \mathrm{II}_1 & \mathrm{II}_2 \end{array} \\ \begin{array}{c} \mathrm{I}_1 \\ \mathrm{I}_2 \end{array} & \begin{pmatrix} 3 & 2 \\ 0 & 4 \end{pmatrix} \end{array}.$$

There is no domination, so using (2.42) and (2.43) gives:

$$v_{\mathrm{I}} = \frac{3.4 - 0.2}{3 + 4 - 0 - 2} = 2.4 \tag{3.21}$$

and

$$\bar{\mathbf{x}} = \left(\frac{4 - 0}{3 + 4 - 0 - 2}, \frac{3 - 2}{3 + 4 - 0 - 2}\right) = (\tfrac{4}{5}, \tfrac{1}{5}). \tag{3.22}$$

In the game with II's payoff, it is useful to write the matrix so that II's strategies are given by the rows of the matrix. This is because II is the 'maximiser' in this game, and all through Chapter 2, the maximiser was the row player. We are used to looking for domination when the game is in this form. Thus, we get:

$$\begin{array}{c c} & \begin{array}{cc} \mathrm{I}_1 & \mathrm{I}_2 \end{array} \\ \begin{array}{c} \mathrm{II}_1 \\ \mathrm{II}_2 \end{array} & \begin{pmatrix} 2 & 3 \\ 1 & 4 \end{pmatrix} \end{array}. \tag{3.23}$$

Here I_1 dominates I_2, and then II_1 dominates II_2 in the reduced game. Thus, $v_{\mathrm{II}} = 2$, and the maximin strategy is II_1. So the maximin values are (2.4,2) and the maximin–maximin pair is $\bar{\mathbf{x}} = (\frac{4}{5},\frac{1}{5})$, $\bar{\mathbf{y}} = (1,0)$.

3.6 SOLUTION CONCEPTS OF NON-ZERO-SUM GAMES

Thus, we can find the equilibrium pairs and maximin–maximin pairs for 2×2 games, and methods are being developed of finding them for all games. However, the 'maximin–maximin' idea of a solution no longer makes much sense, because it assumes a player is both trying to minimise his opponent's payoff and maximise his own payoff. These two objectives could be diametrically opposed, as the payoff matrix of Example 3.1 showed. Thus, equilibrium pairs are considered the more acceptable solution concept, but the difficulty with them is that, usually, there are far too many of them in a game. Several authors have suggested choosing, as a solution to the game, a subset of the equilibrium pairs with particular properties. We shall briefly describe three of these suggestions and point out the difficulties with them.

3.6.1 Solutions in the Nash Sense

Definition 3.2. A game has a solution in the **Nash sense** if every pair of equilibrium pairs is **interchangeable**, i.e. if $(\mathbf{x}_1,\mathbf{y}_1)$, $(\mathbf{x}_2,\mathbf{y}_2)$ are equilibrium pairs so are $(\mathbf{x}_1,\mathbf{y}_2)$ and $(\mathbf{x}_2,\mathbf{y}_1)$. The solution is then the set of equilibrium pairs.

Note that Problem 2.10 shows that all zero-sum games are Nash solvable. Looking at the examples of Section 3.2, Example 3.1 and Prisoners' Dilemma only have one equilibrium pair each (I_2,II_2) and (I_1,II_1), respectively. Thus, they both have solutions in the Nash sense, since they satisfy Definition 3.2 trivially. However, in the Battle of the Sexes (Example 3.2) (I_1,II_1) and (I_2,II_2) are both equilibrium pairs—see Problem 3.1 to find them all—but (I_1,II_2) is certainly not one because it would pay I to change to I_2. Thus, these are not interchangeable and so the game is not solvable in the Nash sense.

So, although it would be nice for equilibrium pairs to have this property, very few do.

3.6.2 Solutions in the strict sense

First two definitions.

Definition 3.3. $(\mathbf{x}_2,\mathbf{y}_2)$ **dominates** $(\mathbf{x}_1,\mathbf{y}_1)$ if $e_1(\mathbf{x}_2,\mathbf{y}_2) \geqslant e_1(\mathbf{x}_1,\mathbf{y}_1)$ and $e_2(\mathbf{x}_2,\mathbf{y}_2) \geqslant e_2(\mathbf{x}_1,\mathbf{y}_1)$ with strict inequality in at least one case.

Definition 3.4. A pair of strategies $(\mathbf{x},\mathbf{y})$ is **Pareto optimal** if it is not dominated. Then we can define the idea of strict solutions of a game by:

Definition 3.5. A game has a solution in the **strict sense** if:

(1) there is an equilibrium pair among the Pareto optimal pairs, and

(2) all Pareto optimal equilibrium pairs are **interchangeable** and have the same payoffs. The solution is then the set of Pareto optimal equilibrium pairs.

In zero-sum games $e_1(\mathbf{x},\mathbf{y}) = -e_2(\mathbf{x},\mathbf{y})$ so no pair of strategies can be dominated. It follows from Lemma 2.3 and Problem 2.10 that all the equilibrium pairs in such games are interchangeable and have the same payoffs, and so zero-sum games always have strict solutions.

Turning to the examples in Section 3.2, Example 3.1 has a solution in the strict sense because (I_2,II_2), the only equilibrium pair, is Pareto optimal. However, in the Battle of the Sexes both (I_1,II_1) and (I_2,II_2) are Pareto optimal equilibrium pairs but they are not interchangeable. So there is no solution in the strict sense. Prisoners' Dilemma also has no solution in the strict sense, but this time because the only equilibrium pair (I_1,II_1) with payoffs $(-9,-9)$ is dominated by (I_2,II_2) with payoffs $(-1,-1)$.

3.6.3 Solutions in the completely weak sense

The idea here is to use domination of strategies to weaken the previous definition sufficiently so that games like Prisoners' Dilemma have a solution. If the strategy sets for the players are X and Y, respectively, we reduce the game by removing dominated strategies as in Section 2.6. First we remove I's strategies which are dominated with respect to I's payoffs when played against Y, leaving X^1, then II's strategies which are dominated with respect to II's payoffs when played against X^1, leaving Y^1. Repeating this procedure we get X^2, Y^2, X^3, Y^3, ... until we cannot remove any more strategies. The resultant strategies set we denote by X^r, Y^r.

Definition 3.6. A game with strategy sets X, Y is **solvable in the completely weak sense** if the reduced game, with strategy sets X^r, Y^r is solvable in the strict sense. The solution set is the strict solution set of the reduced game.

The zero-sum game is solvable in the completely weak sense because, by Lemma 2.1, removing dominated strategies does not change the solution of the game.

In Example 3.1, the reduced strategy sets are I_2 and II_2, respectively, giving a 1×1 game with payoffs (4,4). Thus, this is the solution in the completely weak sense. It was also the solution in the strict sense. For the Battle of the Sexes there is no domination so $X^r = X$, $Y^r = Y$, and since there was no solution in the strict sense, there is none in the completely weak sense. Howver, for Prisoners' Dilemma the reduced set of strategies are $X^r = \{I_1\}$, $Y^r = \{II_1\}$ and so the game is the 1×1 one $(-9,-9)$, which obviously has a solution in the strict sense. Thus, $\{I_1,II_1\}$ is the solution of Prisoners' Dilemma in the completely weak sense.

These three suggestions for a solution of the game have the disadvantage that not all games have such a solution. Other suggestions like the perfect equilibria of Selten (1975) or the proper equilibria of Myerson (1978) exist in all games, but often do not reduce the number of equilibrium pairs by much.

3.7 COOPERATIVE GAMES

In the Prisoners' Dilemma, it is obvious that if both players can cooperate on the 'keep silent, keep silent' strategy they will do better than the 'confess, confess' solution of the non-cooperative game. Again when discussing the Battle of the Sexes, we suggested that tossing a coin to decide whether both go to the theatre or the football match was the best thing to do, but that this required the players to cooperate. We think of cooperation as discussing the game before hand with the other player, and then deciding on a joint strategy for both players. The players agree that player I will play $\mathbf{x} \in X$, say, and II will play $\mathbf{y} \in Y$ and it is assumed that this agreement will be binding on both players. Obviously this assumption of enforceable agreements is often not realistic, and indeed the troubles that arise in industrial relations or in international treaties occur because there is no way one player can be forced to keep his word. However, that doesn't stop us looking at the interesting and important question of what are the strategies the players are likely to decide on when they are cooperating. The possibility that the agreement might be broken should not prevent one seeking the most acceptable agreement.

Recall the Battle of the Sexes (Example 3.2) where the payoff matrix was:

$$\begin{array}{c} \\ \mathrm{I}_1 \\ \mathrm{I}_2 \end{array} \begin{array}{c} \begin{array}{cc} \mathrm{II}_1 & \mathrm{II}_2 \end{array} \\ \begin{pmatrix} (1,4) & (0,0) \\ (0,0) & (4,1) \end{pmatrix} \end{array} . \tag{3.24}$$

If I plays $\mathbf{x} = (x, 1 - x)$ and II plays $\mathbf{y} = (y, 1 - y)$, then the payoffs are:

$$e_1(\mathbf{x},\mathbf{y}) = 5xy + 4 - 4x - 4y, \qquad e_2(\mathbf{x},\mathbf{y}) = 5xy + 1 - x - y. \tag{3.25}$$

In the non-cooperative game the possible payoffs are given by (3.25) as x and y vary from 0 to 1. This is given by Fig. 3.4.

If the players cooperate, then the suggestion is to toss a coin and either both go to the theatre or both go to the football. This corresponds to playing $(\mathrm{I}_1,\mathrm{II}_1)$ or $(\mathrm{I}_2,\mathrm{II}_2)$ each with probability ½ and gives an expected payoff of $\frac{1}{2}(1,4) + \frac{1}{2}(4,1) = (2\frac{1}{2},2\frac{1}{2})$. This is not in the region of non-cooperative payoffs.

For cooperative games, since **jointly randomised strategies** are allowed, the payoff region will be larger than that of the corresponding non-cooperative game. If (u_1,v_1), (u_2,v_2) are payoffs in the non-cooperative game, then in the cooperative game, by playing the joint strategy which chooses the strategies that give (u_1,v_1) with probability α, and those that give (u_2,v_2) with probability $1 - \alpha$, we get the expected payoff $\alpha(u_1,v_1) + (1 - \alpha)\ (u_2,v_2)$ Thus, the payoff region R for the

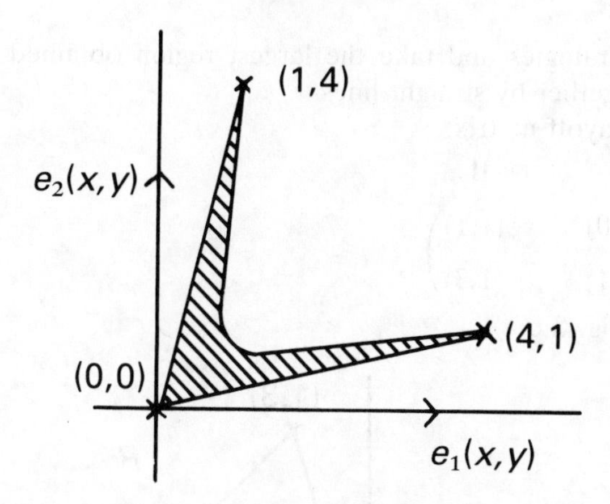

Fig. 3.4—Payoff region of 'Battle of the Sexes'.

cooperative game is the convex closure of the payoff region of the non-cooperative game, i.e. the smallest region that contains all lines which start and end in the non-cooperative region (see Fig. 3.5).

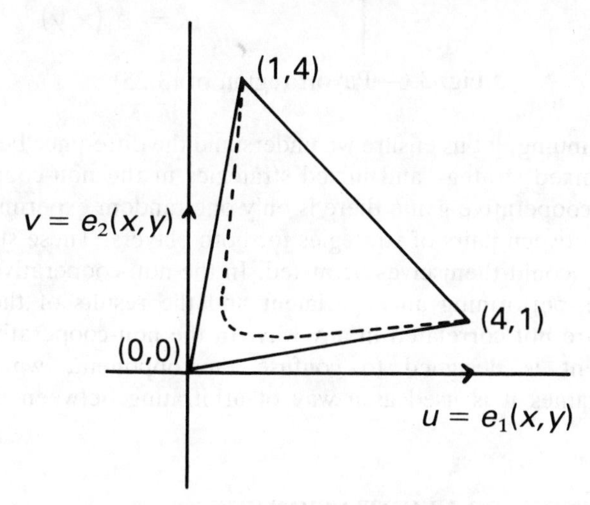

Fig. 3.5—Cooperative payoff region of 'Battle of the Sexes'.

Notice in Fig. 3.5 that the vertices (corners) of the region are the payoffs of the pure strategies. It is always the case that the vertices of R are the payoffs of pure strategies playing one another. However, not all the payoffs of pure strategy versus pure strategy need be vertices of R. This gives a quick and easy way of finding R, the payoff region of the cooperative game. Just plot the payoffs corresponding to both players playing pure strategies and take the largest region obtained by joining these points together by straight lines.

So, for the payoff matrix:

$$\begin{array}{cc} & \begin{array}{cc} \mathrm{II}_1 & \mathrm{II}_2 \end{array} \\ \begin{array}{c} \mathrm{I}_1 \\ \mathrm{I}_2 \end{array} & \begin{pmatrix} (0,0) & (1,1) \\ (3,1) & (1,3) \end{pmatrix} \end{array}, \tag{3.25}$$

R is given by Fig. 3.6.

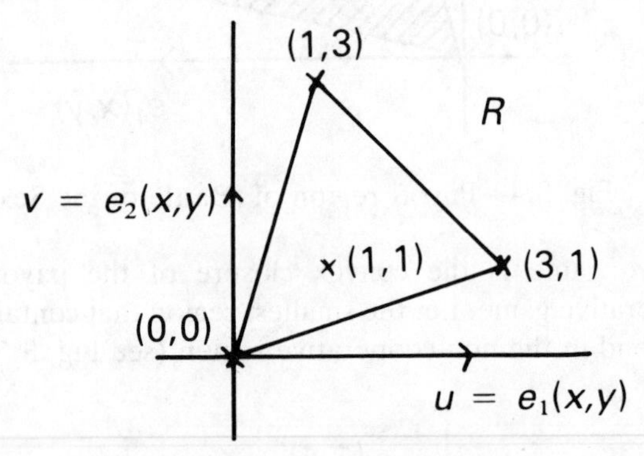

Fig. 3.6—Payoff region of (3.25).

Before continuing, let us ensure we understand the difference between a cooperative mixed strategy and mixed strategies in the non-cooperative game. In the cooperative game there is only one random experiment and this chooses between pairs of strategies for both players. These strategies for the players could themselves be mixed. In the non-cooperative game, each player is performing an experiment and the results of these two experiments are not correlated in any way. In the non-cooperative case, the experiment is designed to confuse the opponent, whereas in cooperative games it is used as a way of arbitrating between different outcomes.

3.8 BARGAINING OR NEGOTIATION SET

As soon as we allow the two players to cooperate, the problem shifts from the actual playing of the game to the preplay negotiation. Which of the possible payoffs in the cooperative game are the players likely to agree on?

To answer this we recall the definition of domination (Definition 3.3) of pairs of strategies, but write it in terms of the payoffs of these strategies.

Definition 3.3′. A pair of payoffs (u,v) in a cooperative game is **jointly dominated by** (u',v') if $u' \geqslant u$, $v' \geqslant v$ and $(u',v') \neq (u,v)$.

Definition 3.4′. A pair of payoffs (u,v) is **Pareto optimal** if it is not jointly dominated.

Clearly, the players in the game should only be interested in Pareto optimal payoffs since otherwise they can both do as well and one can do better than the present payoffs. Moreover, player I can always ensure himself a payoff

$$v_{\mathrm{I}} = \max_{x \in X} \min_{y \in Y} e_1(\mathbf{x},\mathbf{y})$$

just by not cooperating with II, while II can similarly ensure himself

$$v_{\mathrm{II}} = \max_{y \in Y} \min_{x \in X} e_2(\mathbf{x},\mathbf{y}).$$

Thus, whatever the outcome of the bargaining session, I ought to receive at least v_{I} and II, v_{II}, otherwise they can do better by ignoring their opponent and playing their maximin strategy.

Von Neumann and Morgernstern (1944) suggested that any solution of a cooperative game must be part of the **bargaining** set B (sometimes called the **negotiation** set), where

$$B = \{(u,v) \mid u \geqslant v_{\mathrm{I}}, \geqslant v_{\mathrm{II}}, (u,v) \text{ Pareto optimal in } R\}. \tag{3.26}$$

One of the difficulties with bargaining problems is the assumptions that the payoffs are the players' utilities. Thus, player I is happier with (4,1) than (1,10) and player II less happy. The temptation is to think of the two players' utilities on the same scale, so that a change of +9 for II more than compensates for one of −3 for I. Since utility functions are not defined uniquely and the scale can be varied, we must resist this tendency to compare different people's utilities. If we do succumb to interpersonal comparisons we will warn the reader.

The choice still remains which point of B is likely to be decided on, and in the next three sections we examine this problem.

3.9 NASH'S BARGAINING AXIOMS

Nash (1950a) considered a slightly more general bargaining problem than which point of the bargaining set should be chosen. He said, suppose there is a transaction between two people in which the resulting feasible payoffs to the players form a closed bounded convex set P. (The payoff region of a finite cooperative game is always closed, bounded and convex.) There is also a special payoff $(u_0,v_0) \in P$ called the **status quo** point, which is the outcome the participants will receive if they can't agree on the transaction. Nash then listed some axioms, which he felt an independent arbitrator would follow in arriving at a satisfactory arbitration solution (u^*,v^*).

The idea is that if the participants agree to the axioms as general principles then they can apply the arbitration procedure that satisfies the axioms in all situations without recourse to a real arbitrator. We define an arbitration procedure, ψ, as a map from (u_0,v_0) and P to some point (u^*,v^*) in P, i.e.

$$\psi((u_0,v_0),P) = (u^*,v^*). \tag{3.27}$$

The axioms Nash suggested for such arbitration procedures, ψ, are the following.

N1 (as least as good as status quo): $u^* \geqslant u_0$, $v^* \geqslant v_0$.
N2 (feasibility): $(u^*,v^*) \in P$.
N3 (Pareto optimality): If $(u,v) \in P$, and $u \geqslant u^*$, $v \geqslant v^*$, then $u = u^*$, $v = v^*$.
N4 (Independence of irrelevant alternatives): If P_1 is a subset of P_2 and if (u^*,v^*) is the solution of the procedure for $((u_0,v_0),P_2)$, then if $(u^*,v^*) \in P_1$ it must also be the solution of the procedure for $((u_0,v_0),P_1)$.
N5 (Invariance under linear transformations): Let P' be obtained from P by the transformation

$$u' = au + b, \qquad v' = cv + d, \qquad a,c > 0, \tag{3.28}$$

then if (u^*,v^*) is the solution of the procedure for $((u_0,v_0),P)$, $(au^* + b, cv^* + d)$ is the solution of the procedure for $((av_0 + b, cv_0 + d),P')$
N6 (Symmetry): If P is symmetric, so that if $(u,v) \in P$, then $(v,u) \in P$, and if $u_0 = v_0$, then $u^* = v^*$.

The first three axioms appear quite reasonable, since the solution must be feasible and if there was another payoff that was as good for both players and strictly better for one, it would be preferable. Also, if the players don't get their status quo values then they won't accept the solution as they are better off if they don't trade. The axiom of irrelevant alternatives (N4) is the one that is most criticised, both in this context and in that of social choice theory (see Sen, 1970).

As an example look at Fig. 3.7. Looking at the payoff region P_2 with status quo point (0,0), then (5,50) seems a fair solution—in the next section

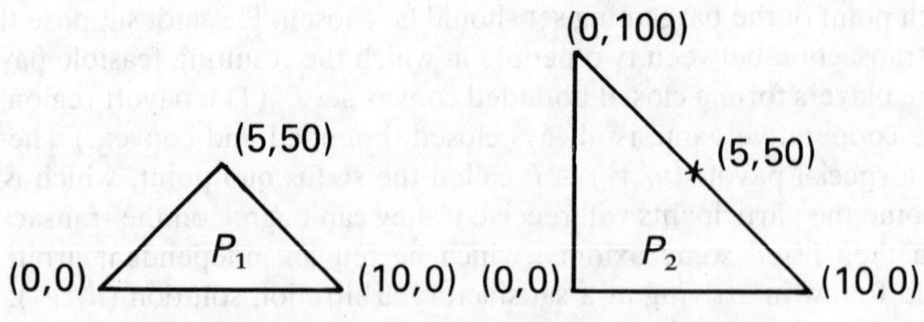

Fig. 3.7—Example of irrelevant alternatives.

we will show it is the arbitration solution. The irrelevant alternative axiom says that if we cut the set of feasible payoffs down to P_1, (5,50) must still be the arbitration solution. This seems very generous to the second player, who now gets his maximum possible payoff, whereas the first only gets half his. It still has merit in that it demands a sort of consistency, but leads to many strange results.

The linear transformation axiom (N5) is readily acceptable, since utilities are only defined uniquely up to linear transformation, and we are forbidding interpersonal comparison of utilities. In monetary terms it says we would make the same transaction whether the price is in pounds or dollars.

The final axiom on symmetry assumes, without stating it, that the two players are roughly of the same size in power and skill in diplomacy. So the only times one might question it, is when one player is on his own while the other might be the rest of the community, or a government or a multi-national company.

Having settled on these axioms, Nash was able to show that there is an arbitration procedure ψ that satisfies these axioms, and it is unique.

Theorem 3.3 (Nash). Consider the following arbitration procedure ψ.

(i) If there exists points $(u,v) \in P$ with $u > u_0$, $v > v_0$, then look at the function

$$f(u,v) = (u = u_0)\,(v - v_0) \tag{3.29}$$

defined over $(u,v) \in P$, $u > u_0$, $v > v_0$. Its maximum occurs at a unique point (u^*,v^*), and define:

$$\psi((u_0,v_0),P) = (u^*,v^*) \text{ in this case.} \tag{3.30}$$

(ii) If there are no points $(u,v) \in P$ with $u > u_0$ and $v > v_0$, then either (a) there are points of the form (u,v_0), $u \geqslant u_0$, or (b) (u_0,v), $v \geqslant v_0$. If u^* is the maximum u-value of points of the first type, and v^* is the maximum v-value of points of the second type, then in (iia) define:

$$\psi((u_0,v_0),P) = (u^*,v_0), \tag{3.31}$$

and in (iib):

$$\psi((u_0,v_0,P) = (u_0,v^*). \tag{3.32}$$

The arbitration function ψ defined by (3.30), (3.31) and (3.32) satisfies N1 to N6 and is the only function to do so.

Expression (3.30) is the important part of the definition of ψ, since it is obvious what to do in cases (iia) and (iib). Notice that since $(u_0,v_0) \in P$, there will always be points satisfying (iia) so one of the three cases holds, and if (iia) and (iib) are both to hold with points (u,v_0), $u > u_0$ and (v_0,v), v

$> v_0$, then by convexity (i) holds. In most theorems it is the existence of a function with certain properties which is the difficult part and the uniqueness is quite easy. Here it is the other way around, but first we must ensure that the definition of ψ in the theorem makes sense, namely that f in (3.29) always has a maximum and it is unique.

Lemma 3.1. If there exist points $(u,v) \in P$ with $u > u_0$, $v > v_0$, then f defined by (3.29) always has a maximum over $(u,v) \in P$, $u > u_0$, $v > v_0$ and the maximum is unique.

Proof. P is a closed bounded set and if we take the intersection with the region $u \geqslant u_0$, $v \geqslant v_0$, which is also closed, that is also closed and bounded. It is a standard theorem in analysis that a continuous function, like f, defined on a closed and bounded set, has a maximum. Moreover, this maximum won't be on the lines (u,v_0) or (u_0,v) since f is zero here and is positive elsewhere in the region. So there is a maximum for f defined on $(u,v) \in P$, $u > u_0$, $v > v_0$.

To prove uniqueness of the maximum suppose there are two points (u_1,v_1), (u_2,v_2) where f takes its maximum value, M say. Since $M > 0$, if $u_1 = u_2$, then $v_1 = v_2$, so we can assume $u_1 < u_2$ and hence $v_1 > v_2$ ($u_1 > u_2$, $v_1 < v_2$ is proved similarly). Since P is convex it contains the point:

$$(u_3,v_3) = \tfrac{1}{2}(u_1,v_1) + \tfrac{1}{2}(u_2,v_2). \tag{3.33}$$

Now

$$\begin{aligned} f(u_3,v_3) &= (\tfrac{1}{2}(u_1+u_2) - u_0)\,(\tfrac{1}{2}(v_1+v_2) - v_0) \\ &= \frac{(u_1-u_0)+(u_2-u_0)}{2} \cdot \frac{(v_1-v_0)+(v_2-v_0)}{2} \\ &= \frac{(u_1-u_0)(v_1-v_0)}{2} + \frac{(u_2-u_0)(v_2-v_0)}{2} \\ &\quad + \frac{(u_2-u_1)(v_1-v_2)}{4} \\ &= \tfrac{1}{2}f(u_1,v_1) + \tfrac{1}{2}f(u_2,v_2) + \frac{(u_2-u_1)(v_1-v_2)}{4} \\ &= M + \frac{(u_2-u_1)(v_1-v_2)}{4} > M, \end{aligned} \tag{3.34}$$

where the last inequality holds because $u_1 < u_2$, $v_1 > v_2$. This contradicts (u_1,v_1), (u_2,v_2) being maximum points unless $u_1 = u_2$ and $v_1 = v_2$, and so the uniqueness is proved.

We now prove the rest of Theorem 3.3.

Proof of Theorem 3.3. First we show ψ defined by (3.30)–(3.32) satisfies Nash's six axioms. Consider first case (i) where (u^*,v^*) is given by (3.30). It follows trivially from the definition of the region we maximise f over, that N1 and N2 hold.

For N3 if $u \geqslant u^* \geqslant u_0$ and $v \geqslant v^* \geqslant v_0$, then:

$$f(u,v) = (u - u_0)(v - v_0) \geqslant (u^* - u_0)(v^* - v_0) = f(u^*,v^*). \quad (3.35)$$

So as (u^*,v^*) maximises $f(u,v)$ this implies $f(u,v) = f(u^*,v^*)$ and thus $u = u^*$, $v = v^*$.

If $P_1 \subset P_2$ and (u^*,v^*) maximises $f(u,v)$ over P_2 it is bound to maximise it over the smaller set P_1, and so N4 holds. N5 goes as follows. If (u^*,v^*) maximises $f(u,v)$ over P, then it must also maximise $ac(u - u_0)(v - v_0)$ over P since $ac > 0$. Thus, (u^*,v^*) maximises $(au + b - (au_0 + b))((cv + d) - (cv_0 + d))$ over P or, alternatively, $(au^* + b, cv^* + d)$ maximises $(u - u'_0)(v - v'_0)$ over P', where $u'_0 = au_0 + b$, $v'_0 = cv_0 + d$.

Finally, for N6, suppose (u^*,v^*) maximises $f(u,v)$ but $u^* \neq v^*$. By the symmetry property of P, (v^*,u^*) is in P, and by convexity so is $\frac{1}{2}(u^*,v^*) + \frac{1}{2}(v^*,u^*) = (u^* + v^*/2, u^* + v^*/2)$. Then

$$f\left(\frac{u^* + v^*}{2}, \frac{u^* + v^*}{2}\right) = \left(\frac{u^* + v^*}{2} - u_0\right)\left(\frac{u^* + v^*}{2} - v_0\right)$$
$$= \frac{u^{*2} + 2u^*v^* + v^{*2}}{4} - (u^* + v^*)u_0 + u_0^2, \quad (3.36)$$

where we have used $u_0 = v_0$. Since $(u^* - v^*)^2 \geqslant 0$ we know $u^{*2} + v^{*2} > 2u^*v^*$ unless $u^* = v^*$ and using this in (3.36) gives:

$$f((u^* + v^*)/2, (u^* + v^*)/2) > u^*v^* - (u^* + v^*)u_0 + u_0^2$$
$$= (u^* - u_0)(v^* - v_0) = f(u^*,v^*). \quad (3.37)$$

This contradicts (u^*,v^*) maximising f, and so $u^* = v^*$.

To prove the uniqueness of the procedure ψ, we assume there is some other procedure $\bar{\psi}$ which also satisfies N1 to N6. In order to be different from ψ there must be some P, (u_0,v_0) where $\psi((u_0,v_0), P) = (u^*,v^*)$ and $\bar{\psi}((u_0,v_0),P) = (\tilde{u},\tilde{v})$ and $(u^*,v^*) \neq (\tilde{u},\tilde{v})$. By using only axioms N1 to N6 we will show $(u^*,v^*) = (\tilde{u},\tilde{v})$, hence proving the uniqueness of ψ.

By N5 if we transform P into the region P' using the transformation:

$$u' = \frac{u - u_0}{u^* - u_0}, \quad v' = \frac{v - v_0}{v^* - v_0}, \quad (3.38)$$

(u_0,v_0) is mapped into (0,0), (u^*,v^*) into (1,1) and $(\tilde{u},\tilde{v})$ into some other point $(\bar{u},\bar{v})$. So by definition of ψ, (1,1) maximises uv over $(u,v) \varepsilon P'$, $u \geqslant 0$, $v \geqslant 0$.

This means that all points $(u,v) \varepsilon P'$ $u \geqslant 0$, $v \geqslant 0$ satisfy $u + v \leqslant 2$, because if there was a point $(1 + x, 1 - x + \varepsilon)$ in P' with $\varepsilon > 0$, then look at the points on the line joining this to (1,1) which, since P' is convex, must be in P'. They are of the form $(1 - \lambda)(1,1) + \lambda(1 + x, 1 - x + \varepsilon)$ for $0 < \lambda < 1$, and evaluating f for these points gives:

$$(1 + \lambda x)(1 + \lambda(\varepsilon - x)) = 1 + \lambda\varepsilon + \lambda^2 x(\varepsilon - x). \tag{3.39}$$

Taking λ sufficiently close to 0, we can ensure $\lambda\varepsilon + \lambda^2 x(\varepsilon - x) > 0$ since $\varepsilon > 0$ and this contradicts the fact that (1,1) maximises uv.

Now look at P_2 the symmetric closure of P', i.e. if (u',v') is in P', then (u',v') and (v',u') are in P_2. It is still true that for all points in P_2, $u + v \leqslant 2$ and so the maximum u for points (u,u) in P_2 is $u = 1$. Any solution procedure satisfying N3 and N6 on $((0,0),P_2)$ must choose (1,1) as its solution, since P_2 is symmetric. Moreover, any solution procedure satisfying N4 will still choose (1,1) for $(P', (0,0))$. Hence, if ψ satisfies N1 to N6 $(\bar{u},\bar{v}) = (1,1)$ and using the transformation (3.37) this gives $(\tilde{u},\tilde{v}) = (u^*,v^*)$. So ψ is unique.

We still have to prove the theorem in case (ii) given by (3.31) and (3.32), but in this case only the procedures defined by (3.31) and (3.32) can be Pareto optimal, feasible and as good as the status quo and so satisfy N1, N2 and N3. Showing ψ also satisfies N4, N5 and N6 in these cases is trivial.

We can now apply this arbitration procedure to choose a point in the negotiation set of cooperative games.

3.10 MAXIMIN BARGAINING SOLUTION

Obviously since the feasible region of payoffs of a cooperative game is convex, we can apply Nash's bargaining model to it, once we have agreed on a status quo point. One obvious suggestion is $u_0 = v_{\mathrm{I}}$, $v_0 = v_{\mathrm{II}}$, which are I and II's maximin values. Notice that it is the maximin values we take, which are what I and II can guarantee themselves, not the payoffs of playing the maximin strategies. The outcome of Nash's bargaining procedure is then called the **maximin bargaining solution**, for obvious reasons. It is also known as the **Shapley solution** because it is related to the Shapley value, which is a solution concept used in n-person games (see Chapter 4, Section 4.8). We will see how to find these bargaining solutions by solving some examples.

Examples 3.4—Consider the matrix:

$$\begin{array}{cc} & \begin{array}{cc} II_1 & II_2 \end{array} \\ \begin{array}{c} I_1 \\ I_2 \end{array} & \begin{pmatrix} (1,2) & (8,3) \\ (4,4) & (2,1) \end{pmatrix} \end{array}. \tag{3.40}$$

First, by looking at the zero-sum games consisting of I and II's payoff, respectively, we can calculate v_I, v_{II}. The matrices of these zero sum games are:

$$\begin{array}{cc} & \begin{array}{cc} II_1 & II_2 \end{array} \\ \begin{array}{c} I_1 \\ I_2 \end{array} & \begin{pmatrix} 1 & 8 \\ 4 & 2 \end{pmatrix} \end{array}, \qquad \begin{array}{cc} & \begin{array}{cc} I_1 & I_2 \end{array} \\ \begin{array}{c} II_1 \\ II_2 \end{array} & \begin{pmatrix} 2 & 4 \\ 3 & 1 \end{pmatrix} \end{array}. \tag{3.41}$$

There is no domination in either matrix and so by (2.42), $v_I = 3\frac{1}{3}$ and $v_{II} = 2\frac{1}{2}$. Now, by drawing a diagram of the payoff region for the cooperative game, we can obtain the negotiation set (Fig. 3.8), where we

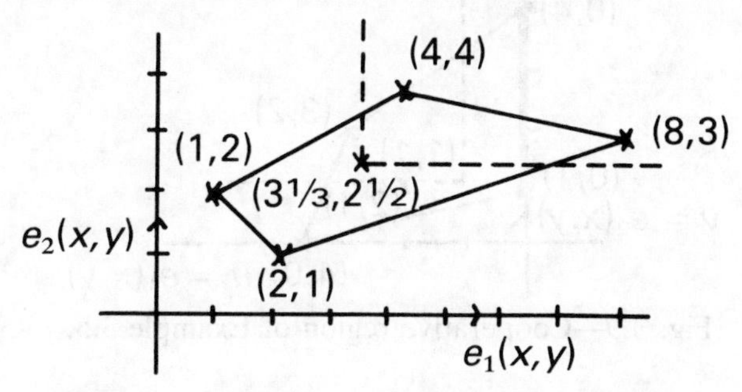

Fig. 3.8—Cooperative region of Example 3.4.

see that the negotiation set is $\{(u,v) \mid u + 4v = 20, 4 \leqslant u \leqslant 8\}$. From the first three of Nash's axioms we know the arbitration solution must be in the negotiation set. So we want to maximise $(u - 3\frac{1}{3})(v - 2\frac{1}{2})$ for (u,v) in this set. Since these (u,v) satisfy $u + 4v = 20$, we can also rewrite this as maximise $(u - 3\frac{1}{3})(5 - \frac{1}{4}u - 2\frac{1}{2})$, for $4 \leqslant u \leqslant 8$. Finding where the derivative is 0 gives:

$$\frac{d}{du}(u - 3\tfrac{1}{3})(5 - \tfrac{1}{4}u - 2\tfrac{1}{2}) = 5 - \tfrac{1}{4}u - 2\tfrac{1}{2} - \tfrac{1}{4}(u - 3\tfrac{1}{3}) = 0,$$

therefore

$$\tfrac{1}{2}u = 5 - 2\tfrac{1}{2} + \tfrac{10}{12} = 3\tfrac{1}{3}, \tag{3.42}$$

therefore

$$u = 6\tfrac{2}{3}.$$

It is trivial that the turning point, if it exists, must be a maximum since we are looking at a function of the form $(u - 3\frac{1}{3})(2\frac{1}{2} - \frac{1}{4}u) = -\frac{1}{4}u^2 + 3\frac{1}{3}u - 8\frac{1}{3}$ which only has one turning point that is a maximum. This holds no matter what the specific values involved in the analysis. So $(6\frac{2}{3}, 3\frac{1}{3})$ is the unique solution to Nash's procedure.

Example 3.5 — This game has a negotiation set which consists of two piecewise linear parts. The payoff matrix is:

$$\begin{array}{cccc} & \mathrm{II}_1 & \mathrm{II}_2 & \mathrm{II}_3 \\ \mathrm{I}_1 & \multicolumn{3}{c}{\begin{pmatrix} (2,1) & (3,2) & (0,4) \\ (0,1) & (4,0) & (2,1) \end{pmatrix}} \\ \mathrm{I}_2 & & & \end{array}, \tag{3.43}$$

and the maximin values are $v_{\mathrm{I}} = 1$, $v_{\mathrm{II}} = 1$. By drawing the payoff region for the cooperative game (Fig. 3.9), we see the negotiation set is those

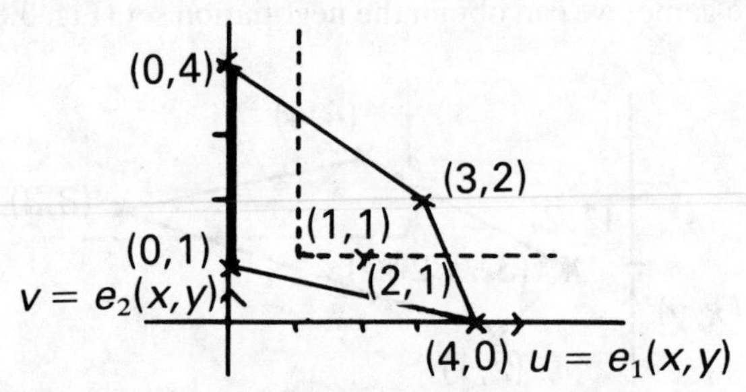

Fig. 3.9—Cooperative region of Example 3.5.

points on the lines joining (4,0) to (3,2) and (0,4) to (3,2) which also satisfy $u \geqslant v_{\mathrm{I}} = 1$ and $v \leqslant v_{\mathrm{II}} = 1$. So the maximin bargaining solution must be in the set:

$$\{(u,v)|2u + 3v = 12, 1 \leqslant u \leqslant 3\} \cup \{(u,v)|2u + v = 8, 3 \leqslant u \leqslant 3\tfrac{1}{2}\}. \tag{3.44}$$

First we find which (u,v) maximises $(u-1)(v-1)$ on $2u + 3v = 12$, $1 \leqslant u \leqslant 3$. This is the same as maximising $(u - 1)(4 - \frac{2}{3}u - 1)$ which has a turning point at

$$d/du(u-1)(3 - \tfrac{2}{3}u) = 3 - \tfrac{2}{3}u - \tfrac{2}{3}(u-1) = 0. \tag{3.45}$$

This gives $u = \frac{11}{4}$, and $v = \frac{13}{6}$ which, since it is in the region $1 \leqslant u \leqslant 3$, is the maximum point on the first set of (3.44). Turning to the second set we want to maximise $(u - 1)(v - 1)$ or $(u - 1)(8 - 2u - 1)$ over it.

$$d/du(u-1)(7 - 2u) = 7 - 2u - 2(u-1) = 0 \tag{3.46}$$

tells us the turning point is $u = 2\frac{1}{4}$, which is not within the region $3 \leqslant u \leqslant 3\frac{1}{2}$ in which we are interested. Thus, the maximum is at one of the end points and, on checking, we see it is at (3,2). However, as this is also part of the other set, the overall maximum is at $u = 2\frac{3}{4}$, $v = 2\frac{1}{6}$, which is the **maximin** bargaining solution.

Example 3.6. Consider the payoff matrix:

$$\begin{array}{c} \\ \mathrm{I}_1 \\ \mathrm{I}_2 \end{array}\begin{array}{c} \begin{array}{cc} \mathrm{II}_1 & \mathrm{II}_2 \end{array} \\ \begin{pmatrix} (1,4) & (-1,-4) \\ (-4,-1) & (4,1) \end{pmatrix} \end{array} . \tag{3.47}$$

The maximin values are $v_{\mathrm{I}} = v_{\mathrm{II}} = 0$, and the negotiation set is $\{(u,v) | u + v = 5, 1 \leqslant u \leqslant 4\}$ (see Fig. 3.10). From the diagram it is obvious that the payoff region is symmetric, and so by Nash's sixth axiom the solution (u^*,v^*) must have $u^* = v^*$. The only point on the negotiation set which satisfies this condition is $(2\frac{1}{2},2\frac{1}{2})$.

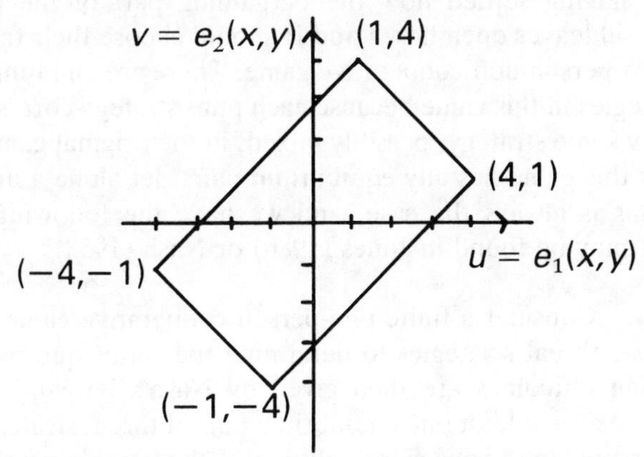

Fig. 3.10—Cooperative region of Example 3.6.

3.11 THREAT BARGAINING SOLUTION

Example 3.6 appears to be a symmetric game between the two players and the maximin bargaining solution reflects that. However, if II decides to play II_1, there is not much I can do. If I plays I_1, then II gets a payoff of 4, the highest possible payoff for II. If I plays I_2, then I gets a payoff of -4, the lowest payoff for I. Thus, II is in a stronger position than I because I has no comparable threat. Yet the maximin bargaining solution gives both the same amount.

Nash (1953) suggested changing the status quo point from the maximin values to some threat outcomes. The idea is that I and II independently threaten to play strategies **x** and **y**, then when each knows what the other threatens they bargain to settle on a pair of payoffs, knowing that if they don't agree they will receive the payoffs $e_1(\mathbf{x},\mathbf{y})$ and $e_2(\mathbf{x},\mathbf{y})$. Thus, they must play their threats if they cannot agree.

Once they have stated their threats, one can think of the bargaining process as a non-cooperative game, where if player I requests u and player II requests v the payoff is (u,v) if $(u,v) \in P$, the payoff region of the game, and $(e_1(\mathbf{x},\mathbf{y}),e_2(\mathbf{x},\mathbf{y}))$ if $(u,v) \notin P$. This game has a large number of equilibrium pairs, but Nash suggests reasons why one particular equilibrium pair should always be chosen. This is the solution one obtains by applying the Nash bargaining scheme with a status quo point $(e_1(\mathbf{x},\mathbf{y}),e_2(\mathbf{x},\mathbf{y}))$. It is probably more honest just to assume that the players agree with Nash's axioms and so automatically choose this solution, rather than to try to justify choosing it from among all other equilibrium pairs of the bargaining game. This solution is called the **threat bargaining solution** or sometimes the **Nash bargaining solution**.

However, having settled how the bargaining part of the problem is solved, this still leaves open how I and II should choose their threats. This is really a two-person non-cooperative game. There are an infinite number of pure strategies in this game because each pure strategy corresponds to a threat to play some strategy, possibly mixed, in the original game. It is not obvious that this game has any equilibrium pairs, let alone a unique pair. However, this is always the case, and we have the following theorem, whose proof may be found in Jones (1980) or Nash (1953).

Theorem 3.4. Consider a finite two-person cooperative game where the players choose threat strategies to determine the status quo payoffs, and the bargaining outcomes are then given by Nash's bargaining scheme. There always exists at least one equilibrium pair of threat strategies and all equilibrium pairs lead to the same solution of the bargaining game.

Thus, the only problem is to find these optimal threats. This is relatively simple in the case when the Pareto optimal set is of the form $\{(u,v)\,|\,au + v = b,\ c_2 \geqslant u \geqslant c_1\}$, as follows. Suppose I threatens strategy **x** and II threatens **y**, then I will end up with u that maximises

$$(u - e_1(\mathbf{x},\mathbf{y}))(b - au - e_2(\mathbf{x},\mathbf{y})). \tag{3.48}$$

At a turning point

$$d/du(u - e_1(\mathbf{x},\mathbf{y}))(b - au - e_2(\mathbf{x},\mathbf{y})) = b - au - e_2(\mathbf{x},\mathbf{y}) - a(u - e_1(\mathbf{x},\mathbf{y})) = 0.$$

Therefore

$$u = \frac{1}{2a}(b + ae_1(\mathbf{x},\mathbf{y}) - e_2(\mathbf{x},\mathbf{y})), \tag{3.49}$$

$$= \frac{1}{2a}(b + \sum_{i=1}^{n}\sum_{j=1}^{m} x_i(ae_{ij}^1 - e_{ij}^2)y_j. \tag{3.50}$$

So, to make u as large as possible, I wants to choose x to maximise

$$\sum_{i=1}^{n}\sum_{j=1}^{m} x_i(ae_{ij}^1 - e_{ij}^2)y_j \tag{3.50}$$

against any $\mathbf{y}$. Similarly, since substituting (3.50) into $au + v = b$ gives

$$v = ½\left(b - \sum_{i=1}^{n}\sum_{j=1}^{m} x_i(ae_{ij}^1 - e_{ij}^2)y_j\right), \tag{3.51}$$

II chooses $\mathbf{y}$ so as to minimise

$$\sum_{i=1}^{n}\sum_{j=1}^{m} x_i(ae_{ij}^1 - e_{ij}^2)y_j, \quad \text{for any } \mathbf{x}.$$

Thus, in finding their optimal threats, I and II appear to be playing the zero-sum game with payoff matrix $ae_1(\, , \,) - e_2(\, , \,)$. The optimal strategies $\mathbf{x}^*$ and $\mathbf{y}^*$ of this game are the optimal threats and so the status quo point will be $(e_1(\mathbf{x}^*,\mathbf{y}^*),e_2(\mathbf{x}^*,\mathbf{y}^*))$. Notice also that if w^* is the solution to the zero-sum game $ae_1 - e_2$, we can substitute this in (3.50) and (3.51) and the threat bargaining solution will be:

$$u^* = \frac{1}{2a}(b + w^*), \qquad v^* = ½(b - w^*). \tag{3.52}$$

Thus, we can either solve $ae_1 - e_2$ to get the optimal threats, or more quickly just use (3.52). However, we must make sure that the solution satisfies $c_1 \leqslant u^* \leqslant c_2$, otherwise the threat bargaining solution must be one of the end points of the negotiation set. We will now find the threat bargaining solution to the examples dicussed above.

Example 3.6

$$\begin{array}{c} \\ I_1 \\ I_2 \end{array} \begin{array}{c} \begin{array}{cc} II_1 & II_2 \end{array} \\ \begin{pmatrix} (1,4) & (-1,-4) \\ (-4,-1) & (4,1) \end{pmatrix} \end{array}$$

Since from Fig. 3.10 the negotiation set is $u + v = 5, 4 \geqslant u \geqslant 1$, by previous analysis we look at the zero-sum game $1.e_1 - e_2$, which gives a payoff matrix

$$\begin{array}{cc} & \begin{array}{cc} II_1 & II_2 \end{array} \\ \begin{array}{c} I_1 \\ I_2 \end{array} & \begin{pmatrix} -3 & 3 \\ -3 & 3 \end{pmatrix} \end{array}. \tag{3.54}$$

II_1 dominates II_2 so the solution is $\mathbf{x}^* = (x,1-x)$, $\mathbf{y}^* = (1,0)$ and $w^* = -3$. Substituting in (3.52) gives:

$$u^* = \frac{1}{2.1}(5-3) = 1, \qquad v^* = \tfrac{1}{2}(5-(-3)) = 4. \tag{3.55}$$

Just to check this, take any of the optimal strategies of (3.54), say $\mathbf{x}^* = (0,1)$, $\mathbf{y}^* = (1,0)$ as the optimal threats. Then the status quo payoff $(e_1(\mathbf{x}^*,\mathbf{y}^*), e_2(\mathbf{x}^*,\mathbf{y}^*)) = (-4,-1)$. Applying Nash's bargaining scheme we want to maximise $(u-(-4))(v-(-1))$, where $u + v = 5$, which is equivalent to:

$$\text{maximise } (u+4)(5-u+1) \text{ on } 1 \leqslant u \leqslant 4. \tag{3.56}$$

At a turning point

$$d/du(u+4)(6-u) = 6-u-(u+4) = 0, \tag{3.57}$$

therefore $u = 1$ and so $v = 4$. Thus, the threat bargaining solution moves the payoff (1,4) as far as possible to II's advantage along the negotiation set. This confirms our suggestion that the strategy II_1 is a very strong threat for II.

Example 3.4

This had payoff matrix

$$\begin{pmatrix} (1,2), & (8,2) \\ (4,4), & (2,1) \end{pmatrix}, \tag{3.58}$$

a Pareto optimal set $\tfrac{1}{4}u + v = 5$, $4 \leqslant u \leqslant 8$, and the maximin bargaining solution was $(6\tfrac{2}{3}, 3\tfrac{1}{3})$. To find the threat solution we look at the game $\tfrac{1}{4}e_1 - e_2$, with payoff matrix

$$\begin{array}{cc} & \begin{array}{cc} II_1 & II_2 \end{array} \\ \begin{array}{c} I_1 \\ I_2 \end{array} & \begin{pmatrix} -1\tfrac{3}{4} & -1 \\ -3 & -\tfrac{1}{2} \end{pmatrix} \end{array}. \tag{3.59}$$

II_1 dominates II_2, and then I_1 dominates I_2 so the solution is (I_1, II_1) with value $w^* = -1\tfrac{3}{4}$. Substituting this in (3.52) with $a = \tfrac{1}{4}$, $b = 5$ gives $u^* = 6\tfrac{1}{2}$, $v^* = 3\tfrac{3}{8}$, which is within the negotiation set $4 \leqslant u \leqslant 8$. Thus, the threat solution $(6\tfrac{1}{2}, 3\tfrac{3}{8})$ with threats (I_1, II_1) is slightly more in II's favour than the maximin solution $(6\tfrac{2}{3}, 3\tfrac{1}{3})$.

If the negotiation set consists of more than one linear piece, then finding the threat solution can become very difficult. To help understand what occurs, notice the following result.

Lemma 3.2. If (u^*, v^*) is the solution of Nash's bargaining scheme with status quo point (u_0, v_0), and the negotiation set at (u^*, v^*) is a straight line with (u^*, v^*) not at an end point, then the slope of the line joining (u_0, v_0) to (u^*, v^*) is the negative of the slope of the negotiation set at (u^*, v^*).

Proof. Suppose(u^*, v^*) lies on $au + v = b$, $c_1 < u < c_2$, then Nash's scheme requires it to maximise $(u - u_0)(v - v_0)$ over this set, i.e. maximise $(u - u_0)(b - au - v_0)$. Since it is in the interior of the set $c_1 < u < c_2$, it must be a turning point so the derivative of $(u - u_0)(b - v_0 - au)$ must be zero at (u^*, v^*), i.e.

$$\text{So } (b - v_0 - au_0^*) - a(u^* - u_0) = 0. \tag{3.60}$$

$$u^* = \frac{b - v_0 + au_0}{2a}, \qquad v^* = \frac{b + v_0 - au_0}{2}, \tag{3.61}$$

$(v^* - v_0)/(u^* - u_0) =$ a, whereas the slope of $au + v = b$ is $-a$.

So, for a game where the negotiation set is piecewise linear, ABC in Fig. 3.11 say, take any point D on this set and draw the line DE through D with slope the negative of the pareto optimal boundry at D. Any point F on DE if taken as the threat leads to D as the threat bargaining solution. For threat points like J, lying on BG, BH or between them, the Nash procedure will give B as the solution. However, when we work backwards and use the Nash procedure to find the best threats, it will only work if the best threat, like F, corresponds to a solution D on the interior of some line segment. In this case we apply the analysis leading to (3.52) first to AB,.

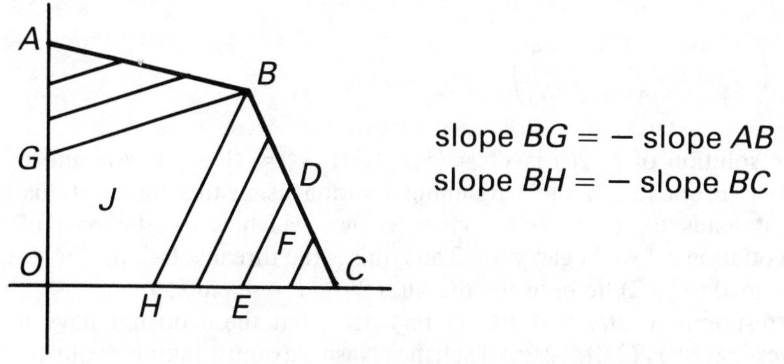

Fig. 3.11—Nash bargaining scheme.

where it leads to a solution not on AB, and then to BC, where it leads to D. However, if the optimal threat was J, applying the procedure to AB leads to a solution not on AB and the same result ocurs for BC. We then have to identify the threat strategies that lead to threats in $OGBH$, and look carefully at those to find the optimal threat.

We close by finding the threat solution for Example 3.5, where the negotiation set has two linear parts (see Fig. 3.9) and the payoff matrix is:

$$\begin{array}{c} \\ \mathrm{I}_1 \\ \mathrm{I}_2 \end{array} \begin{array}{c} \begin{array}{ccc} \mathrm{II}_1 & \mathrm{II}_2 & \mathrm{II}_3 \end{array} \\ \begin{pmatrix} (2,1) & (3,2) & (0,4) \\ (0,1) & (4,0) & (2,1) \end{pmatrix} \end{array} . \tag{3.62}$$

The Pareto optimal set is in two parts $\{(u, v)|2u + v = 8, 3 \leqslant u \leqslant 4\}$ and $\{(u, v)|\tfrac{2}{3}u + v = 4, 0 \leqslant u \leqslant 3\}$. (This is bigger than the negotiation set (3.44).)

If the threat solution is on the first part, then the corresponding threat is given by the solution of $2e_1 - e_2$, i.e.

$$\begin{bmatrix} 3 & 4 & -4 \\ -1 & 8 & 3 \end{bmatrix}.$$

The solution of this is $\mathbf{x}^* = (\tfrac{4}{11}, \tfrac{7}{11})$, $\mathbf{y}^* = (\tfrac{7}{11}, 0, \tfrac{4}{11})$ and $w^* = \tfrac{5}{11}$. Substituting this into (3.52) with $a = 2$ gives $u^* = 2\tfrac{5}{44}$, $v^* = 3\tfrac{17}{22}$, which is not in the range $3 \leqslant u \leqslant 4$ allowed for this line. Thus, the most suitable threat solution cannot lead to a bargaining solution on the interior of this line.

Turning to the other part of the Pareto optimal set, if the solution is on this line $\tfrac{2}{3}u + v = 4$, then the corresponding threat strategies are obtained by solving $\tfrac{2}{3}e_1 - e_2$:

$$\begin{pmatrix} \tfrac{1}{3} & 0 & -4 \\ -1 & \tfrac{8}{3} & \tfrac{1}{3} \end{pmatrix}.$$

The solution of (3.26) is $\mathbf{x}^* = (\tfrac{4}{17}, \tfrac{13}{17})$, $\mathbf{y}^* = (\tfrac{13}{17}, 0, \tfrac{4}{17})$ and $w^* = -\tfrac{35}{51}$. From (3.52) the bargaining solution using this threat status quo point leads to $u^* = 2\tfrac{33}{68}$, $v^* = 2\tfrac{35}{102}$, which is on the part of the negotiation set we began with. Thus, this is the threat solution. The threats that lead to (3.2) lie between the lines $2u - v = 4$ and $\tfrac{2}{3}u - v = 0$, which corresponds to BH and BG in Fig. 3.11, but these do not have to be investigated as $(2\tfrac{33}{68}, 2\tfrac{35}{102})$ satisfies Nash's axioms, taking as our threats $(\tfrac{4}{17}, \tfrac{13}{17})$ and $(\tfrac{13}{17}, 0, \tfrac{4}{17})$.

3.12 FURTHER READING

There is much more literature on non-zero-sum games than zero-sum games, since there is no obvious solution concept for the former. Moreover, many of the most interesting conflict problems are non-zero-sum. Again the basic ideas are in von Neumann and Morgenstern (1947) and are explained well by Luce and Raiffa (1957), though our proof of the existence of equilibrium pairs was given by Nash (1950b). To actually find all the equilibrium pairs in a bimatrix game is still a topic which generates much research; see Winkels (1979).

The theory of cooperative games is even more interesting because it involves some of the most important conflict situations that can arise. Even if you only allow two pure strategies to each player, Rapoport and Guyer (1966) described 78 different types of game you could have. By far the most famous one is the Prisoners' Dilemma, and for more details of the theory and applications of this problem the reader should look at Rapoport and Chammah (1965), Rapoport (1974) or Brams and Straffin (1979). For an amusing tour around the applications of other cooperative games, Hamburger's book *Games as Models of Social Phenomena* (1979) is an easy read.

The last part of the chapter dealt with bargaining theory, which in some sense has outgrown this chapter because all the ideas used can be applied when there are more than two players. The maximin bargaining solution is closely related to the Shapley value which we introduce in the next chapter on n-person games, whereas Harsanyi (1963) generalises the threat bargaining model to n-person games. Raiffa (1953) and Braithwaite (1955) suggested schemes that tried to avoid inter-person comparison of utility and yet satisfied all of Nash's axioms, except that of irrelevant alternatives, Both, however, involve changes of utility scale, which it can be argued are implicit comparisons of utility. For further details in this area, see Luce and Raiffa (1957, ch. 6) or Harsanyi (1977).

Problems for Chapter 3

3.1. Use the swastika method to find *all* the equilibrium pairs in (a) Example 3.1, (b) Example 3.2 (Battle of the Sexes) and (c) Example 3.3 (Prisoners' Dilemma), where the respective payoff matrices are:

$$\text{(a)} \begin{pmatrix} (2,2) & (3,3) \\ (1,1) & (4,4) \end{pmatrix}, \quad \text{(b)} \begin{pmatrix} (1,4) & (0,0) \\ (0,0) & (4,1) \end{pmatrix}, \quad \text{(c)} \begin{pmatrix} (-9,-9) & (0,-10) \\ (-10,0) & (-1,-1) \end{pmatrix}$$

3.2. Find the maximin-maximin pairs and all equilibrium pairs of the two games:

(a) $\begin{pmatrix} (2,1) & (4,3) \\ (6,2) & (3,1) \end{pmatrix}$, (b) $\begin{pmatrix} (4,-30) & (10,6) \\ (12,8) & (5,4) \end{pmatrix}$.

3.3. Suppose there are only two countries with nuclear weapons. Set up the problem of whether or not each should disarm as a 2×2 non-zero-sum game. Under what circumstances does this game have the same difficulties as Prisoners' Dilemma?

Whooping cough vaccine can cause brain damage in children, and is only given to babies more than 6 months old. The disease itself can also kill, and babies under 6 months old are most at danger. Consider the problem of whether or not to vaccinate your children as a two-player non-zero-sum game, where the players are yourself and 'a typical parent in the rest of society'. Take into account that vaccination of the population reduces the chance of an epidemic and so lessens the chances of non-vaccinated babies catching the disease.

Finally, look at the problem of whether employees should ask for high or low wage claims. Think of two different employees and set it up as a 2×2 two-person non-zero-sum game. How could this game be thought of as another example of Prisoners' Dilemma?

3.4 In a two-person non-zero-sum game $e_1(\mathrm{I}_1,\mathrm{II}_j) < e_1(\mathrm{I}_2,\mathrm{II}_j)$ for all j, $1 \leqslant j \leqslant n$. Prove that if $(\mathbf{x}^*,\mathbf{y}^*)$ is an equilibrium pair and $\mathbf{x}^* = (x^*_1,x^*_2,\ldots,x^*_n)$, then $x^*_1 = 0$. (A strategy in an equilibrium pair cannot choose a strictly dominated strategy with positive probability.) Use this result to find all the equilibrium pairs in the game with payoff matrix:

$$\begin{array}{cc} & \mathrm{II} \\ \mathrm{I} & \begin{pmatrix} (2,4) & (0,2) & (3,0) \\ (1,1) & (3,3) & (2,2) \end{pmatrix}. \end{array}$$

Find the maximin–maximin pair for this game and show it is not an equilibrium pair.

3.5. Consider the 3×3 two-person non-zero-sum game with payoff matrix:

$$\begin{pmatrix} (3,2) & (3,0) & (2,2) \\ (1,0) & (3,3) & (0,3) \\ (0,2) & (0,0) & (3,2) \end{pmatrix}$$

Find all the equilibrium pairs of the game. Use the same idea as in the 2×2 case, but instead of drawing a diagram to find what pairs satisfy both sets of conditions, you have to check this by inspection.

By looking at the pair (I_2,II_2), what does this tell us about the similar result to that proved in Problem 3.4, but with strict domination replaced by domination?

3.6. Find the solutions in (i) the Nash sense, (ii) the strict sense and (iii) the completely weak sense to the two games in Problem 3.2 and the one game in Problem 3.5.

3.7. Find the negotiation set, the maximin bargaining solution and the threat bargaining solution of the two games in Problem 3.2.

3.8. Sometimes it appears a player would prefer to play a game without cooperating with the other player. The payoff matrix for a two-person non-zero-sum game is:

$$\begin{array}{cc} & \begin{array}{cc} II_1 & II_2 \end{array} \\ \begin{array}{c} I_1 \\ I_2 \end{array} & \begin{pmatrix} (3,8) & (4,4) \\ (2,0) & (0,6) \end{pmatrix} \end{array}.$$

Find all its equilibrium pairs when considered as a non-cooperative game. Then find the maximin bargaining solution and the threat bargaining solution when the game is considered as a cooperative game. Which game would II prefer to play?

3.9. Look at the game with payoff matrix:

$$\begin{pmatrix} (a,2) & (3.0) \\ (2,0) & (2,2) \end{pmatrix},$$

where a is arbitrary. Show that the maximin bargaining solution for this game as a function of a is:

$$\begin{array}{rl} (2\tfrac{1}{4},\ 1\tfrac{1}{2}), & \text{if } a \leqslant 2, \\ (\tfrac{3}{4} + \tfrac{3}{4}a,\ 1\tfrac{1}{2}), & \text{if } 2 < a < 3, \\ (a,\ 2), & \text{if } a \geqslant 3. \end{array}$$

What is surprising about this result?

3.10. (Recall Problem 2.6.) Firm I can make either 200 colour TV sets a week or 100 colour and 100 black and white sets, while firm II can make either 50 colour TV sets or 100 black and white sets each week. There is a weekly demand of 200 colour and 100 black and white televisions; colour televisions sell at £200 each, and black and white at £100 each. It costs I £150 to make each colour TV and £60 to make each black and white TV. It costs firm II, £160 to make each colour TV and £85 each black and white TV. If more televisions are made then there is demand, the firms sell the same proportion of what they made (e.g. if I makes 200 colour TVs, II makes 50 colour TVs and demand is 200, then I sells $^{200}/_{250} \times 200 = 160$ and II sells $^{200}/_{250} \times 50 = 40$). Sets not sold are worth nothing, but the firms still have to pay the cost of manufacture. Set this up as a 2 × 2 two-person non-zero-sum game, where both firms are trying to maximise their profit.

Find, to the nearest pound, the profit each firm can ensure itself without cooperation. Describe the negotiation set of 'reasonable' profits if the two firms cooperate. Then find which member of this set is the maximin bargaining solution. Find the threat bargaining solution, and explain why this is to II's advantage.

Chapter 4

N-Person Games

'Too many cooks spoil the broth'—Anon.

4.1 NON-COOPERATIVE GAMES

In this chapter we look at games where the number of players is greater than two, though from time to time we will examine how the ideas we are discussing appear in the two-person game case.

Consider first the case where there is no cooperation between the players. In the two-person case we looked at the ideas of equilibrium pairs and maximin–maximin pairs. The latter seems to be rather far-fetched when there are more than two players, since you require each person to try and maximise his own payoff and minimise each of his opponents' payoffs. Even in the constant-sum case, when the total of the payoffs is the same value no matter what strategies are chosen, this leads to ambiguities in the choice of strategies. However, the idea of an equilibrium pair has an obvious generalisation to n-person.

Definition 4.1. In an n-person non-cooperative game, the n-tuple of strategies $\mathbf{x}_1^*, \mathbf{x}_2^*, \ldots, \mathbf{x}_n^*$, where player i plays the mixed strategy $\mathbf{x}_i^*$, is an equilibrium n-tuple if for all other strategies $\mathbf{y}_1, \mathbf{y}_2, \ldots \mathbf{y}_n$:

$$e_i(\mathbf{x}_1^*, \mathbf{x}_2^*, \ldots, \mathbf{x}_i^*, \ldots, \mathbf{x}_n^*) \geqslant e_i(\mathbf{x}_1^*, \mathbf{x}_2^*, \ldots, \mathbf{y}_i, \ldots, \mathbf{x}_n^*), \qquad 1 \leqslant i \leqslant n. \qquad (4.1)$$

The proof of Nash's theorem sketched in Chapter 3, Section 3.4, then extends to give the following theorem.

Theorem 4.1. Any finite n-person non-cooperative game has at least one equilibrium n-tuple.

Just as in the two-person case, the difficulty is in picking one, or a few, of the many equilibrium n-tuples to choose as a possible solution idea. Nothing constructive has been done on this so let's turn to the much more interesting cooperative games.

4.2 CHARACTERISTIC FUNCTION

In two-person games, non-cooperation between the players corresponds to two coalitions—player 1 by himself and player 2 by himself—whereas cooperation means the formation of the coalition of players 1 and 2 together. (Notice we are now switching notation, and calling the players 1, 2, 48, etc. rather than I, II, XLVIII — for obvious reasons.) However, with n players in the game, $n > 2$, there can be cooperation between some but not all of the players, and so the first thing we have to do in analysing these games is to determine which coalitions of players are likely to form. As in the two-person case, the formation of different sorts of coalitions will lead to vastly different outcomes in the game. When we say players form a coalition, we assume that they act just as one player and so after communicating together, they play a jointly agreed set of strategies with the aim of maximising the sum of the payoffs to the players in the coalition. After this is done comes the problem of sharing out the rewards between the members of the coalition.

So how do we analyse which coalitions will form? Recall that the players are labelled $1,2,3,\ldots,n$, so a coalition S is some subset of $N = \{1,2,\ldots,n\}$. The worst thing that can happen to the coalition S is if the rest of the players, written N–S, form a coalition and try to minimise the payoff that S can obtain. This is then a two-person, S and N–S, non-cooperative game, and so we can calculate the maximum payoff S can ensure itself because this is the 'maximin' value of that game. This gives us the following definition of a **characteristic function** v of a game.

Definition 4.2. The **characteristic function** of an n-person game assigns to each subset S of the players the maximum value $v(S)$ that coalition S can guarantee itself by coordinating the strategies of its members, no matter what the other players do.

It is standard to define the characteristic value of the empty coalition, $\emptyset$, as 0 so:

$$v(\emptyset) = 0. \tag{4.2}$$

What the definition says is that if X_S is the set of strategies available to the players in S and Y_{N-S} is the set of strategies available to the players in N–S, then

$$v(S) = \max_{\mathbf{x} \in X_S} \min_{\mathbf{y} \in Y_{N-S}} \sum_{i \in S} e_i(\mathbf{x},\mathbf{y}), \tag{4.3}$$

where $e_i(\mathbf{x}.\mathbf{y})$ is the payoff to player i if $\mathbf{x}$ and $\mathbf{y}$ are the strategies played by the players. It follows fairly easily from this definition that if S and T are disjoint coalitions, we get:

$$v(S \cup T) \geqslant v(S) + v(T), \quad \text{if } S \cap T = \{\emptyset\}. \tag{4.4}$$

Functions v satisfying (4.4) are called **superadditive**, and it is fairly obvious why this should happen. On the left S and T are working together against N–S–T, whereas on the right S is working against N–S–T and T (in $v(S)$) and T is working against N–S–T and S (in $v(T)$). Below we shall look at some games where the payoffs depend on which coalitions actually form and are not built up, as in (4.3), from the individuals' payoffs which in turn depend only on the strategies the various players play. In such games it can occur that v will no longer be superadditive, but for the time being we will always assume it is.

In some games it doesn't pay to form any sort of coalition, so for all subsets S, T of N we have:

$$v(S \cup T) = v(S) + v(T), \quad \text{if } S \cap T = \{\emptyset\}. \tag{4.5}$$

These games, where v is additive, i.e. satisfies (4.5), are called **inessential** and imply trivially that:

$$v(N) = \sum_{i=1}^{n} v(\{i\}). \tag{4.6}$$

For these it does not matter which coalitions form because it does not make any difference to the game.

Definition 4.3. A game that is not inessential is **essential**.

By thinking about a game in terms of its characteristic function, we have taken yet another step back from the actual playing of the game. Initially we used the extensive form to describe what happens in a game, but when we wanted to use the idea of mixed or randomised strategies we turned to the normal form. When using the normal form, one loses some knowledge about the game. Now to deal with the formation of coalitions we have had to abandon the normal form and with it any knowledge of particular strategies. The only thing that remains is our pessimistic criterion. The definition of a characteristic function is acceptable if the game is **constant sum**, so that every outcome gives a total combined payoff c to the players, because then whenever N–S try to maximise their combined payoffs, they are also minimising the payoff to S. In most games, however, the idea leads to wild underestimates of the strengths of coalitions, which almost borders on paranoia. Think of the two-person game with payoff matrix:

$$\begin{array}{cc} & 2 \\ 1 & \begin{pmatrix} (0,-100) & (1,100) \end{pmatrix} \end{array}. \tag{4.7}$$

Using the characteristic function approach 1 believes 2 will restrict him to 0, even though 2 will suffer a 100 loss rather than a 100 gain in doing so. The characteristic function of this is:

$$v(1) = 0, \qquad v(2) = 100, \qquad v(1,2) = 101. \tag{4.8}$$

In a less mathematical context, if you really believe the characteristic function, then you would never cross a road, because all the car drivers are just waiting to run you over. It is not surprising that such an approach to game theory runs into serious difficulties. Despite these reservations, there is no other approach to n-person games which is as useful as the characteristic function approach for describing succinctly the relative power of various coalitions. Using the doctor's old maxim, and applying it to the characteristic function: 'We prescribe it with care.'

Let's look at two typical examples which we will use throughout this chapter and find their characteristic functions.

Example 4.1 — Oil Market game. Country 1 has oil which it can use to run its transport system at a profit of a per barrel. Country 2 wants to buy the oil to use in its manufacturing industry, where it gives a profit of b per barrel, while Country 3 wants it for food manufacturing where the profit is c per barrel. $a < b \leqslant c$.

The characteristic function for this is:

$v(\emptyset) = 0$, by definition,

$v(1) = a$, because if 2 and 3 form a coalition against 1, they cannot force him to sell the oil so it is worth a to him,

$v(2) = v(3) = v(2,3) = 0$, because any coalition of buyers can't make the seller sell them the oil,

$v(1,2) = b$, because 1 and 2 can use the oil at a profit of b per barrel (1 sells it to 2), and so 3 would have to pay at least b to get it,

$v(1,3) = v(1,2,3) = c$; since 1 and 3 can use l's oil at a profit of c per barrel.

(4.9)

Example 4.2—Lilliput U.N. Security Council. Lilliput has a small version of the U.N. Security Council (the numbers get too hard in the real thing) which has two permanent members, 1 and 2, who have the veto, plus three ordinary members, 3, 4, and 5. For a resolution to be passed (payoff 1 to the coalition who wants it passed), it requires three votes in favour and no vetoes. The payoff is 0 if the resolution is not passed.

The characteristic function just takes values 0 or 1 depending on whether that coalition can get the resolution passed or not. So:

$$\begin{aligned} v(1,2,3) &= v(1,2,4) = v(1,2,5) = v(1,2,3,4) = v(1,2,3,5) \\ &= v(1,2,4,5) = v(1,2,3,4,5) = 1; \end{aligned} \tag{4.10}$$

$v(S) = 0$, all other $S \subset N$.

4.3 STRATEGIC EQUIVALENCE OF CHARACTERISTIC FUNCTIONS

In the oil game it should not matter if the price is given in dollars per barrel or pounds per barrel. It will be the same game, but the numerical values of the characteristic functions will be different. Thus, if

$$v'(S) = c\,v(S), \quad \text{for all } S \subset N, \tag{4.11}$$

we assume we have the same game. Similarly, if player i is paid a_i to play the game for $i = 1,2,\ldots,n$, then this doesn't change the game, but the characteristic function changes from v to v', where

$$v'(S) = v(S) + \sum_{i\in S} a_i. \tag{4.12}$$

Definition 4.4. Two characteristic functions, v and v', are **strategically equivalent** if

$$v'(S) = cv(S) + \sum_{i\in S} a_i. \tag{4.13}$$

This enables us to group games which are strategically equivalent together and pick a typical representative from the group. This is called **normalising** the game and we choose as a representative the game whose one-person coalitions have characteristic function value zero and the coalition of every one has value one.

Lemma 4.1. Every essential game is strategically equivalent to one with $v'(\{i\}) = 0$, $i = 1,2,\ldots,n$, and $v'(N) = 1$.

Proof. Let v be the characteristic function of the game. Define a strategically equivalent function v' by:

$$v'(S) = c\,v(S) - c\sum_{i\in S} v(i), \tag{4.14}$$

where $c^{-1} = v(N) - \sum_{i=1}^{n} v(i)$. It is easy to check this has the required properties.

In our two examples the normalised form of the oil market game is:

$$v'(1) = v'(2) = v'(3) = 0, \qquad v'(1,2) = \frac{b-a}{c-a},$$

$$v'(2,3) = 0, \qquad v'(1,3) = v'(1,2,3) = 1, \tag{4.15}$$

whereas the Security Council game is already in a normalised form. Although it is often useful when proving general results to adopt the normalised form, you lose the obvious interpretations of the original game. ($v(1,2) = b$ is easily understood in the oil market game. $v'(1,2) = (b-a)/(c-a)$ is less transparent.) We will seldom use the normalised form of the characteristic function.

4.4 IMPUTATIONS

There are two important and interrelated questions when analysing n-person games. So far we have discussed what coalitions are likely to form, or rather given a measure to the strengths of the possible coalitions. The second question is: When a coalition does form, how does it share its reward between the individual members of the coalition? It is assumed that the important part of an n-person game is the pre-play negotiations, where coalitions form and the rewards from the game (which can be calculated) shared out. Obviously the distribution of the rewards affects the formation of the coalition because some players might offer a large reward to another player to join a particularly favourable coalition. Each individual would tend to join the coalition that offered and could guarantee him most. So we cannot describe which coalitions form without describing how the payoffs are split between the individual players in the game.

Notice that implicit in this scenario is the assumption that the reward from the game can be split in any way we like, i.e. it is **infinitely divisible**, and also that it is easy to transfer payoffs from one player to another. These are called **side-payments**, and in order to allow them we must suspend all judgements about inter-personal comparison of utility, and think of what is being transferred as money or goods rather than utility.

There is a difference between this idea of a cooperative game and the two-person cooperative game described earlier, in that the players cooperated on their strategies, but kept the payoffs they got from the game. Think of the analogy with robbing a bank. The two-person cooperative game corresponds to the two thieves working together in the robbery, but each keeps the actual money he takes out of the bank. The n-person game is equivalent to the situation where the thieves return to their hideout carrying whatever they took away, put it all in one big heap and share it out according to a predetermined scheme.

Let's return then to the second question we mentioned previously. How are the rewards shared out between the individual players? We call 'reasonable' share outs of the rewards **imputations**.

Definition 4.5. An **imputation** in an n-person game with characteristic function v is a vector $x = (x_1,x_2,\ldots,x_n)$ satisfying:

(i) $\sum_{i=1}^{n} x_i = v(N)$; (4.16)

(ii) $x_i \geqslant v(i)$, for $i = 1,2,\ldots,n$, (4.17)

where x_i is obviously the ith player's reward. (4.16) is a Pareto optimality condition. $v(N)$ is the most the players can get out of the game when they all work together. So, for any possible set of individual rewards x_i, we must have $\sum_{i=1}^{n} x_i \leqslant v(N)$. If this was a strict inequality, then by working together they could always share out the rewards so that everyone got more. (4.17) says that everyone must get as much as they could get if they played by themselves.

Denote the set of all imputations in a game by $E(v)$. In an inessential game, there is only one imputation, but for essential games there are lots. In the oil market game (Example 4.1):

$$E(v) = \{(x_1,x_2,x_3): x_1 \geqslant a, x_2 \geqslant 0, x_3 \geqslant 0, x_1 + x_2 + x_3 = c\}, \quad (4.18)$$

while in the Security Council game:

$$E(v) = \{(x_1,x_2,x_3,x_4,x_5): x_i \geqslant 0, i = 1,2,3,4,5, x_1 + x_2 + x_3 + x_4 + x_5 = 1\}. \quad (4.19)$$

Are some imputations 'better' than others? Look at two imputations x, y. Since

$$\sum_{i=1}^{n} x_i = v(N) = \sum_{i=1}^{n} y_i, \quad (4.20)$$

if for some i $x_i > y_i$, then there must be a j with $y_j > x_j$. So, an imputation can't be better for everyone, but it is possible that for a particular coalition, x is better than y for all its members. This gives us the idea of **domination** of one imputation by another, but to stop coalitions promising their members the world, we always require that a coalition has the 'strength' to provide the better imputation.

Definition 4.6. Let x, y be two imputations, x dominates y over S, $(x >_S y)$ if:

(i) $x_i > y_i$, for all $i \in S$,

(ii) $\sum_{i \in S} x_i \leqslant v(S)$. (4.21)

Condition (4.21) requires that S has enough payoff to ensure its members x.

Definition 4.7. We say x dominates y $(x \succ y)$ if x dominates y for some coalition S.

Notice it is easy to construct situations where $x \succ y$ and $y \succ x$.

4.5 THE CORE

One of the earliest solution concepts was the **core**, first defined explicitly by Gillies (1959). This uses the idea of domination of imputations defined above and says an imputation will persist in the preplay negotiations only if it is not dominated.

Definition 4.8. The **core** of a game v, denoted by $C(v)$, is the set of imputations which are not dominated for any coalition.

Thus, if x is in the core, any coalition which forms, either says x is the best imputation for it, or if it prefers another one y it hasn't the strength $(\sum_{i \in S} y_i > v(S))$ to enforce the change. Notice there can be more than one imputation in the core. There is an alternative algebraic formulation of the core, which is very useful in actually finding the core of a specific game.

Theorem 4.2. x is in the core if and only if

$$\text{(i)} \quad \sum_{i=1}^{n} x_i = v(N), \tag{4.22}$$

and

$$\text{(ii)} \quad \sum_{i \in S} x_i \geqslant v(S), \quad \text{for all } S \subset N \tag{4.23}$$

Proof. Let x satisfy (i) and (ii), then putting S equal to $\{1\}, \{2\}, \ldots, \{n\}$, respectively, in (ii) together with condition (i) shows x is an imputation. To show it is not dominated, suppose the opposite is true, so $y >_S x$. Then $y_i > x_i$ for all $i \in S$, but then using (ii):

$$\sum_{i \in S} y_i > \sum_{i \in S} x_i \geqslant v(S). \tag{4.24}$$

Thus, y cannot dominate x because this fails (4.21) and so x must be undominated.

Conversely, suppose x is in the core, then it is an imputation and so (i) must hold. Suppose for some coalition S, (ii)doesn't hold for x so $\sum x_i < v(S)$. Define ε by:

$$v(S) = \sum_{i \in S} x_i + \#S\,\varepsilon, \tag{4.25}$$

where $\#S$ is the number of players in coalition S.
Define:

$$y_i = \begin{cases} x_i + \varepsilon, & i \in \mathrm{S}, \\ \mathrm{v}\{i\} + (v(N) - v(S) - \sum_{i \notin S} v(\{i\}))/\#N - S, & i \notin S. \end{cases} \tag{4.26}$$

y, defined by (4.26), is an imputation since $\sum y_i = v(N)$ and $y_i \geqslant v(i)$, for all $i = 1,2,\ldots,n$. Moreover, $y_i > x_i$ for all $i \in S$ and $\sum_{i \in S} y_i = v(S)$. Thus, y dominates x on S which contradicts x being in the core, and so x must satisfy (ii) for all coalitions.

The core is a convex set ($x,y \in C(v)$ implies $\lambda x + (1 - \lambda)y \in C(v)$) and is closed (contains its boundary). It has one great disadvantage for a possible 'solution' to n-person games though. In many games the core is empty. This is true in particular for essential constant-sum games, which are the n-person equivalents of zero-sum games.

Lemma 4.2. If v is the characteristic function of an essential constant-sum game, then $C(v) = \{\emptyset\}$.

Proof. Recall that essential means $v(N) > \sum_{i=1}^{n} v(i)$ and constant sum means that $v(N - S) + v(S) = v(N)$ for all coalitions S. We will assume there is an imputation x in the core and derive a contradiction. Using (4.23) and the constant sum property we have:

$$\sum_{i=2}^{n} x_i \geqslant v(2,3,\ldots,n) = v(N) - v(1). \tag{4.27}$$

Since x is an imputation, $x_1 + \sum_{i=2}^{n} x_i = v(N)$ and so subtracting (4.27) gives:

$$x_1 \leqslant v(1). \tag{4.28}$$

The same argument will give $x_i \leqslant v(i)$, for all $i = 2,3,\ldots,n$, and so if the game is essential $\sum_{i=1}^{n} x_i \leqslant \sum v(i) < v(N)$, which contradicts x being an imputation. Thus, the core must be empty.

Let's look at our two examples and use Theorem 4.2 to find the core for these.

Example 4.3—Oil Market game. $v(1) = a$, $v(2) = v(3) = v(2,3) = 0$, $v(1,2) = b$, $v(1,3) = v(1,2,3) = c$. If $\mathrm{x} = (x_1,x_2,x_3)$ is in the core then (4.22) requires $x_1 + x_2 + x_3 = c$, whereas (4.23) gives the following inequalities:

$$\begin{aligned} &x_1 \geqslant a \ (S = \{1\}); && x_2 \geqslant 0 \ (S = \{2\}); \\ &x_3 \geqslant 0 \ (S = \{3\}); && x_1 + x_2 \geqslant b \ (S = \{1,2\}); \\ &x_2 + x_3 \geqslant 0 \ (S = \{2,3\}); && x_1 + x_3 \geqslant c \ (S = \{1,3\}). \end{aligned} \tag{4.29}$$

Notice that $x_1 + x_3 \geqslant c$, $x_2 \geqslant 0$ and $x_1 + x_2 + x_3 = c$ imply $x_2 = 0$, $x_1 + x_3 = c$ and substituting this into $x_1 + x_2 \geqslant b$, gives $x_1 \geqslant b$. Thus:

$$C(v) = \{(x,0,c - x) \colon b \leqslant x \leqslant c\}. \tag{4.30}$$

We can give an interpretation to this result. It assumes 1 and 3 form a coalition and so 1 will sell the oil to 3. 3 pays him x per barrel which must be at least b, otherwise 1 would be better off selling it to 2, and no more than c, so that 3 doesn't pay more than it is worth to him.

Example 4.4—Lilliput U.N. Security Council. $v(1,2,3) = v(1,2,4) = v(1,2,5) = v(1,2,3,4) = v(1,2,3,5) = v(1,2,4,5) = v(1,2,3,4,5) = 1$, $v(S) = 0$ all other S. Applying Theorem (4.2) gives $x_1 + x_2 + x_3 + x_4 + x_5 = 1$

from (4.22) and putting various coalitions into (4.23) we get:

$$x_i \geqslant 0, \qquad i = 1,2,3,4,5;$$

$$x_1 + x_2 + x_3 \geqslant 1, \qquad x_1 + x_2 + x_4 \geqslant 1, \qquad x_1 + x_2 + x_5 \geqslant 1. \tag{4.31}$$

Again $x_1 + x_2 + x_3 + x_4 + x_5 = 1$, $x_1 + x_2 + x_3 \geqslant 1$, $x_4 \geqslant 0$, $x_5 \geqslant 0$ implies $x_4 = 0$, $x_5 = 0$, whereas using $x_1 + x_2 + x_4 \geqslant 1$, $x_3 \geqslant 0$, $x_5 \geqslant 0$ gives $x_3 = 0$ and $x_1 + x_2 = 1$ as well. Thus, the core is:

$$C(v) = \{(x, 1 - x, 0, 0, 0);\ 0 \leqslant x \leqslant 1\}. \tag{4.32}$$

This suggests that all the power resides with the two members with the veto. One way of looking at this is to suppose that in the future each country will be given a percentage of the votes and motions will be passed if more than 50% of the votes are cast for it. However, the percentage of votes given to each country has to be decided under the present voting system. Which distribution of the votes will get passed? (4.32) suggests that only those which give all the votes to 1 and 2 will get through. This seems highly improbable at first, but on second thought we can see how this will happen. Suppose, initially, 1, 2 and 3 decide on a third of the votes each, then 1 and 2 can go to 4, who has no votes at present and offer him $\frac{1}{6}$ of the votes say, if the rest are shared between them. 4 would obviously agree to this, but then 1 and 2 could go to 5 and offer him the same deal for $\frac{1}{12}$ of the votes. He would also accept that, as would 3 for $\frac{1}{24}$ of the votes, and 4 for $\frac{1}{48}$ of the votes the second time around. Continuing this bargaining sequence, 1 and 2 could eventually guarantee themselves as much of the vote as they wanted.

4.6 STABLE SETS

Another solution concept suggested by von Neumann and Morgenstern (1944) is the **stable set** of imputations (also called the von Neumann–Morgenstern solution). This is more complicated than the core, but it was hoped that it would always exist. The idea is that instead of one imputation which every coalition is happy with, there is a set of imputations, so that if we take anything outside the set, there is an imputation inside the set, which some coalition is happier with and has the strength to obtain. Not everyone might be satisfied with this new imputation and so some subset of players might force a change to another imputation outside the set.

However, this new imputation is again dominated by one inside the set and so we will return within the set. Thus, the bargaining process resolves around the set, and we can think of the whole set, as a possible solution. Also, there should be no domination of an imputation within the set by another one within the set, so that they are all as important as one another. Formally we have the following definition.

Definition 4.9. A stable set $S(v)$ of an n-person game with characteristic function v is any set of imputations such that:

(i) if $x, y \in S(v)$, then x must not dominate y or vice versa (internal stability);

(ii) if $z \notin S(v)$, there is an $x \in S(v)$, so that $x \succ z$ (external stability).

Notice that $C(v) \subseteq S(v) \subseteq E(v)$, i.e., the core must be in every stable set. This follows since any imputation outside a stable set must be dominated by (ii) and so the undominated imputations must be within every stable set. The core will be within the intersection of all stable sets, but Lucas (1967) gave an example where it is strictly included in the intersection. In Lucas (1969) he gives an example of non-convex stable sets which are almost impossible to describe.

The first disadvantage of using stable sets as a solution concept is that most games have lots of different stable sets and it is quite difficult to find all of them. For about twenty years, however, a great deal of research effort was put into proving that every game has at least one stable set. However, Lucas (1968) discovered a 10-person game with no stable sets.

Lucas's example depends on a result proved by Gillies (1969) on the connection between **generalised games** and ordinary n-person games. A generalised game is just like an ordinary n-person game except that the characteristic function v is not superadditive. For such games we can define a new function v^* as follows. For any coalition S, let P_S be the collection of all possible ways $P_1, P_2, \ldots, P_n$ of splitting S into mutually disjoint subsets, and then let

$$v^*(S) = \max_{\{P_i \in P_S | 1 \leqslant i \leqslant n\}} \sum_{T \in P_i} v(T). \tag{4.33}$$

The function v^* is superadditive and so we can think of it as defining an ordinary n-person game which has the same set of imputations as the generalised game. Gillies proved that if x and y are imputations, then if x dominates y for some coalition S in the game defined by v, then x dominates y for some coalition S^* in the game defined by v^*. The converse also holds trivially. Thus, a set which is stable in v is also stable in v^* and Lucas showed that the following 10-person generalised game has no stable set:

$$\begin{aligned}
&v(1,2) = v(3,4) = v(5,6) = v(7,8) = v(9,10) = 1,\\
&v(1,3,7) = v(1,5,7) = v(3,5,7) = v(1,3,9) = v(1,5,9) = v(3,5,9) = 2,\\
&v(1,4,7,9) = v(3,6,7,9) = v(2,5,7,9) = 2,\\
&v(1,3,7,9) = v(1,5,7,9) = v(3,5,7,9) = 3, \qquad v(1,3,5,7,9) = 4,\\
&v(1,2,3,4,5,6,7,8,9,10) = 5, \qquad v(S) = 0 \text{ all other } S.
\end{aligned} \tag{4.34}$$

The result of Gillies then proves that the corresponding classical n-person game also has no stable set. So stable sets suffer from the same disadvantage as the core in that they need not exist. However, it also leads to the interesting problem of classifying the games that do not have stable sets.

Let us return to our two standard games and find stable sets for them.

Example 4.5—Oil Market game. Since we know the core must be in the stable set, we want to find what is not dominated by the core. If the core dominates everything, it is a stable set; if not, then the only imputations that can be added to the core to form the stable set are ones that are not dominated by it, because of internal stablility.

First a general comment on domination. An imputation x can never dominate another one y on the coalition of just one player or on the grand coalition of everyone. The first remark follows because if x dominates y on $\{i\}$, $x_i > y_i$, and because y is an imputation $y_i \geqslant v(i)$. Thus, $x_i > v(i)$, which contradicts (4.21) in the definition of domination. The second result about the grand coalition N follows because no imputation can give everyone more than another imputation.

Returning to the Oil Market game, by (4.30) the core is the set $C(v) = \{(x, 0, c - x), b \leqslant x \leqslant c\}$. If we look at $(a, b - a, c - b)$, then since $b - a \nless 0$, $c - b \nless c - x$, imputations in the core only give player 1 more than they get here. By the above comment, the core can not dominate this imputation on $\{1\}$, and so $(a, b - a, c - b)$ is not dominated by the core. What else is not dominated by the core?

Take a typical imputation $y = (y_1, y_2, y_3)$, where $y_1 \geqslant a$, y_2,$y_3 \geqslant 0$ and $y_1 + y_2 + y_3 = c$. We need only to look at coalitions (1,2), (1,3) and (2,3) because of the general comments above. Also, since $y_2 \geqslant 0$, the core does not give 2 more than he gets here and so cannot dominate y on (1,2) or (2,3). This leaves coalition (1,3). For (y_1, y_2, y_3) not to be dominated by an imputation in the core, either $y_1 \geqslant x$ or $y_3 \geqslant c - x$ for all x, $b \leqslant x \leqslant c$. Since x can be c, we cannot guarantee $y_1 \geqslant x$, but as $c - x$ can be at most $c - b$ we want $y_3 \geqslant c - b$.

It helps to understand what is going on by drawing it graphically, as in Fig. 4.1. The set of imputations is the triangular plane ABC, whereas the core is the line DC. The imputations which are dominated by the core are those which have smaller y_1 and y_3 components, i.e. those in $DCBE$, whereas those in ADE are not dominated since $y_3 \geqslant c - b$. The set of imputations ABC actually form a plane so we can draw it as in Fig. 4.2.

Now we cannot add all of AED to the core to form a stable set because some of the imputations on AED dominate other ones on AED. Take a typical point F in AED, as in Fig. 4.3, and draw GFH parallel to ED, so all

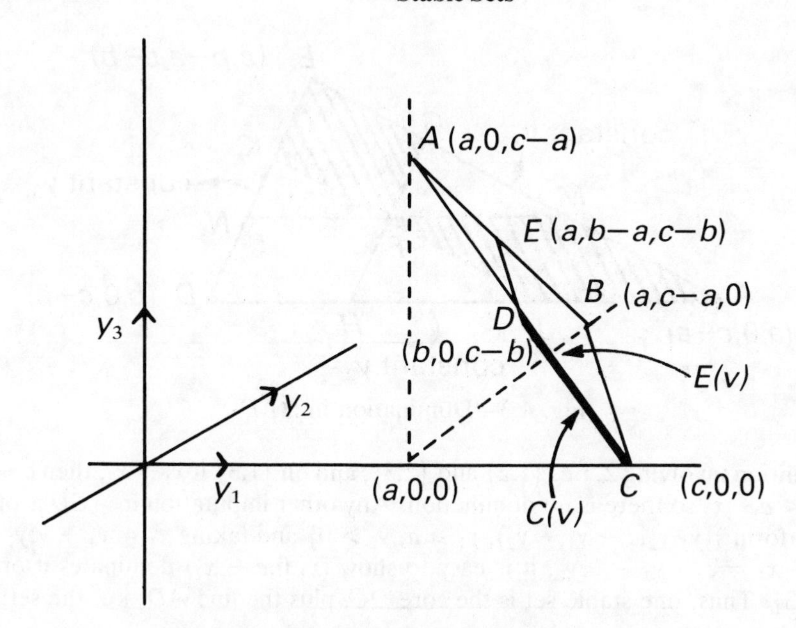

Fig. 4.1—Set of imputations in Example 4.5.

the points on it have the same y_3 component. Repeat the process with JFK, parallel to AE, which has all y_1 values the same in it, and MFN, parallel to AD, which has constant y_2 values. Then F dominates the points inside $GFKE$ on the $\{1,3\}$ coalition, and those inside $MFJA$ on the $\{1,2\}$ coalition. We need to pick a set of points in AED which avoids domination of points in it, by other points in it, but still dominates all the others points in AED.

One possibility is the line AD, i.e. $\{(x,0,c - x),\ a \leqslant x \leqslant b\}$. This is internally stable since $(x_1,0,c - x_1)$ cannot dominate $(x_2,0,c - x_2)$ on any

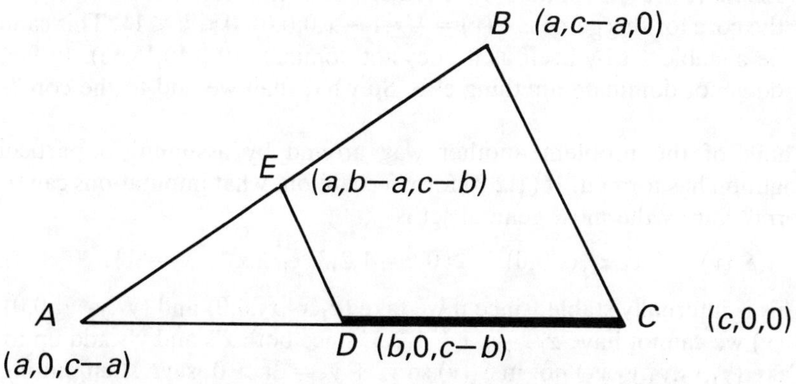

Fig. 4.2—Planar representation of imputation set.

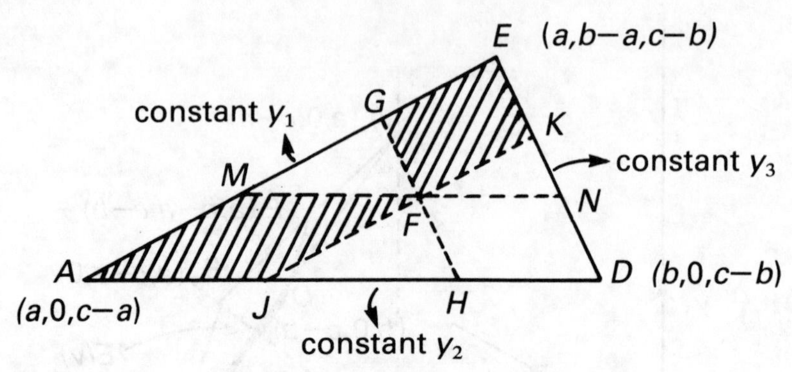

Fig. 4.3—Domination in *AED*.

coalition involving 2, i.e. $\{1,2\}$ and $\{2,3\}$, and on $\{1,3\}$ if $x_1 > x_2$, then $c - x_1 < c - x_2$ so there is no domination. Any other imputation in *AED* is of the form $\{(y_1,y_2,c - y_1 - y_2), y_1 \geqslant a, y_2 > 0\}$ and taking $x_1 = y_1 + ½y_2$, $c - x_1 = c - y_1 - ½y_2$, it is easy to show $(x_1,0,c - x_1)$ dominates it on $\{1,3\}$. Thus, one stable set is the core *DC*, plus the line *AD*, i.e. the set:

$$\{(x,0,c - x), a \leqslant x \leqslant c\} \tag{4.35}$$

Another solution is given by taking the line *ED*, $\{(x,b - x, c - b), a \leqslant x \leqslant b\}$, which again is internally stable, and dominates the rest of *AED* on the coalition $\{1,2\}$. Thus, *ED* plus the core *DC* is another stable set. In fact we can show that for any function f which is non-decreasing in x ($f(x_2) \geqslant f(x_1)$ if $x_2 > x_1$), or non-increasing in x ($f(x_2) \leqslant f(x_1)$ if $x_2 > x_1$), the set

$$\{(a + x,f(x),c - a - x - f(x)), x \geqslant 0, f(x) \geqslant 0, x + f(x) \leqslant b - a\} \tag{4.36}$$

plus the core *DC* form a stable set.

Example 4.6—Lilliput U.N. Security Council. Recall from (4.32) that the core for this game is $C(v) = \{(x,1 - x,0,0,0), 0 \leqslant x \leqslant 1\}$. This cannot be a stable set by itself as it does not dominate $(0,0,⅓,⅓,⅓)$. In fact it does not dominate anything else. So what shall we add to the core?

Think of the problem another way around by assuming a particular coalition has formed. If (1,2,3) form a coalition, what imputations can they form? Surely the most general set is:

$$S_1(v) = \{(x_1,x_2,x_3,0,0); x_i \geqslant 0, i = 1,2,3, x_1 + x_2 + x_3 = 1\}.$$

This is internally stable, since if we take $(x_1,x_2,x_3,0,0)$ and $(y_1,y_2,y_3,0,0)$ in $S_1(v)$ we cannot have $x_i > y_i$, $i = 1,2,3$, since both x's and y's add up to 1. Take (y_1,y_2,y_3,y_4,y_5) not in $S_1(v)$ so $y_4 + y_5 = 3\varepsilon > 0$, say. Then putting $x_i = y_i + \varepsilon$, $i = 1,2,3$, $(x_1,x_2,x_3,0,0)$ dominates this on $\{1,2,3\}$.

A game is called **simple** if for all coalitions S, $v(S) = 0$ or 1. In such games, a **minimum winning coalition** S is one where $v(S) = 1$, and $v(S - \{i\}) = 0$ for any i in S. Take any minimum winning coalition S and let V_S be the set of imputations so that $x_i = 0$ for all $i \notin S$. V_S is always a stable set. The proof is just the justification we gave in the above example, so in Example 4.6 we have stable sets $V_{(1,2,3)}$, $V_{(1,2,4)}$ and $V_{(1,2,5)}$.

4.7 NUCLEOLUS

One of the more recent ideas put forward as a solution to an n-person game is that of the **nucleolus** introduced by Schmeidler (1969). It has two very useful properties:

(a) every game has one and only one nucleolus, and

(b) if the core exists, the nucleolus is part of it.

It is based on the idea of making the most unhappy coalition under it happier than the most unhappy coalition under any other imputation. For any imputation x and any coalition S, let $x(S) = \sum_{i=S} x_i$. Then each coalition looks at $v(S) - x(S)$ and the larger this number the more unhappy the coalition is with that imputation. It is the difference between what they could get by themselves and what they actually get. Define $\theta(x)$ to be the 2^n values $v(S) - x(S)$ for all coalitions S (including N and ϕ) written in decreasing numerical order. If we compare two imputations x and y by looking at the coalition which is unhappiest under each and calculating $v(S) - x(S)$, $v(S') - y(S')$ for these two coalitions, the smaller value is the 'better' imputation; if these numbers are the same, look at the pair of second most unhappy coalitions and compare these and so on. We are ordering $\theta(x)$ and $\theta(y)$ lexicographically (i.e. as in a dictionary) so if $\theta(x) = (\theta(x)_1, \theta(x)_2, \ldots, \theta(x)_{2^n})$, we say:

$$\begin{aligned} \theta(x) < \theta(y), \quad & \text{if } \theta(x)_1 < \theta(y)_1, \\ & \text{or if } \theta(x)_k = \theta(y)_k, \\ & \text{for } k = 1,2,\ldots,i-1 \text{ and } \theta(x)_i < \theta(y)_i. \end{aligned} \tag{4.37}$$

Now we can formally define the nucleolus.

Definition 4.10. The nucleolus $N(v)$ is the smallest imputation under the ordering defined by (4.37), i.e.

$$N(v) = \{x \in E(v) \mid \theta(x) < \theta(y) \text{ for all } y \in E(v)\}. \tag{4.38}$$

To understand the definition let us calculate the nucleolus for our two examples.

Example 4.7—Oil Market game. A typical imputation for this game is:

$$(a + x_1, x_2, c - a - x_1 - x_2). \quad x_1 . x_2 \geqslant 0, \quad x_1 + x_2 \leqslant c - a. \tag{4.39}$$

For (4.39), calculating $v(S) - x(S)$ for all coalitions S we get:

$$v(1) - x(1) = a - (a + x_1) = -x_1; \qquad v(2) - x(2) = 0 - x_2 = -x_2;$$

$$v(3) - x(3) = 0 - (c - a - x_1 - x_2) = x_1 + x_2 - c + a;$$

$$v(1,2) - x(1,2) = b - (a + x_1 + x_2) = b - a - x_1 - x_2;$$

$$v(2,3) - x(2,3) = 0 - (c - a - x_1) = x_1 - c + a;$$
$$v(1,3) - x(1,3) = c - (c - x_2) = x_2;$$

$$v(1,2,3) - x(1,2,3) = c - c = 0; \qquad v(\phi) - x(\phi) = 0. \tag{4.40}$$

Since $x_1, x_2 \geqslant 0$ and $x_1 + x_2 \leqslant c - a$, the only entries in (4.40) which can be positive are x_2 and $b - a - x_1 - x_2$. Obviously, to make these as low as possible we must take $x_2 = 0$ and $x_1 > b - a$. If $x_2 = 0$, the entries in (4.40) become $-x_1, 0, x_1 - c + a, b - a - x_1, x_1 - c + a, 0, 0, 0$. In the values that vary with x_1, we have $b - a - x_1 > -x_1$ so that the largest element is either $b - a - x_1$ or $x_1 - c + a$. As functions of x_1 these look like Fig. 4.4.

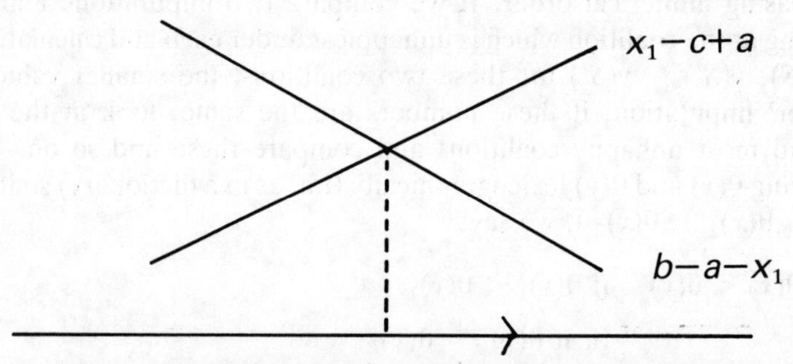

Fig. 4.4—Maximising the minimum of two lines.

To make the largest of these as small as possible, we choose x_1 so that the two functions are the same value, i.e. $b - a - x_1 = x_1 - c + a$. This gives $x_1 = \frac{1}{2}(c + b) - a$, and substituting this and $x_2 = 0$ into (4.39) gives $(\frac{1}{2}(c + b), 0, \frac{1}{2}(c - b))$. This is the nucleolus, and $\theta(x)$ for this imputation is $(0,0,0,0,\frac{1}{2}(b - c), \frac{1}{2}(b - c), \frac{1}{2}(b - c), a - \frac{1}{2}(b + c))$.

Example 4.8—Lilliput U.N. Security Council. If the characteristic function is symmetric for some players (you could interchange them without affecting its value), then the nucleolus will also reflect this symmetry. In the Security Council game, players 1 and 2 are interchangeable, as are 3, 4 and 5. Thus, the only imputations that can be the nucleolus are $\{(x,x,\frac{1}{3}(1-2x),\frac{1}{3}(1-2x),\frac{1}{3}(1-2x)), 0 \leqslant x \leqslant \frac{1}{2}\}$ which reflect this symmetry. The difference, $v(S) - x(S)$, for all coalitions S for such an imputation, leads to the values $\frac{2}{3} - \frac{4}{3}x$, $\frac{1}{3} - \frac{2}{3}x$, 0 or ones which are obviously negative. To minimise the maximum of these, $\frac{2}{3} - \frac{4}{3}x$, we obviously take x as large as we can, i.e. $x = \frac{1}{2}$. Thus, $(\frac{1}{2},\frac{1}{2},0,0,0)$ is the nucleolus.

Notice that in both cases the nucleolus is in the core. This is true in general provided the core exists, because for any $x \in C(v)$, $v(S) - x(S)$ is zero or negative by Theorem 4.2. Thus, all the entries of $\theta(x)$ for an x in the core are zero or negative and when finding the lexicographic minimum of $\theta(y)$ for all y, we must then choose one with entries all zero or negative. But, again by Theorem 4.2, such an imputation must be in the core. This is very useful to remember when finding the nucleolus of a particular game, because in Example 4.7 we could have restricted ourselves to imputations of the form $\{(x,0,c-x), b \leqslant x \leqslant c\}$ immediately.

The proof of the existence of a nucleolus in every game depends on a theorem in mathematical analysis which says a continuous function defined on a closed bounded set always attains its minimum on that set. Since $E(v)$ is closed and bounded, and $\theta(x)_i$, the ith component of $\theta(x)$, is continuous for each i, $1 \leqslant i \leqslant 2^n$, the theorem says $\theta(x)_1$ always attains its minimum. If this is not unique, then think of $\theta(x)_2$ defined on the set of minima of $\theta(x)_1$. This set is also closed and bounded, so $\theta(x)_2$ attains its minima, and we repeat the argument through all the 2^n components of $\theta(x)$. This guarantees that $\theta(x)$ has a lexicographic minimum. The uniqueness follows by showing that if $\theta(x) = \theta(y)$ and $x \neq y$, then $\theta(\frac{1}{2}(x+y)) < \theta(x)$ (see Problem 4.10).

4.8 SHAPLEY VALUE

The previous solution concepts were equilibrium outcomes of the game, in that they considered one or a set of imputations that are likely to persist during the preplay negotiations—the core because it is undominated, the stable set because it dominates everything outside it, and the nucleolus because it minimises the maximum unhappiness. Shapley (1953) looked at what each player could reasonably expect to get before the game has begun. He put forward three axioms, which he felt $\phi_i(v)$, player i's expectation in a game with characteristic function v, should satisfy.

S1 $\phi_i(v)$ is independent of the labelling of the players. If π is a permutation of $1,2,\ldots,n$ and πv is the characteristic function of the game, with the players numbers permuted by π, then

$$\phi_{\pi(i)}(\pi v) = \phi_i(v). \tag{4.41}$$

S2 The sum of the expectations should equal the maximum available from the game, so

$$\sum_{i=1}^{n} \phi_i(v) = v(N). \tag{4.42}$$

S3 If u, v are the characteristic functions of two games, $u + v$ is the characteristic function of the game of playing both the games together. ϕ should satisfy

$$\phi_i(u + v) = \phi_i(u) + \phi_i(v). \tag{4.43}$$

S1 and S2 are quite acceptable, but S3 is decidedly odd. What exactly is meant by playing two games together? Is it that having chosen a coalition you must stay with it through both games? In which case it is not at all obvious why what you expect, in these two games with this restriction, is the same as if you are allowed to join different coalitions in each game. However, given these assumptions, Shapley proved the following theorem.

Theorem 4.3. There is only one function which satisfies S1, S2 and S3, namely:

$$\phi_i(v) = \sum_{S:i\in S} \frac{(\#S - 1)!(n - \#S)!}{n!} (v(S) - v(S - \{i\})) , \tag{4.44}$$

where the summation is over all coalitions S which contain player i and $\#S$ is the number of players in coalition S. $\phi_i(v)$ is called the **Shapley value.**

We can interpret (4.44) in a rather neat way. The idea is that players arrive at the game in random order. If, when player i arrives, $S - \{i\}$ are the players who have already arrived, he gets the extra amount he brings to the game, namely $v(S) - v(S - \{i\})$. The probability that he actually arrives after $S - \{i\}$ have arrived and before the remaining $N - S$ come, is the number of sequences starting with $S - \{i\}$, in some order, then player $\{i\}$, followed by $N - S$ in some order, divided by the total number of different ways people can arrive. This is $(\#S - 1)!(n - \#S)!/n!$

This seems a much more reasonable explanation of (4.44) than trying to justify S3, and make it quite acceptable as the value each player will expect from the game. Let us now look at the Shapley value for our standard game.

Example 4.9—Oil Market game. Recall $v(1) = a$, $v(2) = v(3) = v(2,3) = 0$, $v(1,2) = b$, $v(1,3) = v(1,2,3) = c$. To find $\phi_1(v)$ we sum over the coalitions $S = \{1\},\{1,2\},\{1,3\}$ and $\{1,2,3\}$, and substituting into (4.44) gives:

$$\phi_1(v) = \frac{0!2!}{3!}(a-0) + \frac{1!1!}{3!}(b-0) + \frac{1!1!}{3!}(c-0) + \frac{2!0!}{3!}(c-0)$$

$$= \tfrac{1}{2}c + \tfrac{1}{3}a + \tfrac{1}{6}. \qquad (4.45)$$

For $\phi_2(v)$ we sum over $\{2\}$, $\{1,2\}$, $\{2,3\}$ and $\{1,2,3\}$ so:

$$\phi_2(v) = \frac{0!2!}{3!}(0-0) + \frac{1!1!}{3!}(b-a) + \frac{1!1!}{3!}(0-0) + \frac{2!0!}{3!}(c-c)$$

$$= \tfrac{1}{6}b - \tfrac{1}{6}a. \qquad (4.46)$$

For $\phi_3(v)$ we sum over $\{3\}$, $\{1,3\}$, $\{2,3\}$ and $\{1,2,3\}$. So:

$$\phi_3(v) = \frac{0!2!}{3!}(0-0) + \frac{1!1!}{3!}(c-a) + \frac{1!1!}{3!}(0-0) + \frac{2!0!}{3!}(c-b)$$

$$= \tfrac{1}{2}c - \tfrac{1}{6}a - \tfrac{1}{3}b. \qquad (4.47)$$

As you would expect, since $c \geqslant b \geqslant a$, then $\tfrac{1}{2}c + \tfrac{1}{3}a + \tfrac{1}{6}b \geqslant \tfrac{1}{2}c - \tfrac{1}{6}a - \tfrac{1}{3}b \geqslant \tfrac{1}{6}b - \tfrac{1}{6}a$ and so $\phi_1(v) \geqslant \phi_3(v) \geqslant \phi_2(v)$.

Example 4.10 Lilliput U.N. Security Council. This has characteristic function $v(1,2,3) = v(1,2,4) = v(1,2,5) = v(1,2,3,4) = v(1,2,3,5) = v(1,2,4,5) = v(1,2,3,4,5) = 1$ with $v(S) = 0$ for all other S. By the symmetry of v, it is obvious that $\phi_1(v) = \phi_2(v)$ and $\phi_3(v) = \phi_v v = \phi_5(v)$. For $\phi_1(v)$, we sum (4.44) over all coalitions with player 1 in it, but the only coalitions where there is a difference between $v(S)$ and $v(S - \{1\})$, are $\{1,2,3\}$, $\{1,2,4\}$, $\{1,2,5\}$, $\{1,2,3,4\}$, $\{1,2,3,5\}$, $\{1,2,4,5\}$ and $\{1,2,3,4,5\}$. Thus:

$$\phi_1(v) = 3\,\frac{2!2!}{5!}(1-0) + 3\,\frac{3!1!}{5!}(1-0) + \frac{4!0!}{5!}(1-0) = \frac{9}{20}. \qquad (4.48)$$

Similarly for $\phi_3(v)$, the only coalition S which contains 3 and where $v(S)$ and $v(S-3)$ are different, is $\{1,2,3\}$:

$$\phi_3(v) = \frac{2!2!}{5!}(1-0) = \frac{1}{30}. \qquad (4.49)$$

Thus, $\phi_1(v) = \phi_2(v) = 0.45$, $\phi_3(v) = \phi_4(v) = \phi_5 v = 0.0333$, and so players 1 and 2 appear fourteen times more important than 3, 4 and 5.

4.9 OTHER SOLUTION CONCEPTS

We have in no way exhausted the suggestions that have been made of useful solution concepts in n-person games. In this section we will briefly mention some of the other candidates.

So far all the solutions suggested have been imputations or sets of imputations. Implicit in this is the belief that the grand coalition, N, of everyone, or else a coalition as strong as N, will form so that $\sum_{i=1}^{n} x_i = v(N)$. But since several of the solution concepts need not exist, for some games, what does happen in these games? Once we start thinking along these lines, we are led to drop the requirement that the solutions be imputations, and look more closely at what coalitions are actually forming. We only require that the rewards to the players are Pareto optimal for these, rather than for the best possible coalition N. We are led then to the idea of individually rational payoff configurations, instead of imputations.

Definition 4.11. An **individually rational payoff configuration** (i.r.p.c.) consists of a set of disjoint coalitions $B_1, B_2, \ldots, B_m$ whose union is every player, and a payoff vector $x = (x_1, x_2, \ldots, x_n)$ so that :

$$\text{(i)} \quad \sum_{i \in B_j} x_i = v(B_j), \quad \text{for } j = 1,2,\ldots,m; \tag{4.50}$$

$$\text{(ii)} \quad x_i \geqslant v(i). \tag{4.51}$$

Thus, the only difference from an imputation is that the Pareto optimality condition (4.50) is over the particular coalitions rather than over N, as in (4.16).

In the idea of a bargaining set, introduced by Aumann and Maschler (1964), one looks to see which i.r.p.c. will remain using the following analysis.

Consider two players, k and l, in the same coalition B_j with rewards x = $(x_1, \ldots, x_n)$. Player k thinks of all the coalitions he can join, but with player l excluded. Suppose there is such a coalition C where all its members could get more than at present. In symbols, this means there is a vector $\mathbf{y}$ = $(y_1, \ldots, y_n)$ with $\sum_{i \in C} y_i = v(C)$, $y_k > x_k$ and $y_i \geqslant x_i$ for all other i in C. We then say player k has an **objection** against player l, in that he can turn to player l and ask why should he stay in B_j with him, when he can do better and the other players who make up C can do as well by joining it.

In the face of this attack, player l might do likewise and think of all coalitions he can join but which exclude k. He looks for one D and a set of rewards $\mathbf{z} = (z_1, \ldots, z_n)$ so that $\sum_{i \in D} z_i = v(D)$,. $z_i \geqslant x_i$ for all i in D and $z_i \geqslant y_i$ for all i in $C \cap D$. If such a set exists, l is said to have a **counter-objection** because he can turn to k and say: 'We can form a set D without you, so everyone does as well as at present, and those in D who you need to form C do as well in my coalition D as they do in your coalition C.'

If player l has no counter-objection against k's objection, the objection is said to be justified and we have the following definition of a **bargaining set.**

Definition 4.12. The bargaining set M is the set of all individually rational payoff configurations in which no player has a justified objection against any other member of the same coalition.

Obviously, calculating the bargaining set takes some time since one must look at each possible coalition configuration in turn, and for each coalition in it examine each pair of players to see if there are justified objections. Look at the Oil Market game yet again.

Example 4.11—Oil Market game. Looking at coalition structure $\{1\}$, $\{2\}$, $\{3\}$, the only i.r.p.c. is $\mathbf{x} = (a,0,0)$, and since there are no two-person coalitions there can be no objections. For $\{12\}$, $\{3\}$, the i.r.p.c. are $\{x = (x, b - x, 0),\ b \geqslant x \geqslant a\}$ and it is easy to see 1 has an objection against 2 by offering 3, $c - y$ say, where $c > y > x$. 2 has no counter-objection to this except in the case $(b,0,0)$, where he is already getting 0, and so his counter-objection is to work by himself and still get 0. For $\{1\}$, $\{23\}$, the only i.r.p.c. is $(a,0,0)$ and again although both 2 and 3 have objections against each other, by offering to work with 1 their partner has a counter-objection since by working by himself he will still get 0, which is what he gets now. For both ($\{13\}$, $\{2\}$) and $\{1,2,3\}$, an i.r.p.c. of the form $(x,0,c - x)$, $b \leqslant x \leqslant c$, gives rise to no justified objections, but the other i.r.p.c.'s do. Thus, the bargaining set is:

$$\begin{aligned} M = \ & \{1\}, \{2\}, \{3\}: && (a,0,0); \\ & \{12\}, \{3\}: && (b,0,0); \\ & \{1\}, \{23\}: && (a,0,0); \\ & \{13\}, \{2\}: && (x,0,c - x),\ b \leqslant x \leqslant c; \\ & \{123\}: && (x,0,c - x),\ b \leqslant x \leqslant c. \end{aligned} \tag{4.52}$$

Another way of concentrating on a special class of outcomes is the **kernel** suggested by Davis and Maschler (1965). In this we again start with an individually rational payoff configuration $(B_1,\ldots,B_m,x)$, and suppose a set of players D think about forming a coalition. For each such coalition we define $e(D) = v(D) - \sum_{i \in D} x_i$ as the **excess**, just as in the definition of a nucleolus. Here of course the larger the excess, the more likely this potential coalition is to form. Now consider two players, k and l, belonging to the same coalition B_j at present and denote by $T_{k,l}$ all the coalitions that can be formed which include k, but not l. Define the maximum excess of k over l as $S_{k,l}$:

$$S_{k,l} = \max_{D : D \in T_{k,l}} e(D). \tag{4.53}$$

This number is supposed to represent k's hope of the best advantage he can get by leaving l and joining a new coalition. Similarly, we can define $S_{l,k}$ as the excess of l over k in exactly the same way. Obviously, either $S_{l,k} > S_{k,l}$, $S_{k,l} > S_{l,k}$ or $S_{k,l} = S_{l,k}$. We say player k outweighs player l if:

$$S_{k,l} > S_{l,k} \quad \text{and} \quad x_l > v\{l\}, \tag{4.54}$$

and vice versa if l outweighs k. Notice there are two conditions in (4.54). Firstly, player k must hope to get more by abandoning l than l gets by abandoning k, and also l must be getting more than he would get if he left the present coalition and worked on his own.

If player k does not outweigh player l, nor player l outweigh k, we say k and l are in equilibrium. This can happen in one of three ways:

$$S_{k,l} = S_{l,k}; \tag{4.55}$$

$$S_{k,l} > S_{l,k} \quad \text{and} \quad x_l = v(l); \tag{4.56}$$

$$S_{l,k} > S_{k,l} \quad \text{and} \quad x_k = v(k). \tag{4.57}$$

If all players are in equilibrium we have the following definition.

Definintion 4.13. The **kernel**, K, is the set of all individually rational payoff congifurations such that every pair of players in the same coalitions of the configuration is in equilibrium.

Davis and Maschler (1965) showed that the kernel exists for all games and is a subset of the bargaining set. Using this we can find the kernel of the Oil Market game.

Example 4.12. (4.52) gives us the elements of the bargaining set, so because of the above remark we need only concentrate on those i.r.p.c. For ({1}, {2}, {3}, $(a,0,0)$) since there are no coalitions with two players, the definition of a kernel is trivially satisfied. With ({12}, {3}, $(b,0,0)$ we need only look at whether 1 or 2 outweigh each other. $S_{1,2} = c - b > S_{2,1} = 0$, but since $x_2 = v\{2\} = 0$, (4.56) is satisfied and the players are in equilibrium. Similarly, for ({1}, 3}, $(a,0,0)$) $S_{3,2} = c - a > S_{2,3} = b - a$ but $x_2 = v\{2\} = 0$ again means the players are in equilibrium. For ({13}, {2}, $(x,0,c - x)$, $b \leqslant x \leqslant c$) we have $S_{1,3} = b - x$ and $S_{3,1} = x - c$. At $x = ½(b + c)$, $S_{1,3} = S_{3,1}$ so (4.55) holds, but for all other values we do not get the two equalities we require for (4.56) or (4.57). Thus, the only i.r.p.c. in the kernel is ({13}, {2}, $(½(b + c),0,½(c - b))$). A similar analysis works for the general coalition, giving the kernel K as:

$$\begin{aligned} K = \;&\{1\}, \{2\}, \{3\}, && (a,0,0); \\ &\{12\}, \{3\}, && (b,0,0); \\ &\{1\}, \{23\}, && (a,0,0); \\ &\{13\}, \{2\}, && (½b + ½c,0,½c - ½b); \\ &\{123\}, && (½b + ½c,0,½c - ½b). \end{aligned} \tag{4.58}$$

4.10 FURTHER READING

Once again, the foundations of this chapter are to be found in von Neumann and Morgenstern's book (1947), and a good survey of the early years is found in Luce and Raiffa (1957). More recently, there is an excellent survey of the period in the sixties, when one by one the conjectures about n-person games were disproved, by Lucas (1971), who found most of the counter-examples. The book by Rapoport (1970) is also an easy introductory guide to the main concepts, and Shubik (1982) gives a comprehensive account of the material.

The area of cooperative n-person games is still the most active area of research in game theory. New solution concepts are still appearing in the characteristic form of the game described in this chapter. Roth (1976) introduced the ideas of **subsolutions** and the **supercore**, which have strong connections with stable sets and the core, but have the extra property that every game has a subsolution.

The Shapley value was the beginning of the value theory approach to determining a unique imputation for each game. The value is usually arrived at by applying bargaining axioms, which have some ideas of equity and reasonableness involved in them. As was mentioned in Chapter 3, if we allow transfer of utility, then the Shapley value applied to two-person games gives the maximin–maximin bargaining solution. The n-person analogue of the threat bargaining solution was introduced by Harsanyi (1959), and in his book (1977) he looks at the whole problem in more detail. Banzhof in 1968 introduced a different value function for voting games, like the Lilliput U.N. Security Council one, by changing the weighting given to the various coalitions.

One of the criticisms of the characteristic function approach is that there is no real connection between the imputations and the possible coalition structure apart from the implicit assumption that a coalition as strong as $\{1,2,\ldots,n\}$ will form. Luce (1954) suggested one should look at pairs (x,P), where x is an imputation and $\mathbf{P}$ a partition of the n players, so that $\sum_{i\in S} x_i \geqslant v(S)$ for all coalitions S in $\psi(\mathbf{P})$. (This is a weaker condition than (4.23) for the core but stronger than (4.50) for i.r.p.c.) The important point that Luce added is that only certain coalition structures $\psi(\mathbf{P})$ can be formed from $\mathbf{P}$ by players changing coalitions. These are given by the function ψ and a pair (x,P) is called ψ-stable if it doesn't pay any of the allowed coalitions to form.

Another one of the assumptions made in the characteristic function form is that side-payments can be made between the players. If one drops this assumption, one also avoids the problem of transferable utility. The only difference is that the characteristic function $v(S)$, of a coalition S, is a set of vectors $\mathbf{x}$, so that S can guarantee all its members their respective components in **x**. For a good survey of this area see Aumann (1967). You can also extend the value theory based on Shapley's value to these games without side-payments. This λ-transfer value has been considered by Harsanyi (1963) and Shapley (1969).

An alternative formulation to the characteristic function for games with side-payments, was the partition function introduced by Thrall and Lucas (1963). The partition function F is defined on the set of all partitions $P = (B_1, B_2, B_n)$ of $\{1,2,\ldots,n\}$. Each coalition B_i in P is assigned a real number $F(B_i)$, which is what that coalition will receive if that particular partition forms. It is then easy to define imputations, domination of imputations, and stable sets in the obvious way, but again there are games which have no stable sets in this context.

PROBLEMS OF CHAPTER 4

4.1. A two-person non-zero-sum game has payoff matrix:

$$\begin{pmatrix} (4,1) & (1,0) \\ (-1,2) & (2.3) \end{pmatrix}.$$

Using the techniques of Chapter 3, show the negotiation set of this game is $\{(u,v) | u + v = 5, 2 \leqslant u \leqslant 4\}$ and the maximin–maximin bargaining solution is $(1\frac{1}{4}, \frac{9}{4})$.

Consider the same game as an n-person game ($n = 2$), and show its characteristic function is $v(1) = 1\frac{1}{2}$, $v(2) = 1$, $v(1,2) = 5$. Hence, find all the imputations. Why isn't the set of imputations equal to the negotiations set for this problem? Show that the Shapley value is equal to the maximin bargaining solution. What is the core and the nucleolus for this game?

4.2. Prove that in all two-person games the core will be the whole set of imputations and that this is the only stable set.

4.3. Show that in any two-person game, which when considered as an n-person game, has characteristic function $v(1) = a$, $v(2) = b$, $v(1,2) = c$, the nucleolus and the Shapley values are both $(\frac{1}{2}a + c - b), \frac{1}{2}(b + c - a))$. Show if the negotiation set is $u + v = c$ this is also the outcome of the maximin bargaining solution.

Consider a three-person game with characteristic function $v(1) = v(2) = v(3) = 0$, $v(1,2) = a$, $v(1,3) = b$, $v(2,3) = c$, $v(1,2,3) = 1$. Show that the nucleolus for such a game is $((a + b - 2c + 1)/3, (a + c - 2b + 1)/3, (b + c - 2a + 1)/3)$. What is the Shapley value? Under what conditions on a, b and c are the Shapley value and the nucleolus the same? Consider a general three-person game with $v(1) = a_1$, $v(2) = a_2$, $v(3) = a_3$, $v(1,2) = a_{12}$, $v(1,3) = a_{13}$, $v(2,3) = a_{23}$, $v(1,2,3) =$

a_{123}. By applying a transformation which makes it strategically equivalent to the above game, describe the condition on the a's which mean the nucleolus and the Shapley value are the same.

4.4. For the game $v(1) = 4$, $v(2) = v(3) = 0$, $v(1,2) = 5$, $v(1,3) = 7$, $v(2,3) = 6$, $v(1,2,3) = 10$, find (a) the Shapley value; (b) the core; (c) the nucleolus; and (d) prove $\{(4,6 - x,x), 0 \leqslant x \leqslant 6\}$ is a stable set.

4.5. Consider the game with characteristic function $v(1) = 1$, $v(2) = 2$, $v(3) = 3$, $v(1,2) = 3$, $v(1,2) = 10$, $v(2,3) = 6$, $v(1,2,3) = 12$. Find (a) the set of imputations; (b) the core; (c) the nucleolus; (d) the bargaining set; and (e) the kernel.

4.6. On Wimbledon Common, Bungo, the Womble, has found a mattress which he values at £2. Wellington has not got one, but would like just one mattress, which he would value at £6. Orinoco, who likes his comforts, also has found a mattress which he thinks is worth £4, but he would not mind a second one which he would value at £3. Set up the buying and selling as a three-womble game, and show that its characteristic function is:

$v(1) = 2$, $v(2) = 0$, $v(3) = 4$, $v(1,2) = 6$, $v(1,3) = 7$, $v(2,3) = 6$, $v(1,2,3) = 10$.

Find the Shapley values of the game, and describe the set of imputations. What is the subset of these that form the core of the game?
Show that imputations $\{(2 + x,4 - x,4), 0 \leqslant x \leqslant 4\}$ form a stable set.

4.7. In the local park there are two ponds, a large one and a small one. Four ducks, called Arthur, Buck, Chuck and Donald, live on them. Three-quarters of the bread crumbs are thrown into the large pond by the people who use the park, and only a quarter of them are thrown into the small pond. Each duck has to choose one of the ponds to live on and they can form gangs to frighten other ducks off their pond. Arthur can frighten off Buck and also Chuck and Donald individually, and even Chuck and Donald together. All other gangs of two or three ducks can frighten off any individual duck on his own. Buck can frighten off Chuck, and also Donald while Chuck can frighten off Donald. When gangs of two ducks meet, Arthur and Buck can frighten off Chuck and Donald, Arthur and Chuck frighten off Buck and Donald, but Buck and Chuck frighten off Arthur and Donald. Set this up as a four-player game, where the payoff is the proportion of bread crumbs each duck gets. Find the characteristic function, the set of imputations, and the core of the game. Which, if any, of the following sets of imputations is a stable set?

$$S_1 = \{(\tfrac{1}{4}, x, \tfrac{3}{4} - x, 0),\quad 0 \leqslant x \leqslant \tfrac{3}{4}\};$$
$$S_2 = \{(\tfrac{1}{4} + x, \tfrac{1}{2} - x, 0, \tfrac{1}{4}),\quad 0 \leqslant x \leqslant \tfrac{1}{2}\};$$
$$S_3 = \{(\tfrac{1}{2} - x, x, x, \tfrac{1}{2} - x),\quad 0 \leqslant x \leqslant \tfrac{1}{2}\}.$$

If Donald flies away, vowing to become a movie star, write down the characteristic function of the remaining three-player game. Why is it obvious that this game has no core? Describe one stable set for the three-player game.

4.8. Consider the jury system where one requires ten of the twelve jurors to vote a man guilty, for him to be found guilty. Describe the characteristic function of this game where $v(S) = 1$, if, when the members of S believe him guilty, he will be found guilty. Show there is no core and that the Shapley value and the nucleolus are both $(\tfrac{1}{12}, \tfrac{1}{12}, \ldots, \tfrac{1}{12})$.
If the judge believes a defendent is innocent he will direct the jury to find him so, which they have to do; only if he believes him guilty will he let the jury decide for themselves. In this 13-person game, what is the characteristic function? Show that the core is $(1,0,0,0,\ldots,0)$ where player 1 is the judge, and that the Shapley value is $(\tfrac{18}{78}, \tfrac{5}{78}, \ldots, \tfrac{5}{78})$. What is the nucleolus? What are the stable sets?

4.9. Voting tactics in a committee of n people can be thought of as a game, where the payoff is 1 to any coalition S who can pass a motion they agree with and 0 if they cannot get it passed by the committee. Player 1 is the chairman of the committee and there are two types of voting rules for chairmen:

(a) he may not vote unless there is a tie, and then he has the casting vote; or
(b) he votes like any other member of the committee, and if there is a tie, he has the casting vote.

Explain why, if n is odd, then the Shapley value of the chairman is the same as any other committee member no matter which rule is used. Suppose n is even, find the Shapley value of the chairman first under rule (a) and then under rule (b).
A committee consisting of five people $\{1,2,3,4,5\}$ chaired by 1, votes to decide which of two subcommittees shall decide on any motion. Subcommittee A consists of $\{1,2,3\}$ and is chaired by 1; subcommittee B consists of $\{2,4\}$ and is chaired by 2. All committees operate under rule (b). Write down all the winning coalitions S (i.e. the coalitions where $v(S) = 1$). Hence, find the Shapley values of all the players.

4.10. If x and y are two imputations, $x \neq y$, and θ is defined as in (4.37), prove that if $\theta(x) = \theta(y)$, then $\theta(\frac{1}{2}(x + y)) < \theta(x)$. Hence, prove the nucleolus of a game is unique.

CHAPTER 5

Market games and oligopoly

'Grace is given by God, but knowledge is bought in the market'—A.H. Clough

5.1 EDGEWORTH MARKET GAMES

One of the earliest applications of game theory was in mathematical economics to describe the bargaining involved in trading. The simplest such model, suggested by Edgeworth (1881), in fact predates game theory by half a century, but still can be interpreted as a game. Edgeworth supposed that there were only two commodities to be traded, A for apples, say, and B for bread. Assume there are M apple traders, where trader i starts with resources $(a_i,0)$, $i = 1,2,\ldots,M$, meaning he has a_i apples and 0 bread. Similarly, assume there are N bread traders, starting with resources $(0,b_j)$, $j = M + 1, M + 2,\ldots,M + N$, respectively. The utility of trader i is $u_i(a,b)$ if he has a apples and b bread, and we assume each of the utility functions u_i is **concave**. This is an unnecessarily strong assumption, but we have to assume something that means the traders actually prefer to trade and so prefer some combination of the two goods to more extreme outcomes. u is **concave** if $u(\lambda\mathbf{x} + (1 - \lambda)\mathbf{y}) \geqslant \lambda u(\mathbf{x}) + (1 - \lambda)u(\mathbf{y})$ for $0 \leqslant \lambda \leqslant 1$, so in this case we want:

$$u_i(\lambda a_1 + (1 - \lambda)a_2,\lambda b_1 + (1 - \lambda)b_2) \geqslant \lambda u_i(a_1,b_1) + (1 - \lambda)u_i(a_2,b_2). \tag{5.1}$$

This is an easy condition to use. The problem is to find out which trades take place if traders each try to maximise their utility.

5.2 [1,1]-MARKET GAME

This is the game where there is only one trader of each type, and the A-trader starts with $(a,0)$, the B-trader with $(0,b)$. It has become standard to describe the results in terms of a box diagram—the Edgeworth box—as follows. Firstly we can plot the utility of trader A on a graph by drawing his indifference curves (just like the contour lines on a map). At all the points

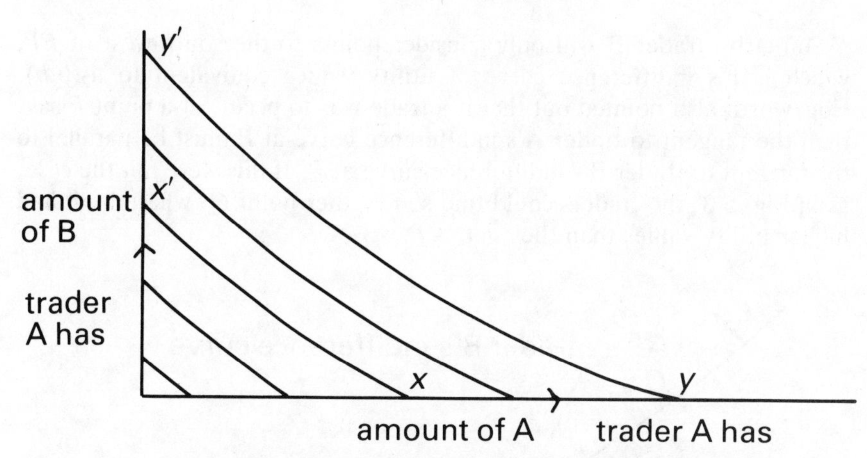

Fig. 5.1—Indifference curves of trader A.

on the curve XX' in Fig. 5.1 he has the same utility and this is less than his utility at all the points on the curve YY'. Thus, trader A obviously tries to go as far to the north-east as he can, whereas the concavity of his utility function is expressed in the way his indifference curve actually curves.

A similar diagram will represent the utility curves of the B-trader, but if the A trader has (x,y) after the trade, notice that the B-trader will have $(a - x, b - y)$. Thus, we can represent both traders' indifference curves on the same diagram by rotating the diagram for B through 180° and identifying its origin with the point (a,b) on the diagram for A. This leads to the Edgeworth box, Fig. 5.2, whose length is a and height is b. Each point in the box represents a possible result of the trading where A has (x,y) and B has $(a - x, b - y)$. Obviously, trader A will only consider points to the north-east of ED as these are the points where his utility is at least $u_1(a,0)$—the utility he gets if he does not trade.

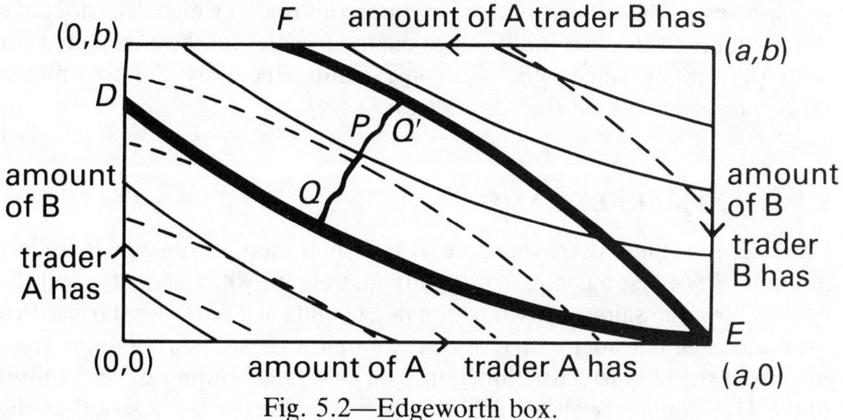

Fig. 5.2—Edgeworth box.

Similarly, trader B will only consider points to the south-west of EF, which is his indifference curve of utility values equivalent to $u_2(0,b)$. Edgeworth also pointed out that if a trade was to occur, at a point P say, then the tangent to trader A's indifference curve at P must be parallel to the tangent to trader B's indifference curve at P. If this were not the case, as in Fig. 5.3, the traders could find some other point Q, where both had higher utility values than they get at P.

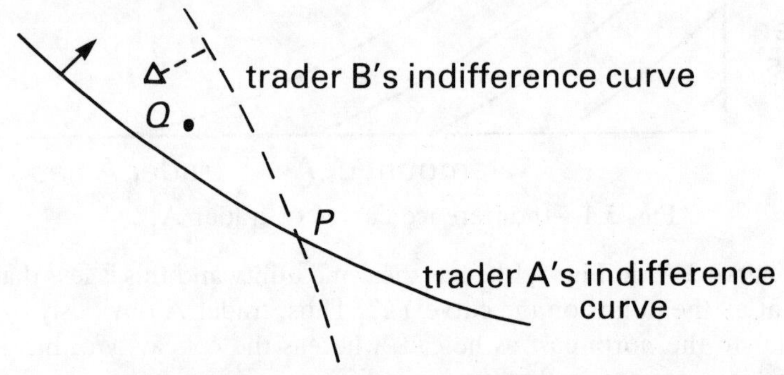

Fig. 5.3—Non-tangential indifference curves.

Edgeworth suggested that the solution of the trading will be on the curve QQ' between ED and EF, where the points on QQ' have the property that the tangents to the two indifference curves are parallel. This is called the contract curve.

If we think of this model as a two-person non-zero-sum game, we can see that the contract curve is in fact the negotiation set (Chapter 3, Section 3.8). Firstly, any point to the north-east of ED ensures that trader A gets at least his maximin value $u_1(a\ 0)$, and similarly those to the south-west of EF ensure trader B at least his maximin value $u_2(0,b)$. The argument of the previous paragraph shows that any point in this region which does not have the tangents to the two indifference curves parallel can be dominated for both players by some other outcome. Thus, the only Pareto optimal outcomes are those on the contract curve QQ'.

5.3 [M,N]-MARKET GAME

Consider the game where there are M-type A-traders and N-type B-traders as an $M + N$-person game. To simplify matters we will assume that all the traders have the same utility function $u(x,y)$ and each A-trader starts with a of A and each B-trader with b of B. If we follow the ideas in Chapter 4, we must find the characteristic function v, of this game. Suppose S is a subset of the $M + N$ players with s_1 type A-traders and s_2 type B-traders, then the best S can ensure itself is the highest sum of the utilities of its members that can be obtained when they trade with each other. Obviously, as in the Oil Market game example of Chapter 4, the worst the remaining players can do to S is not to trade with them. Notice the characteristic function formulation implies side-payments take place. It is infeasible to have pieces of apple and slices of breads changing hands, so obviously the side-payments will be in a divisible commodity like money. However, we transfer the utility of these side-payments into the equivalent utility of apples and bread. Now:

$$v(S) = \max_{x_1,\ldots,x_{s_1+s_2},y_1,\ldots,y_{s_1+s_2}} \sum_{i=1}^{s_1+s_2} u(x_i,y_i), \tag{5.2}$$

where

$$\sum_{i=1}^{s_1+s_2} x_i = s_1 a, \qquad \sum_{i=1}^{s_1+s_2} y_i = s_2 b.$$

Since the utility function u is concave, (5.1) implies that:

$$u(x_1,y_1) + u(x_2,y_2) \leqslant 2u(\tfrac{1}{2}(x_1 + x_2),\tfrac{1}{2}(y_1 + y_2)). \tag{5.3}$$

i.e. the sum of the utilities is maximised when both players get the same amount. We can repeat this result for any number of players just by applying (5.1) over and over again and get:

$$\sum_{i=1}^{s_1+s_2} u(x_i,y_i) \leqslant (s_1 + s_2)u(s_1a/(s_1 + s_2),s_2b/(s_1 + s_2)). \tag{5.4}$$

So:

$$v(S) = (s_1 + s_2)u(s_1a/(s_1 + s_2),s_2b/(s_1 + s_2)). \tag{5.5}$$

Notice that in the [1,1] game, which we looked at earlier, the characteristic function is:

$$v(1) = u(a,0), \qquad v(2) = u(0,b) \quad \text{and} \quad v(1,2) = 2u\left(\frac{a}{2}, \frac{b}{2}\right), \tag{5.6}$$

and so the set of imputations is:

$$E(v) = \{(u(a,0) + pc,u(0,b) + (1 - p)c), \qquad 0 \leqslant p \leqslant 1\}, \tag{5.7}$$

where

$$c = 2u(a/2,b/2) - u(a,0) - u(0,b).$$

We could think of p as the price of the goods. It reflects the number of units of B exchanged for one unit of A. Remember (Problem 4.2) that in a two-person game, the core is equal to the set of imputations so (5.7) represents the core. It corresponds to just one point on the contract curve

QQ'—the point $(\frac{1}{2}a, \frac{1}{2}b)$—and the range of possible utilities to the players arise because they are allowed to make side-payments to one another so as to agree on that share out. This exemplifies again the discussion in (4.4) on the difference between two-person cooperative games and the same games thought of as n-person games.

5.4 [1,N]-MARKET GAMES

Having worked out the characteristic function for the general game, let us look at some special cases to see if the results describe any of the economic phenomena we observe in reality. The first situation to look at is the case of a monopolist. Suppose there is only one A-trader, but N B-traders. What happens to the core in these games? We would suspect that the monopolist could charge as high a price as he wants, provided it is still worth the others trading, and the following result tends to confirm this belief.

Theorem 5.1. The imputation $x^* = ((N+1)u(a/N+1, Nb/N+1) - Nu(0,b), u(0,b), \ldots, u(0,b))$ lies in the core.

Proof. Using Theorem 4.2, $\mathbf{x}^* = (x_1^*, x_2^*, \ldots, x_{N+1}^*)$ can only be in the core if, for $S = \{1\} \cup \{K\}$, where K is some k of the B-traders,.

$$x_1^* + \sum_{i \in K} x_i \geqslant v(S). \tag{5.8}$$

This corresponds to proving:

$$(N+1)u(a/N+1, Nb/N+1) - (N-k)u(0,b) \geqslant (k+1)u(a/k+1, kb/k+1), \tag{5.9}$$

or

$$(N+1)u(a/N+1, Nb/N+1) \geqslant (N-k)u(0,b) + (k+1)u(a/k+1, kb/k+1). \tag{5.10}$$

But (5.10) holds because the concavity of u says the best way of maximising the sum of the utility of $N+1$ people is when they all get the same. For coalitions S which don't contain 1 it is trivial that $\sum_{i \in S} x_i^* = v(S)$ and so x^* is in the core.

A harder result, which we will not prove here, says that for any $\varepsilon > 0$, no matter how small, there is an n_ε, so that for $N \geqslant n_\varepsilon$, the only imputations, x, in the core of the game $[1,N]$ have $x_1 > x_1^* - \varepsilon$. This means that as N gets bigger, the core is getting smaller and smaller until in the limit it will consist of x^* alone.

5.5 [N,N]-MARKET GAME

If we look at the games where the number of A-traders and B-traders are the same, we can show that there is a stable set analogous to the set of imputations in the [1,1]-market game. Essentially each trader of the same commodity gets the same amount and the set is parameterised by p, which can be thought of as the market price.

Theorem 5.2. In the $[N,N]$-market game there is a stable set of the form:

$$\{(u(a,0) + pc,\ldots,u(a,0) + pc,u(0,b) + (1-p)c,\ldots,u(0,b) + (1-p)c) \quad 0 \leqslant p \leqslant 1\}, \tag{5.11}$$

where

$$c = 2u\left(\frac{a}{2},\frac{b}{2}\right) - u(a,0) - u(0,b).$$

Proof. Internal stability is trivially true, since coalitions of one type of trader cannot guarantee each person more than $u(a,0)$ or $u(0,b)$, respectively. If the coalition has both types of traders in it, then as p varies, one type of trader has a better utility value and the other has a worse utility value so there cannot be any domination on that coalition. To prove external stability, take an imputation y not of the above form. Then there must be a pair of traders, i and j, where i is an A-trader and j is a B-trader so that:

$$y_i + y_j < 2u(a/2,b/2), \tag{5.12}$$

since the sum of the $2N$ values $y_i, i = 1,\ldots,2N$ is $2Nu(a/2,b/2)$. Given this, we can now choose a p so that:

$$\frac{y_i - u(a,\,0)}{2u((a/2,\,b/2)) - u(a,\,0) - u(0,\,b)} < p < 1 - \frac{y_j - u(0,\,b)}{2u((a/2,\,b/2)) - u(a,\,0) - u(0,\,b)} \tag{5.13}$$

Then for this p we have $u(a,0) + pc > y_i$ and $u(0,b) + (1-p)c > y_j$, where c is defined in (5.11). Thus, the imputation with this p in the stable set dominates y on $\{i,j\}$.

For these $[N,N]$-games, we can also ask what happens to the core, and this depends on whether the economy has **increasing** or **decreasing returns of scale.**

Definition 5.1. For a coalition S, and an $i \notin S$, we let $S' = S \cup \{i\}$. Then if for all such S and i

$$v(S')/\#S' \geqslant v(S)/\#S, \tag{5.14}$$

the economy is said to have increasing returns of scale.

This corresponds to the economy where big is beautiful since each person who joins a trading group increases the average wealth. In this case we get:

Theorem 5.3. In the $[N,N]$-market game with increasing returns of scale there is always a core.

Proof. It is enough to show that the imputation

$$x^* = \left(\frac{v(2N)}{\#2N}, \frac{v(2N)}{\#2N}, \ldots, \frac{v(2N)}{\#2N}\right) \tag{5.15}$$

is in the core. This follows trivially from Theorem 4.2 since for any coalition S of $\#S$ people, (5.14) implies:

$$\sum x_i^* = \#S \frac{v(2N)}{\#2N} \geqslant v(S). \tag{5.16}$$

Obviously we would define an economy with (5.14) replaced by:

$$v(S')/\#S' < v(S)/\#S, \tag{5.17}$$

as one with **decreasing returns of scale** and it is an easy matter to show that in such a case no core can exist (see Problem 5.3).

5.6 DUOPOLY AND OLIGOPOLY

The previous few sections dealt with a game model of an economic situation that was akin to bartering. The two types of traders essentially swapped goods with one another and the problem was to find out how they cooperated. Although this gives interesting insights into how trading will occur, it does not really model how the marketplace works.

One of the problems that does occur time and again in economic theory is to describe competition among a few firms, who are producing the same or closely related products. If there are only two firms, this is called **duopoly** theory; if more than two, **oligopoly** theory. These producing firms can separately decide on the price they will charge for their product (or alternatively how much of it they will make), and this then determines the demand for the product (or alternatively the price that the buyers will pay for it). The producing firms are then thought of as players in a game, where the rewards to them are their profits, and their object is to maximise these profits. By thinking of such a model as a non-cooperative or a cooperative game we can look at what we mean by market equilibrium, collusion between the firms, and different forms of price fixing.

Let us try to explain how these duopoly and oligopoly models fit in with the $[M,N]$-market games described earlier. Essentially duopoly is a

$[2,\infty]$-market game and oligopoly an $[M,\infty]$-market game, where the first type of traders are the producing firms, who have the particular product they want to sell. The second type are the buyers, or consumers, who have money which they exchange for the product. However, there are so many consumers that we no longer consider them as individuals and we represent their requirements by one utility function $u(p_1,p_2,\ldots,p_M,q_1,q_2,\ldots,q_M)$, where p_i is the price the ith producing firm sets on its product and q_i is the amount of that product bought by consumers. The assumption is that the consumers are told the prices $p_1,p_2,\ldots,p_M$ and then choose the quantities $q_1,\ldots,q_m$ so as to maximise their utility function. This reduces the consumers to a set of price–demand equations which connect the demand q_i for firm i's product, with the prices $p_1,p_2,\ldots,p_M$ set by the firms, i.e.

$$q_i = f_i(p_1,p_2,\ldots,p_M). \tag{5.18}$$

With these equations we can now work out the producers' utility which of course is simply their profit. So for firm i it is:

$$e_i(p_1,p_2,\ldots,p_M) = p_iq_i - c_i(q_i) = p_if_i(p_1,\ldots,p_M) - c_i(f_i(p_1,\ldots,p_M)), \tag{5.19}$$

where c_i is the production cost function for i.

It is equally valid, and perhaps nearer the market game ethos, to say that the producers decide on the quantities, $q_1,q_2,\ldots$, etc. that they will produce and then the consumers' utility function determines the prices p_i they will pay for these products. The profit for a firm can then be worked out as a function of the quantities produced. We will stick with the former price formulation model however.

Obviously, we have an M-person non-zero-sum game and our object is to show how some of the economic solutions suggested for this model can be reinterpreted in a game theory context. Some of them will correspond to ideas from non-cooperative games, others with ideas from cooperative games. It is worth noting that oligopoly theory is a halfway house between monopoly theory and pure competition. In monopoly theory there is only one producing firm, who then selects his price so as to maximise $pf(p) - c(p)$, whereas in pure competition it is assumed there are so many producing firms that essentially demand is unlimited and so there is a fixed price $\bar{p}$ for a particular firm's product, independent of how much it produces. Again there is no game theory involved because the firm just has to decide on the quantity q it will produce by maximising $\bar{p}q - c(q)$.

5.7 COURNOT EQUILIBRIUM

Models of oligopolistic competition date from even farther back than market games, and in one of the first papers, Cournot (1838) suggested a possible solution to the model which, in our notation, reads as follows.

Definition 5.2. A **Cournot equilibrium** is a vector of prices $p^c = (p_1^c, p_2^c, \ldots, p_M^c)$ so that for all the firms, $i = 1, \ldots, M$:

$$e_i(p_1^c, p_2^c, \ldots, p_i^c, \ldots, p_M^c) = \max_{p_i} e_i(p_1^c, p_2^c, \ldots, p_i, \ldots, p_M^c). \tag{5.20}$$

Thus, given the other firms' prices as fixed, the ith firm's Cournot equilibrium price maximises its profit and this condition holds for all the firms. This definition obviously corresponds to the idea of an equilibrium n-tiple in n-person non-cooperative games. Notice, however, that these games have infinite numbers of pure strategies for each player—the prices they choose—and so we cannot appeal to Nash's theorem (Theorem 4.1) directly. However, there is an implicit upper bound on the price a firm will charge, corresponding to no demand at that price, so we can say the price chosen must be bounded above and below (by zero). Thus, the strategy set for each firm is closed, bounded and convex and if we add some reasonable conditions on the functions f_i of (5.18)—namely that they are decreasing in p_i and increasing in p_j, j not equal to i—this is enough to guarantee the existence of a Cournot equilibrium. For more details on the existence conditions of Cournot equilibrium see Friedman (1977).

As an example look at the following duopoly model.

Example 5.1. Assume there are two firms, $i = 1,2$ and that the price–demand functions are:

$$q_1 = f_1(p_1, p_2) = \max(1 + \tfrac{1}{3}p_2 - \tfrac{1}{2}p_1, 0); \tag{5.21}$$

$$q_2 = f_2(p_1, p_2) = \max(1 + \tfrac{1}{4}p_1 \; \tfrac{1}{2}p_2, 0), \tag{5.22}$$

and for simplicity we assume $c_1(q_1) = c_2(q_2) = 0$. Notice that because of the implicit constraints, $q_1 \geqslant 0$, $q_2 \geqslant 0$, we can assume that $0 \leqslant p_1 \leqslant 2 + \tfrac{2}{3}p_2$, and $0 \leqslant p_2 \leqslant 2 + \tfrac{1}{2}p_1$. For prices in these ranges, the profit functions are:

$$e_1(p_1, p_2) = p_1 + \tfrac{1}{3}p_1p_2 - \tfrac{1}{2}p_1^2,$$
$$e_2(p_1, p_2) = p + \tfrac{1}{4}p_1p_2 - \tfrac{1}{2}p_2^2. \tag{5.23}$$

Just as in the swastika method for finding equilibrium pairs we can show that in a Cournot equilibrium, p_1^c must maximise $e_1(p_1, p_2^c)$ so it must satisfy $\mathrm{d}e_1(p_1, p_2^c)/\mathrm{d}p_1 = 0$, provided the solution is a maximum and lies within the price limits above. Similarly, the Cournot equilibrium requires $\mathrm{d}e_2(p_1^c, p_2)/\mathrm{d}p_2 = 0$ and on doing the differentiation we get:

$$\mathrm{d}e_1\,(p_1, p_2)/\mathrm{d}p_1 = 1 + \tfrac{1}{3}p_2 - p_1 = 0, \tag{5.24}$$

$$\mathrm{d}e_2\,(p_1, p_2)/\mathrm{d}p_2 = 1 + \tfrac{1}{4}p_1 - p_2 = 0. \tag{5.25}$$

Solving (5.24) and (5.25) gives $p_1 = \tfrac{16}{11}$ and $p_2 = \tfrac{15}{11}$ which satisfies the price constraints, and it is easy to show from the form of the functions in

(5.23) that they give maxima. Hence, the Cournot equilibrium is $p_1 = 16/11$, $p_2 = 15/11$ and the corresponding profits are $e_1(16/11, 15/11) = 1.06$, and $e_2(16/11, 15/11) = .93$.

5.8 OTHER SOLUTION CONCEPTS IN DUOPOLY

In this section we will concentrate on the game where there are only two firms. In the Cournot equilibrium the corresponding prices satisfy the maximisation conditions $de_i(p_1,p_2)/dp_i = 0$, for $i = 1,2$—equations (5.24) and (5.25) in the previous example. We can usually rewrite these equations to express the price that firm 1 will choose given firm 2's price and vice versa, i.e.

$$p_1 = g_1(p_2), \qquad p_2 = g_2(p_1). \tag{5.26}$$

In Example 5.1, this corresponds to $p_1 = 1 + \frac{1}{3}p_2$ for firm 1 and $p_2 = 1 + \frac{1}{4}p_1$ for firm 2 (see Fig. 5.4). Suppose firm 1 had to give its price before firm 2 gives its price. Then, obviously, firm 2 will choose its price so as to maximise its profit given p_1. It will choose $p_2 = g_2(p_1)$. If firm 1 were to recognise this is going to happen he will realise that his profit is now purely a function of the price he sets, namely $e_1(p_1,g_2(p_1))$. Thus, he should choose his price p_1 so as to maximise $e_1(p_1,g_2(p_1))$, and announce his price.

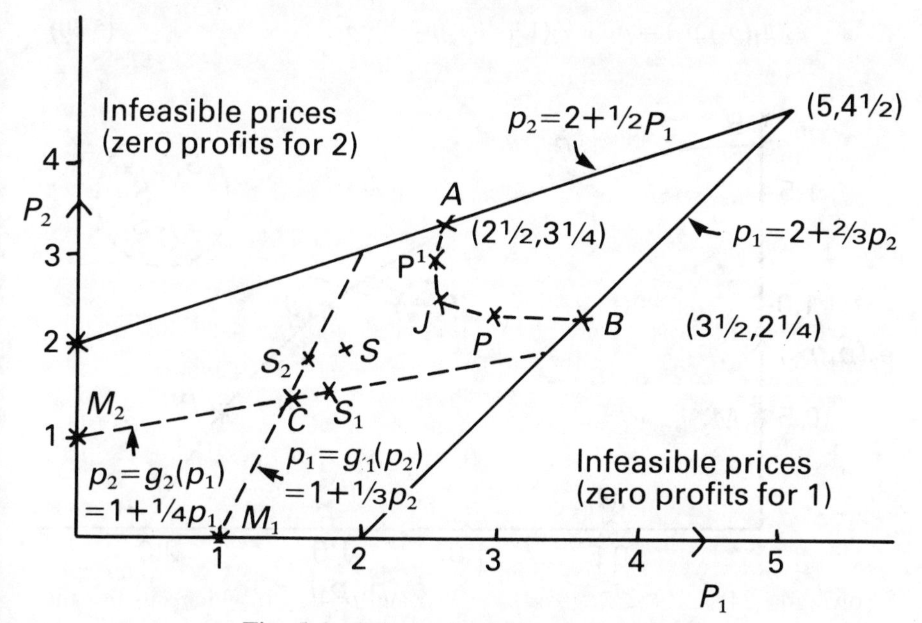

Fig. 5.4—Prices of solution concepts.

This type of strategy was suggested by Stackleberg in 1934, and he calls the firm who announces its price the **leader** and the one who obeys its Cournot reaction function, g, the **follower**. In our example we get the following results.

Example 5.1 (continued). The reaction functions are given by (5.24) and (5.25), namely:

$$p_1 = g_1(p_2) = 1 + \tfrac{1}{3}p_2, \tag{5.27}$$

$$p_2 = g_2(p_1) = 1 + \tfrac{1}{4}p_1. \tag{5.28}$$

Hence, if firm 1 is the leader, he maximises

$$e_1(p_1,g_2(p_1)) = p_1 + \tfrac{1}{3}p_1\,(1 + \tfrac{1}{4}p_1) - \tfrac{1}{2}p_1^2 \tag{5.29}$$

which, on using $de_1/dp_1 = 0$, gives $p_1 = \tfrac{8}{5}$. Again, it is easy to check that this maximises (5.29) and on substituting into (5.28) we get that if firm 1 sets its price at $\tfrac{8}{5}$, then firm 2 will choose its price to be $p_2 = \tfrac{7}{5}$ and the corresponding profits are $e_1(\tfrac{8}{5},\tfrac{7}{5}) = 1.067$, $e_2(\tfrac{8}{5},\tfrac{7}{5}) = 0.98$. These are represented by the point S_1 on Fig. 5.4, which shows the prices of the various solution concepts, and in Fig. 5.5, which gives the actual profits of the various solutions.

If we take firm 2 as the leader and assume that 1 will use his Cournot reaction function $p_1 = 1 + \tfrac{1}{3}p_2$, we find that 2 wishes to maximise

$$e_2(g_1(p_2),p_1) = p_2 + \tfrac{1}{4}(1 + \tfrac{1}{3}p_2)p_2 - \tfrac{1}{2}p_2^2. \tag{5.30}$$

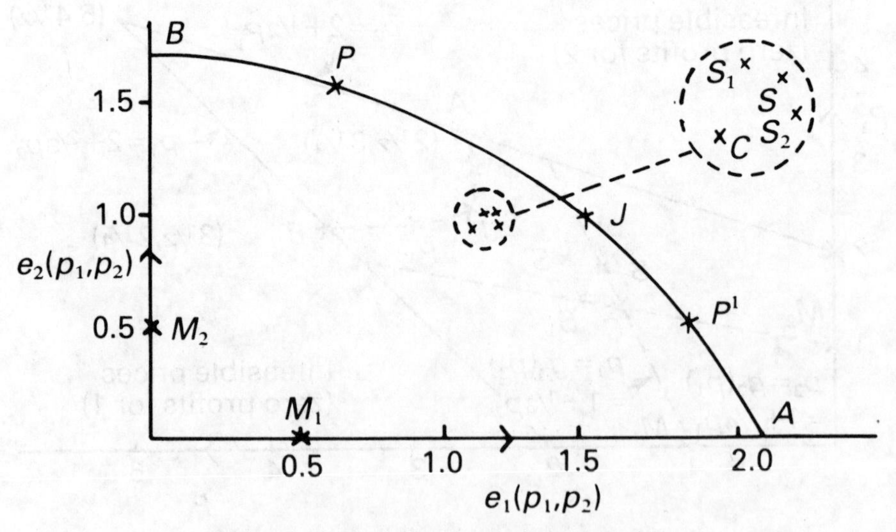

Fig. 5.5—Payoffs of solution concepts.

This leads to a solution $p_2 = 1\frac{1}{2}$ and hence $p_1 = 1\frac{1}{2}$ with corresponding profits $e_1(1\frac{1}{2},1\frac{1}{2}) = 1.125$ and $e_2(1\frac{1}{2},1\frac{1}{2}) = 0.9375$. We represent these outcomes by the points S_2 in Figs. 5.4 and 5.5. Notice that when 1 decides to become the leader and announces his price he does improve on the Cournot equilibrium, but by not as much as firm 2, who it could be argued is less aware of what is going on. Similarly, if firm 2 announces his Stackleberg price $p_2 = 1\frac{1}{2}$, he does improve his payoff as against the Cournot equilibrium, but again not as much as the follower, firm 1. If both firms tried to be leaders in a Stackleberg sense, then firm 1 would choose $p_1 = \frac{8}{5}$ as its price and firm 2 would put $p_2 = 1\frac{1}{2}$. This results in the payoffs $e_1(1.6,1.5) = 1.12$, $e_2(1.6,1.5) = 0.975$, which is shown by point S in the diagrams. Both firms do better when they are leaders, if the supposed follower also tries to be a leader rather than play its Cournot reaction function. Thus, although it pays to announce your price, it is even better when the competing firm also announces its.

When we think of duopoly as a cooperative game, we can ask what solutions are likely to occur. One that is often quoted in economic theory is the **joint maximisation of profit**. It is used as a benchmark to measure the possible collusion between the firms. The two firms choose their prices p_1,p_2 so as to maximise $e_1(p_1,p_2) + e_2(p_1,p_2)$. This is nothing else than the characteristic function value $v(1,2)$ for the game when considered as an n-person game. However, such a point is just one of many Pareto optimal pairs of prices. The negotiation set was the basic solution concept in two-person cooperative games, and so one might well ask what is the negotiation set for duopoly games. This will involve calculating all the Pareto optimal points, not just the joint maximisation of profit, and also finding out the maximin values for this game. Let us return to our example to make these calculations more specific.

Example 5.1 (continued). To find joint maximisation of profit we need to maximise

$$e_1(p_1,p_2) + e_2(p_1,p_2)$$
$$= p_1 + \tfrac{1}{3}p_1p_2 - \tfrac{1}{2}p_1^2 + p_2 + \tfrac{1}{4}p_1p_2 - \tfrac{1}{2}p_2^2, \qquad (5.31)$$

and on setting the first derivatives of (5.31) equal to zero we get:

$$\frac{d}{dp_1}(e_1 + e_2) = 1 + \frac{7}{12}p_2 - p_1 = 0;$$
$$\frac{d}{dp_2}(e_1 + e_2) = 1 + \frac{7}{12}p_1 - p_2 = 0, \qquad (5.32)$$

which gives $p_1 = p_2 = 12/5$ as the joint maximisation of profit prices. The actual profits divide out as $e_1(12/5, 12/5) = 1.44$ and $e_2(12/5, 12/5) = 0.96$, and are represented in Figs. 5.4 and 5.5 by the point J.

To find the maximin values and prices for the two firms we need to find $\max_{p_1} \min_{p_2} e_1(p_1,p_2)$ and $\max_{p_2} \min_{p_1} e_2(p_1,p_2)$, respectively. However, an examination of the form of $e_1(p_1,p_2)$ shows immediately that the worst firm 2 can ever do to firm 1's profits is to set its price p_2 equal to zero. This makes obvious economic sense and in that case firm 1's profits will be $e_1(p_1,0) = p_1 - \frac{1}{2}p_1^2$. This is maximised at $p_1 = 1$ which ensures a profit of 0.5. This is the point M_1 in the previous figures. Similarly, firm 2's maximin profit is also 0.5 corresponding to prices $p_1 = 0$ and $p_2 = 1$. This is M_2 in the diagrams and these two points represent the highest profits the firms can guarantee themselves no matter what the price strategy of their opponent. Lastly, we want to find the Pareto optimal payoffs to the firms in order to find the negotiation set. These will be pairs of prices (p_1^*, p_2^*) so that there is no other pair (p_1,p_2) with

$$e_i(p_1,p_2) \geqslant e_i(p_1^*,p_2^*), \qquad i = 1,2, \tag{5.33}$$

and strict inequality in at least one component. One way of finding these pairs is to solve the problem:

$$\begin{aligned} &\text{maximise} \quad e_1(p_1,p_2) \\ &\text{subject to} \quad e_2(p_1,p_2) \geqslant c, \\ &\qquad\qquad p_1,p_2 \text{ feasible}, \end{aligned} \tag{5.34}$$

for all values c. Thus, in Example 5.1 we need to solve:

$$\begin{aligned} &\text{maximise} \quad p_1 + \tfrac{1}{3}p_1p_2 - \tfrac{1}{2}p_1^2 \\ &\text{subject to} \quad p_2 + \tfrac{1}{4}p_1p_2 - \tfrac{1}{2}p_2^2 \geqslant c, \\ &\qquad\qquad 1 + \tfrac{1}{4}p_1 - \tfrac{1}{2}p_2 \geqslant 0, \\ &\qquad\qquad 1 + \tfrac{1}{3}p_2 - \tfrac{1}{2}p_1 \geqslant 0, \\ &\qquad\qquad p_1,p_2 \geqslant 0. \end{aligned} \tag{5.35}$$

This is a non-linear maximisation problem with constraints and the standard way of solving it is to use Lagrange multipliers, which is outside the scope of this book. However, as c varies the solutions lie on the curve AJB (for those who want to practise their Lagrange multipliers the curve is $\frac{1}{4}p_1^2 + \frac{1}{3}p_2^2 - p_1p_2 + \frac{2}{3}p_2 + \frac{3}{4}p_1 - 1 = 0$). The Pareto optimal set is that part of the curve where both firms get at least their maximin values of 0.5, i.e. the curve PJP'.

5.9 QUANTITY MODEL, SYMMETRIC GAMES AND OLIGOPOLY THEORY

The model we have been looking at in the previous few sections takes the price of the product as the variable that the firms choose. It also differentiates between the products produced by the two firms, since the demand functions $f_1(p_1,p_2)$ and $f_2(p_1.p_2)$ are different. As we mentioned earlier, we could build a similar model where it is the quantities produced, q_1 and q_2, which are the decision variables and we invert the consumer's utility function the other way to get equations which will determine the price p_i for the ith firm's product as a function of q_1 and q_2, i.e. $p_i = h_i(q_1,q_2)$, $i = 1,2$. We can then find the profit functions and from that the Cournot equilibrium and the other solution concepts. Surprisingly, they do not always give the same results as the price model (see Problem 5.5).

If, however, we remove the differentiation between the two firms' products, this will no longer happen. In this symmetric game, we take the price–quantity equations of the two firms to be the same, and so there is only one price which determines the total quantity made, or vice versa, depending on which model is chosen. In this case the two models will give the same results.

Although we initially described the problem in terms of several firms, our example concentrated on the solution concepts when there are only two firms. How many of the ideas discussed there go across to oligopoly theory when there are several firms? The definition of Cournot equilibrium was given in this context, and to find it in the case of M competing firms means solving M equations rather than the two of (5.24) and (5.25) in the example. The Stackleberg 'leader–follower' strategies have more than one extension to oligopoly models. One way is to split the firms into two groups, one who will announce their price and the others who will use some sort of reaction function to choose a price given this partial information on prices. An alternative generalisation is to assume that the firms will announce their prices in a known order, say $1,2,\ldots,M$. Then assume that the last firm will set its price by using the Cournot reaction function $p_M = g_M(p_1,p_2,\ldots,p_{M-1})$. The M-1th firm will then set its price taking into account, $p_1,p_2,\ldots,p_{M-2}$, and the function g_M so as to maximise its profit. This gives a reaction function $p_{M-1} = g_{M-1}(p_1,\ldots,p_{M-2})$ which can be used by the M-2th firm in setting its price, and so on.

The cooperative concepts of joint maximisation of profit and Pareto optimal sets have obvious generalisations to the oligopoly case, though the actual calculation of Pareto optimal sets becomes very complicated. The n-person solution concepts, discussed in Chapter 4, of the core, the nucleolus, stable sets and the Shapley value, can also be calculated for these oligopoly models, but they have received less attention than the ideas with more direct economic interpretations.

5.10 FURTHER READING

Market games, duopoly and oligopoly theory are usually considered as part of mathematical economics and there is a tremendous literature at all levels on these topics. Here we will try to point out a few of the authors who treat these problems primarily from a game theoretic point of view.

The market games which are covered in the first half of this chapter originated with the work of Edgeworth in 1881. However, the reinterpretation of these market problems as games in the sense we understand them was made by Shubik (1959a). A more abstract treatment of the model was given by Shapley (1959) in the same volume. These two authors combined in several other papers to describe the connection between this approach to price formulation and other ways (Shapley and Shubik, 1967), and also to reinterpret results from game theory in economic form (Shapley and Shubik, 1969). In this latter paper they formally defined what was a market game, and research into such games is still continuing, e.g. Billera (1981).

The history of duopoly and oligopoly theory pre-dates game theory by an even larger margin than market games. Cournot (1838) constructed a model of N competing firms selling identical products, and allowed them to fix the quantity they produced. He identified what we now call the Cournot equilibrium for this model. Half a century later, Bertrand (1883) reworked this model using price rather than quantity produced as the firms' strategic variables. There were many refinements of these models over the next fifty years, and Hotelling (1929) was probably the first to differentiate between the firms' products. Once von Neumann and Morgenstern's book was published and Nash (1951) had extended the non-cooperative equilibrium to n-person games, there has been a steady stream of research work on applying game theory concepts and ideas to oligopoly models. Foremost among the workers in this area have been the three S's—Shubik, Shapley and Selten. An understanding of the development of game theory to these competitive market situations can be got by reading Shubik's book (1959b) written just after the start of this development, together with more recent ones by Friedman (1977) and Shubik and Levitan (1980). Friedman's book develops the dynamical aspects of price setting in oligopoly theory (we shall return to this in Chapter 7), while Shubik's later book adds inventory and advertising considerations to the basic oligopoly problem. A good economic introduction to the connection between game theory and economics in general is given by Bacharach (1977).

Problems for Chapter 5.

5.1. In the [1,1] market game, the type 1 trader starts with a and the type 2 trader starts with b. If each has concave utility function $u(\ ,\ )$ show that the Shapley value and the nucleolus of the game is:

$$(u(\tfrac{1}{2}a,\tfrac{1}{2}b) + \tfrac{1}{2}u(a,0) - \tfrac{1}{2}u(0,b),$$
$$u(\tfrac{1}{2}a,\tfrac{1}{2}b) + \tfrac{1}{2}u(0,b) - \tfrac{1}{2}u(a,0)).$$

If in fact $u(x,y) = (1 + x)(1 + y)$ and $a = b = 1$, show that the negotiation set (or contract curve) is $\{(x,x)|\sqrt{2} - 1 \leq x \leq 2 - \sqrt{2}\}$. What is the maximin bargaining solution?

5.2. In the game [1,2] where 1 begins with $(a,0)$, 2 and 3 begin with $(0,b)$, show that the core is of the form:

$$\left\{\left(3u\left(\frac{a}{3}, \frac{2b}{3}\right) - x_2 - x_3, x_2, x_3\right)\right.$$
$$\left. u(0, b) \leq x_i \leq 3u\left(\frac{a}{3}, \frac{2b}{3}\right) - 2u\left(\frac{a}{2}, \frac{b}{2}\right)\right\}.$$

Find the Shapley value and the nucleolus of this game.

5.3. Prove that the core is empty for the n-person game v which describes an economy with decreasing returns of scale, so for $S \subset S'$, $v(S)/\#S > v(S')/\#S'$.

5.4. In a [2,2] market game, all the players have utility function $e(x,y) = (1 + x)(2 + y)$. If type-A traders start with (1,0) and the B-type traders start with (0,1), find the set of imputations. Find the core, the Shapley value and the nucleolus of this game. (Recall stable sets are given for this game by Theorem 5.2.)

5.5. Consider the quantity model of the problem in Example 5.1. Show that if firm 1 makes quantity q_1 and firm 2 makes a quantity q_2, then the prices are given by $p_1 = \max\{0,5 - 3q_1 - 2q_2\}$, and $p_2 = \max\{0,4\tfrac{1}{2} - 1\tfrac{1}{2}q_1 - 3q_2\}$. Prove that the Cournot equilibrium is $q_1 = 7/11$, $q_2 = 13/22$ with profits 1.20 and 1.047. Find the Stackleberg solutions with firm 1 as the leader, then with firm 2 as the leader. What happens if both try to be leaders? Find the joint maximisation of profit quantities and what are the maximin solutions.

5.6. In a price model of duopoly the quantities q_1,q_2 produced by the two firms are functions of their prices p_1,p_2. If $q_1 = \max\{0,10 + 2p_2 - p_1\}$, $q_2 = \max\{0,20 + p_1 - 3p_2\}$, find the Cournot equilibrium, the two Stackleberg solutions, the joint maximisation of profits (be careful) and the maximin strategies and profits.

5.7. Consider the three-firm duopoly model where the quantity produced by each firm q_i, $i = 1,2,3$, is related to the prices p_1, p_2, p_3 by $q_1 =$

$\max\{0,10 + p_3 - p_1\}$, $q_2 = \max\{0,10 + p_3 - p_2\}$, $q_3 = \max\{0,20 + p_1 + p_2 - 4p_3\}$. Find the Cournot equilibrium, joint maximisation of profits and maximin values for this model. If firm 1 first announces its price, then 2, then 3, find the Stackleberg solution for this type of leader follower model.

If you set this up as three-person game, prove that the characteristic function of the game is $v(1) = v(2) = 25$, $v(3) = 25$, $v(1,2) = 50$; $v(1,3) = v(2,3) = 100$ and $v(1,2,3) = 250$. Hence, find the set of imputations, the core, the Shapley value and the nucleolus of this game.

CHAPTER 6

Metagames

'He is a fool that thinks not that another thinks'—Herbert

6.1 OBJECTIVES OF METAGAMES

As researchers started to experiment by playing games it became apparent that the outcomes actually arrived at were not necesssarily those predicted by the theory. Nigel Howard (1966a,1966b,1971) was interested in why the actual outcomes of non-cooperative games like Prisoners' Dilemma might vary from the equilibrium pairs. For Prisoners' Dilemma (Examples 1.4 and 3.3) the equilibrium pair is confess–confess, but actual results show that don't confess–don't confess is also common. Howard assumes that each player tries to predict which strategies his opponents will choose in order to plan his own. This leads to the idea of an actual stable outcome, where each player correctly predicts all the other players' strategies, and so the outcome.

Howard identified such outcomes as equilibria of games based on the actual game being played, and called these enlarged games **metagames**. These are games where the players' strategies are really reaction functions to the other players' strategies and, surprise, surprise, the idea was first looked at by von Neumann and Morgenstern (1944) who called them majorant and minorant games. The difficulty in such games is that once we allow reaction functions to players choices, then why not reaction functions to reaction functions, and a whole infinite regression looms large. Howard's achievement was to show this will not happen.

The only assumption in finding the actual stable outcomes is that each player, having predicted his opponent's strategies, will choose the outcome he prefers most. Thus, it is sufficient that each player be able to make a comparison between any two outcomes. Howard doesn't specify more about the basic game, but since almost all the examples looked at are models of one-off games, it is natural to concentrate only on pure strategies. This also means that the actual numerical payoffs no longer matter, but only the ordering of the payoffs. Thus, if we are looking at pure strategies only, the two payoff matrices of (6.1) lead to the same dominance relations:

$$\begin{pmatrix} (-9,-9), & (0,-10) \\ (-10,0), & (-1,-1) \end{pmatrix} \qquad \begin{pmatrix} (2,2) & (4,1) \\ (1,4) & (3,3) \end{pmatrix} \tag{6.1}$$

We will give the definitions in terms of n-person games but our examples are of two-person games where it is easy to see what is happening.

6.2 METAGAMES AND METAEQUILIBRIA

Consider a game G with n players, where player i has a set X_i of pure strategies, $i = 1,2,\ldots,n$. First of all we are going to redefine the idea of equilibrium in n-person non-cooperative games (Definition 4.1) so that it extends easily to metagames.

Definition 6.1. A **rational outcome** for player i is an n-tiple of strategies $x_1^*,x_2^*,\ldots,x_n^*$, where

$$e_i(x_1^*,x_2^*,\ldots,x_i^*,\ldots,x_n^*) \geqslant e_i(x_1^*,x_2^*,\ldots,x_i,\ldots,x_n^*), \quad \text{for all } x_i \in X_i. \tag{6.2}$$

We denote the set of **rational outcomes** for player i in game G by $R_i(G)$.

We can now give a reformulation of the definition of equilibrium n-tiple; see Definition 4.1.

Definition 6.2. An **equilibrium** in an n-person game G is an outcome which is rational for all players. So the set of equilibria, $E(G)$. satisfies:

$$E(G) = \bigcap_{i=1}^{n} R_i(G). \tag{6.3}$$

Now to consider which outcomes are stable, look at the metagame $1G$. In this game each of the players, except player 1, chooses his strategy in the basic game G, and then player 1 chooses his strategy in the basic game in the knowledge of these choices. By looking at such games we are able to find out what 1 would choose to do if he could predict the other players' strategies, before choosing his own. These are the rational outcomes for 1 in this game, and we label the outcomes $R_1(1G)$. We can then find the equilibrium $E(1G)$ for this game, which are the outcomes that will occur if all the players believe they are in this type of game and predict correctly their opponents' strategies.

Before looking at more complicated metagames, let us look at what happens if we apply these ideas to Prisoners' Dilemma. In this case G is:

$$\begin{array}{lcc} & \text{c: confess} & \text{d: don't confess} \\ \text{c: confess} & (-9,-9) & (0,-10) \\ \text{d: don't confess} & (-10,0) & (-1,-1) \end{array}, \tag{6.4}$$

although, as we explained earlier, we could just as well take the other matrix in (6.1). The sets of pure strategies are $X_1 = X_2 = \{c,d\}$, and it is easy to check that the rational outcomes for 1 are $R_1(G) = \{(c,c),(c,d)\}$ since if 2 plays c, 1 prefers the -9 of confess to the -10 of don't confess, whereas if 2 plays d, 1 prefers the 0 of confess to the -1 of don't confess. Similarly, $R_2(G)$ is $\{(c,c),(d,c)\}$ since if 1 confesses, 2 prefers confessing to not confessing and if 1 does not confess, 2 again prefers confessing. Using this to find the equilibrium pairs, we confirm the result of Example 3.3 that

$$E(G) = R_1(G) \cap R_2(G) = (c,c). \tag{6.5}$$

In the game $1G$, 2's strategies are still the set $X_2 = \{c,d\}$, but 1 now chooses his strategy knowing 2's choice. Thus, his strategies are reaction functions to 2's choice, and form the set $F_1 = \{f | f: X_2 \to X_1\}$. There are four such functions:

$f_1(c) = c$, $f_1(d) = c$, which is always confess,

$f_2(c) = d$, $f_2(d) = d$, which is never confess,

$f_3(c) = c$, $f_3(d) = d$, which is do the same as 2 did,

$f_4(c) = d$, $f_4(d) = c$, which is do the opposite of 2.

The payoff matrix for the game 1G is then:

$$\begin{array}{cc} & \quad 2 \\ & \begin{array}{cc} c & d \end{array} \\ 1 & \begin{array}{c} f_1 \\ f_2 \\ f_3 \\ f_4 \end{array} \begin{pmatrix} (-9,-9) & (0,-10) \\ (-10,0) & (-1,-1) \\ (-9,-9) & (-1,-1) \\ (-10,0) & (0,-10) \end{pmatrix}, \end{array} \tag{6.6}$$

and it is then easy to find that the rational outcomes and equilibria are:

$$R_1(1G) = \{(f_1,c),(f_3,c),(f_1,d),(f_4,d)\}, \tag{6.7}$$

$$R_2(1G) = \{(f_1,c),(f_2,c),(f_3,d),(f_4,c)\}, \tag{6.8}$$

$$E(1G) = R_1(1G) \cap R_2(1G) = \{(f_1,c)\}. \tag{6.9}$$

Notice the outcomes which are rational for 1 are those which maximise 1's payoff in a column, whereas the ones which are rational for 2 maximise 2's payoff along a row. The equilibrium outcomes of $E(1G)$, (f_1,c) actually corresponding to an outcome in the basic game, because if 2 chooses c, and 1 then chooses his reaction function f_1, he actually plays c in G since $f_1(c) = c$. Thus, the outcome (c,c) in G occurs when (f_1,c) is played in $1G$. This gives us our final definition of this section—that of a **metaequilibrium outcome.**

Definition 6.3. An outcome $(x_1^*,x_2^*,\ldots,x_n^*)$ in the original game G which arises from an equilibrium in a game $k_1k_2\ldots k_rG$, where r can be any non-negative integer and $k_1,k_2,\ldots,k_r$ any sequence of the players, including repetitions, is called a **metaequilibrium**. $\hat{E}(G)$ denotes the set of all metaequilibria. $\hat{E}(k_1k_2\ldots k_rG)$ are the metaequilibria arising from $k_1k_2\ldots k_rG$, and $k_1k_2\ldots k_r$ is called the **title** of the game.

Thus, we know that (c,c) is in $\hat{E}(G)$, because it was in $E(G)$ and because it was the derived outcome of the equilibria in $E(1G)$. To find all the elements of $\hat{E}(G)$ we must look at the other metagames based on G. We can construct $2G$ in exactly the same way as $1G$, and because of the symmetry between the players in G we get a symmetrical result, namely:

$$R_1(2G) = \{(c,f_1),(c,f_3),(d,f_1),(d,f_4)\}; \tag{6.10}$$

$$R_2(2G) = \{(c,f_1),(c,f_2),(d,f_3),(c,f_4)\}; \tag{6.11}$$

$$E(2G) = R_1(2G) \cap R_2(2G) = \{(c,f_1)\}. \tag{6.12}$$

Again the metaequilibrium in G corresponding to (c,f_1) is (c,c).

We could now think of $1G$ and $2G$ (and up to nG in the n-person case) as the basic games and look at the metagames built on them. Let us concentrate on $21G$. This corresponds to the game where 1 announces which strategy he will play in the game $1G$—which you recall is of the form 'if 2 does this, I will do that'—and then 2 decides on which strategy he will play in response to this. This is modelling the situation when 2 tries to predict 1's strategy, and he also has to allow for the fact that 1 is trying to predict his (2's) strategy. The rational outcomes in this game tell you what the various players would do given their predictions in this case, and the equilibria (or metaequilibria in the original game G) are the outcomes, when everyone predicts correctly their opponents strategies. In $21G$, 1's strategy set is still $F_1 = \{f: X_2 \to X_1\}$ but 2's is now $G_2 = \{g: F_1 \to X_2\}$. There are sixteen functions in G_2 each of which can be defined by a vector like (c,d,c,c) which gives the responses to the functions f_1, f_2, f_3 and f_4, i.e. $g(f_1) = c$, $g(f_2) = d$, $g(f_3) = c$, $g(f_4) = c$. Just this once we will write out the payoff matrix of $21G$ (see page 133). The $R_1(21G)$ outcomes are marked with a superscript 1 and the $R_2(21G)$ are marked with a superscript 2. It is easy to see that:

$$
\begin{array}{ll}
 & \qquad\qquad\qquad\qquad\qquad\qquad 2 \\
 & \begin{array}{cccccc} g_1 & g_2 & g_3 & g_4 & g_5 & g_6 \\ (c,c,c,c) & (c,c,c,d) & (c,c,d,c) & (c,d,c,c) & (d,c,c,c) & (c,c,d,d) \end{array} \\
\begin{array}{ll} & f_1\text{: c always} \\ 1 & f_2\text{: d always} \\ & f_3\text{: same as 2} \\ & f_4\text{: opposite to 2} \end{array} & \left(\begin{array}{cccccc}
(-9,-9)^{12} & (-9,-9)^2 & (-9,-9)^2 & (-9,-9)^2 & (0,-10)^1 & (-9,-9)^2 \\
(-10,0)^2 & (-10,0)^2 & (-10,0)^2 & (-1,-1)^1 & (-10,0)^2 & (-10,0)^2 \\
(-9,-9)^1 & ((-9,-9) & (-1,-1)^{12} & (-9,-9) & (-9,-9) & (-1,-1)^2 \\
(-10,0)^2 & (0,-10)^1 & (-10,0)^2 & (-10,0)^2 & (-10,0)^2 & (0,-10)^1
\end{array}\right. \\
 & \begin{array}{cccccc} g_7 & g_8 & g_9 & g_{10} & g_{11} & g_{12} \\ (c,d,c,d) & (c,d,d,c) & (d,c,c,d) & (d,c,d,c) & (d,d,c,c) & (c,d,d,d) \end{array} \\
\begin{array}{l} f_1 \\ f_2 \\ f_3 \\ f_4 \end{array} & \begin{array}{cccccc}
(-9,-9)^2 & (-9,-9)^2 & (0,-10)^1 & (0,-10)^1 & (0,-10)^1 & (-9,-9)^2 \\
(-1,-1) & (-1,-1)^1 & (-10,0)^2 & (-10,0)^2 & (-1,-1) & (-1,-1) \\
(-9,-9) & (-1,-1)^{12} & (-9,-9) & (-1,-1)^2 & (-9,-9) & (-1,-1)^2 \\
(0,-10)^1 & (-10,0)^2 & (0,-10)^1 & (-10,0)^2 & (-10,0)^2 & (0,-10)^1
\end{array} \\
 & \begin{array}{cccc} g_{13} & g_{14} & g_{15} & g_{16} \\ (d,c,d,d) & (d,d,c,d) & (d,d,d,c) & (d,d,d,d) \end{array} \\
\begin{array}{l} f_1 \\ f_2 \\ f_3 \\ f_4 \end{array} & \left.\begin{array}{cccc}
(0,-10)^1 & (0,-10)^1 & (0,-10)^1 & (0,-10)^1 \\
(-10,0)^2 & (-1,-1) & (-1,-1) & (-1,-1) \\
(-1,-1)^2 & (-9,-9) & (-1,-1)^2 & (-1,-1)^2 \\
(0,-10)^1 & (0,-10)^1 & (-10,0)^2 & (0,-10)^1
\end{array}\right).
\end{array}
\tag{6.13}
$$

$$E(21G) = R_1(21G) \cap R_2(21G) = \{(f_1,g_1),(f_3,g_3),(f_3,g_8)\}. \tag{6.14}$$

Since $g_1(f_1) = \mathrm{c}$ and $f_1(c) = \mathrm{c}$, (f_1,g_1) corresponds to (c,c) yet again, but as $g_3(f_3) = \mathrm{d}$ and $f_3(d) = \mathrm{d}$, (f_3,g_3) corresponds to (d,d) in the basic Prisoners' Dilemma game. So does (f_3,g_8) if you work it out. Thus, (d,d) is now a metaequilibrium and $\hat{E}(G)$ contains both (c,c) and (d,d).

What are the strategies that make (d,d) a metaequilibria and so an actual stable outcome? f_3 corresponds to 1 saying 'I will do what 2 does', whereas g_3 is the strategy where 2 says 'I will confess unless 1 chooses to do as I do, in which case I will be silent'. If the players recognise these as the intentions of their opponent, then the 'don't confess' pair of strategies is the reasonable outcome. This exemplifies the usefulness of metagame analysis, in that it identifies what outcomes can be expected from a game, and the type of strategy that makes such outcomes likely. Thus, if the prisoners were allowed to talk to one another and each can convince the other that the above strategies are what he intends to 'follow', then the 'don't confess–don't confess' outcome will follow—at least in the long run. Obviously, such an analysis is useful in any multi-person decision problem, where formal bargaining sessions occur, such as international agreements or management–worker negotiations.

By symmetry, if we do the same analysis for $12G$ (see Problem 6.1) we again get both (c,c) and (d,d) as metaequilibria.

However, we still don't know what $\hat{E}(G)$ is, or so it seems. To do this we must look at $121G$ which corresponds to 1 saying: 'When I predict 2's strategy, I must allow for the fact that he is allowing for the fact that I am predicting his own strategy.' As for $1212121G$ or $2222G$, would you like to work out what is going on here? But apparently we must find the equilibria in all these games to find the set of metaequilibria. However, don't get disheartened because in the next section we will show that no more metaequilibria exist beyond these in $12G$ and $21G$.

6.3 METARATIONALITY THEOREM

It is fairly obvious from the previous section that the metagames are much larger than the original game they are based on. Thus, writing down the payoff matrix for them might be quite a considerable task, especially for n-person games, let alone finding their rational outcomes. The following theorem saves us all the trouble of doing this, because it identifies all the outcomes in the basic game which correspond to rational outcomes for a particular player in the metagame. We call such outcomes in the basic game, **metarational** and denote them by $\hat{R}_i(k_1k_2...k_rG)$, etc.

Theorem 6.1. In the metagame $k_1k_2...k_rG$, let F_i be the set of players whose number follows the last time i appears in $k_1k_2...k_r$, or is all the players who appear in $k_1k_2...k_r$ if i does not appear. Let P_i be the set of players who are not in F_i, and who precede the last appearance of player i in $k_1k_2...k_r$. Let U_i be the players who are not in P_i or F_i or $\{i\}$. An outcome $\bar{x} = (\bar{x}_{F_i},\bar{x}_{P_i},\bar{x}_{U_i},\bar{x}_i)$ in G is **metarational** for i in $k_1k_2...k_rG$, if

$$\min_{x_{P_i}} \max_{x_i} \min_{x_{F_i}} e_i(x_{F_i},x_{P_i},\bar{x}_{U_i},x_i) \leqslant e_i(\bar{x}). \tag{6.15}$$

If the game G has only two players, 1 and 2, then the theorem says that in the game $1G$. $(\bar{x}_1,\bar{x}_2)$ is metarational for 1 if

$$\max_{x_1} e_1(x_1,\bar{x}_2) \leqslant e_1(\bar{x}_1,\bar{x}_2), \tag{6.16}$$

since $F_1 = P_1 = \{\emptyset\}$, $U_1 = 2$ for $1G$. $(\bar{x}_1,\bar{x}_2)$ is metarational for 2 if

$$\max_{x_2} \min_{x_1} e_2(x_1,x_2) \leqslant e_2(\bar{x}_1,\bar{x}_2), \tag{6.17}$$

since $F_2 = \{1\}$, $P_2 = U_2 = \{\emptyset\}$ for $1G$. In the metagame $21G$ based on a two-player G, Theorem 6.1 says $(\bar{x}_1,\bar{x}_2)$ is metarational for 1 if

$$\min_{x_2} \max_{x_1} e_1(x_1,x_2) \leqslant e_1(\bar{x}_1,\bar{x}_2), \tag{6.18}$$

since $P_1 = \{2\}$, $F_1 = U_1 = \{\emptyset\}$ for $21G$. $(\bar{x}_1,\bar{x}_2)$ is metarational for 2 if

$$\max_{x_2} \min_{x_1} e_2(x_1,x_2) \leqslant e_2(\bar{x}_1,\bar{x}_2) \tag{6.19}$$

as in (6.16). Similar results hold for $2G$ and $12G$, namely that in $2G$ $(\bar{x}_1,\bar{x}_2)$ is metarational for 1 if

$$\max_{x_1} \min_{x_2} e_1(x_1,x_2) \leqslant e_1(\bar{x}_1,\bar{x}_2) \tag{6.20}$$

(since $F_1 = \{2\}$, $P_1 = U_1 = \{\emptyset\}$. $(\bar{x}_1,\bar{x}_2)$ is metarational for 2 if

$$\max_{x_2} e_2(\bar{x}_1,x_2) \leqslant e_2(\bar{x}_1,\bar{x}_2) \tag{6.21}$$

$(F_2 = P_2 = \{\emptyset\}, U_2 = \{1\})$.
Finally, for $12G$, we have the following requirements. $(\bar{x}_1,\bar{x}_2)$ is metarational for 1 if

$$\max_{x_1} \min_{x_2} e_1(x_1,x_2) \leqslant e_1(\bar{x}_1,\bar{x}_2) \tag{6.22}$$

$(F_1 = \{2\}, P_1 = U_1 = \{\emptyset\})$. $(\bar{x}_1,\bar{x}_2)$ is metarational for 2 if

$$\min_{x_1} \max_{x_2} e_2(x_1,x_2) \leqslant e_2(\bar{x}_1,\bar{x}_2) \tag{6.23}$$

$(F_2 = U_2 = \{\emptyset\}, P_2 = \{1\})$.

Notice that in each case the maximisation is over the strategies for the player whose metarational outcomes we are considering. The minimisation is over the other players' strategies and precedes or follows the maximisation depending on whether they precede or follow that player in the title.

Let us check that the metarational outcomes given by (6.16)–(6.19) agree with those we worked out from the payoff matrices (6.6) and (6.13) of the Prisoners' Dilemma. For $1G$ (6.16) says $(\bar{x}_1,\bar{x}_2)$ is metarational for 1 if it is the highest payoff to 1 against that specific $\bar{x}_2$, i.e. highest in a column. Going back to the payoff matrix of the original game G, (6.4), this gives $\hat{R}_1(1G) = \{(c,c),(c,d)\}$. For player 2's metarational outcomes (6.17) requires $\max_{x_2} \min_{x_1} e_2(x_1,x_2)$. Now $\min_{x_1} e_2(x_1,c) = -9$ and $\min_{x_1} e_2(x_1,d) = -10$. So, $\max_{x_2} \min_{x_1} e_2(x_1,x_2)$ is the maximum of -9 and -10. This gives $\hat{R}_2(1G) = \{(c,c),(d,c),(d,d)\}$, where $\hat{R}_i$ are the metarational outcomes. These results agree with $R_1(1G)$ and $R_2(1G)$ in (6.7) and (6.8).

For $21G$, the condition for metarational outcomes for player 2 is the same as for $1G$ so $\hat{R}_2(21G) = \hat{R}_2(1G) = \{(c,c),(d,c),(d,d)\}$. For player 1, (6.18) requires us to find $\min_{x_2} \max_{x_1} e_1(x_1,x_2)$. For $x_2 = c$ we have $\max_{x_1} e_1(x_1,c) = -9$ from (6.4) again, whereas for $x_2 = d$, $\max_{x_1} e_1(x_1,d) = 0$. So $\min_{x_2} \max_{x_1} e_1(x_1,x_2)$ is -9 and hence $\hat{R}_1(21G) = \{(c,c),(c,d),(d,d)\}$. If you want to spend a pleasant few minutes, you can check these results agree with the 1,2 superscripts on matrix (6.13).

The discerning reader might have realized the proof of Theorem 6.1 is a long time coming, and given my track record in previous chapters might suspect I will omit it. Quite right. It is a long, but not too difficult, exercise in mathematical logic, involving manipulation of the quantifiers 'for all' and 'there exists' and is based on the reflexivity of preference relations ($\geqslant$) and the axiom of choice. Those who wish to study it further can find it in Howard's book (1971, pp. 89–96), and those who read there will bask in the realisation of my unforgiveable sin.

As important as the theorem is, its real strength is only realised when we add the following corollary to it.

Theorem 6.2. Given a metagame $k_1k_2...k_rG$, then if all except the last (nearest to G) appearance of a player in the title $k_1k_2...k_r$ is deleted, the resultant game, $k_1'k_2'...k_s'G$, has the same metarational outcomes as the original metagame.

Proof. The result follows immediately from Theorem 6.1 once it is realised that the sets F_i, P_i and U_i will be the same for both metagames.

Theorem 6.2 means that the game $121G$ or even $1212121G$ gives rise to the same metarational outcomes as $21G$, and $2222G$ gives the same result as $2G$. Thus, if the basic game has only two players, we need only look at the metagames $1G$, $2G$, $12G$ and $21G$ to find all possible metarational outcomes. For n-player basic games, we need only go as far as the $n!$ nth level metagames which are all possible permutations on the numbers 1 to n. Thus, there is no infinite hierarchy of metagames we have to examine in order to find all metarational outcomes.

Our objective is to find the metaequilibria $\hat{E}(G)$, but in this section we have concentrated on the metarational outcomes $\hat{R}_i(k_1k_2...k_rG)$. You might think that the metaequilibria for any metagame are the intersections of the metarational outcomes for all players. This is true if the underlying game has only two players, so that:

$$\hat{E}(12G) = \hat{R}_1(12G) \cap \hat{R}_2(12G) \quad \text{and}$$
$$\hat{E}(21G) = \hat{R}_1(21G) \cap \hat{R}_2(21G).$$

For games with three or more players, this need not be true and so you can have outcomes which are metarational for all the players, but are not metaequilibria since the players have different rational outcomes in the metagame, but they reduce to the same outcome in the basic game. Howard (1968), however, proved that such outcomes will be metaequilibria for even higher order metagames. So even for n-player ($n > 2$) basic games if we look at the intersection of the metarational outcomes for the nth level games, they will all be metaequilibria for higher level games.

6.4 EXAMPLES OF METAGAME ANALYSIS

We now look at two international conflicts, which on the surface appear to have much in common, but resulted in completely different outcomes. We see what the metagame analysis would suggest as the stable outcomes in each case.

Example 6.1—Falkland's War. In 1982, Argentina successfully invaded the Falkland Islands, and deported the British administration. The

British government then prepared a Task Force to retake the islands. The options open to the two 'players' then were that the British could abandon their plans to retake the islands or continue with the reinvasion. The Argentine government could withdraw from the islands or maintain their presence there. It seems reasonable that the two governments ranked the four possible outcomes as follows:

		Argentine island forces	
		withdraw (w)	maintain (m)
U.K. invasion	abandon (a)	(4,2)	(1,4)
	continue (c)	(3,1)	(2,3)

(6.24)

where 4 is preferred to 3 to 2 to 1. In reality, the invasion continued and the Argentine forces stayed on the island, with the resultant conflict. Could this have been avoided? Was there another stable outcome?

The only equilibrium pair in G is (c,m), the actual outcome of the conflict. Using (6.16) to (6.23) we have:

$$\hat{R}_1(1G) = \{(a,w),(c,m)\}, \qquad \hat{R}_2(1G) = \{(a,m),(c,m)\}; \tag{6.25}$$

$$\hat{R}_1(2G) = \{(a,w),(c,w,),(c,m)\}, \qquad \hat{R}_2(2G) = \{(a,m),(c,m)\}; \tag{6.26}$$

$$\hat{R}_1(12G) = \{(a,w),(c,w),(c,m)\}, \qquad \hat{R}_2(12G) = \{(a,m),(c,m)\}; \tag{6.27}$$

$$\hat{R}_1(21G) = \{(a,w),(c,w),(c,m)\}, \qquad \hat{R}_2(21G) = \{(a,m),(c,m)\}. \tag{6.28}$$

since

$$\max_{x_1} \min_{x_2} e_1(x_1,x_2) = \min_{x_1} \max_{x_2} e_1(x_1,x_2) = 2$$

and

$$\min_{x_1} \max_{x_2} e_2(x_1,x_2) = \max_{x_2} \min_{x_1} e_2(x_1,x_2) = 3.$$

From (6.25) to (6.28) it is obvious that (c,m) is the only metaequilibrium in each metagame, and so $\hat{E}(G) = $ (c,m). Thus, the actual outcome is the only stable one in this situation.

Example 6.2—Cuban Missile Crisis. In 1962 the U.S. government discovered that the U.S.S.R had missiles stationed in Cuba. They made preparations for an invasion of Cuba to remove the missiles. Thus, the U.S. had two courses of action—to abandon its invasion or to continue it, while the U.S.S.R. could withdraw its missiles or maintain them. Notice these are essentially the same options as the two players had in the Falklands War, but this time the preferences are:

		U.S.S.R missiles	
		withdraw (w)	maintain (m)
U.S. invasion plans	abandon (a)	(3,3)	(2,4)
	continue (c)	(4,2)	(1,1)

(6.29)

The only equilibria in this basic game are (c,w) and (a,m) which are essentially complete victory for either side, but the actual outcome was (a,w). Let's find what are the stable outcomes:

$$\hat{R}_1(1G) = \{(c,w),(a,m)\}, \qquad \hat{R}_2(1G) = \{(a,w),(c,w),(a,m)\}; \tag{6.30}$$

$$\hat{R}_1(2G) = \{(a,w),(c,w),(a,m)\}, \quad \hat{R}_2(2G) = \{(a,m),(c,w)\}; \tag{6,31}$$

$$\hat{R}_1(12G) = \{(a,w),(c,w),(a,m)\}, \quad \hat{R}_2(12G) = \{(a,w),(c,w),(a,m)\}; \tag{6.32}$$

$$\hat{R}_1(21G) = \{(a,w),(c,w),(a,m)\}, \quad \hat{R}_2(21G) = \{(a,w),(c,w),(a,m)\}; \tag{6.33}$$

since

$$\max_{x_1} \min_{x_2} e_1(x_1,x_2) = \min_{x_2} \max_{x_1} e_1(x_1,x_2) = 2$$

and

$$\max_{x_2} \min_{x_1} e_2(x_1,x_2) = \min_{x_1} \max_{x_2} e_2(x_1,x_2) = 2$$

also. Thus,

$$\hat{E}(1G) = \hat{E}(2G) = \{(a,m),(c,w)\}; \qquad \hat{E}(12G) = \hat{E}(21G) = \{(a,m),(c,w),(a,w)\}. \tag{6.34}$$

So the actual outcome is again stable since it is a metaequilibrium for $12G$ and $21G$. If we write out the payoff matrices, we find the strategies that actually make (a,w) stable are when the U.S. threatens to continue the invasion if missiles are maintained, but will abandon the invasion if they are withdrawn, whereas the U.S.S.R. will maintain the missiles unless the U.S. plays this strategy.

6.5 SYMMETRIC METAEQUILIBRIA

On examining the metarational outcomes in the metagames discussed so far, you might notice that if an outcome is metarational for G, it is metarational for $1G$, $2G$, $12G$ and $21G$, and ones which are metarational for $1G$ are also metarational for $21G$. Thus, a metarational outcome is metarational for all higher order metagames based on that game. This result is always true as the following theorem shows.

Theorem 6.3. If $\bar{x} = (\bar{x}_1,\bar{x}_2,\ldots,\bar{x}_n)$ is metarational for i in the game $k_2k_3\ldots k_rG$, it is also metarational for i in the game $k_1k_2k_3\ldots k_rG$, no matter what k_1 is.

Proof. In Theorem 6.1, the players divided into three sets, U_i, F_i and P_i, and the condition that $\bar{x}$ is metarational in $k_2k_3\ldots k_rG$ is, from (6.15):

$$\min_{x_{P_i}} \max_{x_i} \min_{x_{F_i}} e_i(x_{F_i},x_{P_i},\bar{x}_{U_i},x_i) \leqslant e_i(\bar{x}). \tag{6.35}$$

On adding k_1 to the title of the game, what happens to the sets? If k_1 has appeared before in the title, then because of Theorem 6.2 $\bar{x}$ will remain metarational. If not and $k_1 = \{i\}$, then because of the definition of F_i there will be no change since $k_2,k_3,\ldots,k_r$ remain in F_i. Thus, (6.35) is the condition for $\bar{x}$ to be metarational for $k_1k_2\ldots k_rG$ as well as $k_2\ldots k_rG$. Lastly, if k_1 was in U_i originally it now enters P_i, and since

$$\begin{aligned}&\min_{x_{k_1}} \min_{x_{P_i}} \max_{x_i} \min_{x_{F_i}} e_i(x_{F_i},x_{P_i},x_{k_1},\bar{x}_{U_i},x_i)\\ &\leqslant \min_{x_{P_i}} \max_{x_i} \min_{x_{F_i}} e_i(x_{F_i},x_{P_i},x_{k_1},\bar{x}_{U_i},x_i),\end{aligned} \tag{6.36}$$

if $\bar{x}$ satisfies (6.35), it also satisfies the equivalent condition for it to be metarational for i in $k_1k_2\ldots k_rG$.

This result implies that we need only look at the games $123\ldots nG$, or $nn - 1\ldots 21G$ and all such permutations of $1,2,\ldots,n$, to find the metaequilibria. These games, where the title is some permutation of $1,2,\ldots,n$, are called **complete** games.

Definition 6.4. An outcome $(x_1^*,x_2^*,\ldots,x_n^*)$ is a **symmetric metaequilibrium** if it is a metaequilibrium in all complete games.

If G has two players the symmetric metaequilibria are $\hat{E}(12G) \cap \hat{E}(21G)$, and for the Prisoners' Dilemma, the Falklands War, and the Cuban missile crisis all the metaequilibria are symmetric. This need not always be the case—look at the following example.

Example 6.3—Warship–Merchantship problem. A warship is trying to sink a merchant ship, where both can go either north or south of an

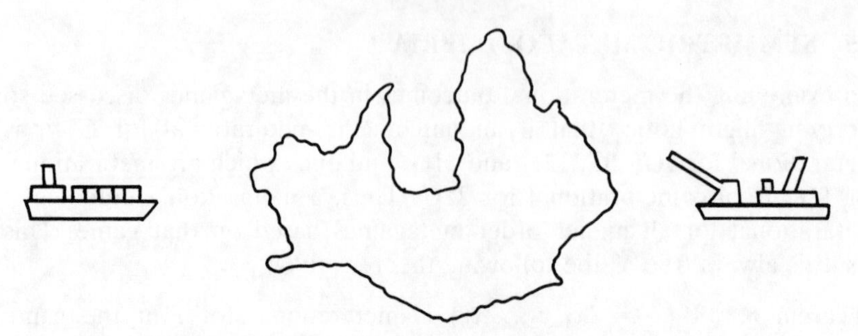

Fig. 6.1—Warship–merchantship game.

island (Fig. 6.1). The payoff matrix is:

$$\begin{array}{ccc} & & \text{Merchant ship 2} \\ & & \begin{array}{cc} \text{N} & \text{S} \end{array} \\ \text{Warship 1} & \begin{array}{c} \text{N} \\ \text{S} \end{array} & \begin{pmatrix} (1,0) & (0,1) \\ (0,1) & (1,0) \end{pmatrix}, \end{array} \tag{6.37}$$

where 1 is victory for that side. There is no equilibrium in G, but on doing the metagame analysis, we get:

$$\hat{R}_1(12G) = \{(\text{N,N}),(\text{S,S}),(\text{N,S}),(\text{S,N})\},$$
$$\hat{R}_2(12G) = \{(\text{N,S}),(\text{S,N})\};$$

$$\hat{R}_1(21G) = \{(\text{N,N}),(\text{S,S})\},$$
$$\hat{R}_2(21G) = \{(\text{N,N}),(\text{S,S}),(\text{N,S}),(\text{S,N})\}.$$

Thus,

$$\hat{E}(12G) = \{(\text{N,S}),(\text{S,N})\}, \qquad \hat{E}(21G) = \{(\text{N,N}),(\text{S,S})\}. \tag{6.38}$$

So all the outcomes are metaequilibria but none is symmetric metaequilibria. This example also shows the psychology behind the various complete games. In $12G$, 2 is assuming he can predict 1's actions, while 1 assumes he can predict 2's policy. If 2 (the merchant ship) says he will go the opposite way to 1, there is nothing 1 can do against that policy. In $21G$, 1 believes he can predict 2's actions and 2 believes he can predict 1's policy. If 1 (the warship) decides to go the same way as he believes 2 will go, there is not much 2 can do against that policy. Most players will tend to think in terms of the metagame where they come last in the title. In this they are predicting the others' actions, whereas they assume the other players are only predicting their policy. However, the symmetric metaequilibria are the ones which are stable for all the players when they think of the game in this way.

6.6 ANALYSIS OF OPTIONS

In order to apply metagame analysis in real situations, one must be able to discuss the concepts involved with non-game theorists. A businessman who brings you in as a consultant will not thank you for telling him the metarational outcomes in 321G. He wants advice on how to proceed with the problem or negotiations at hand. Thus, we must systematically find out his preferences and his beliefs about the other parties' preferences, but in a language he understands. Then he can appreciate what is happening. Also, in most situations people do not want to know all the stable outcomes, but only if the one they are interested in, which could be the status quo, is stable.

In order to satisfy these objectives, several authors have developed a technique for systematically checking what the underlying game is, and whether the particular scenario or outcome of interest is stable. This technique is usually called analysis of options, and is now programmed for microcomputers. You can think of it as 'Metagames without the Mythology' or 'Metagames for Idiots' depending on your standpoint.

The objective of **analysis of options** is to investigate a conflict situation to check whether a particular 'scenario' or outcome is stable in the metagame sense. The only requirement is that the players can always state their preference or indifference between any two scenarios. Usually of course one is working with one player and then one has to use his assessment of the other players' preferences as well as his own preferences. The procedure to go through is as follows.

A. List all the players and all the options or strategies open to each player.
B. Find out which outcome or scenario is to be checked for stability. We call this the **basic scenario**.
C. Choose one of the players, to check whether this scenario will be metarational for him. We do this as follows.
D. Find all the **unilateral improvements** over the basic scenario for this player. These are outcomes which he prefers to the basic scenario and which can be achieved by him changing his strategy but all the other players keeping to their strategy in the basic scenario. If no unilateral improvements, return to C and check on another player.
E. If unilateral improvements exist for a player, find if the other players have **sanctions** against him. A sanction is when the other players change to strategies where no matter what options the player under consideration chooses he will not prefer the outcome to the basic scenario. If there is a sanction against the player, return to C and check on another player.
F. If the player has unilateral improvements, but there are no sanctions against him, we must check if any of the unilateral improvements are

guaranteed improvements. In this case, if the player chooses the option that gave the unilateral improvement, it is a guaranteed improvement if for all choices of options by the other players he still prefers the outcome to the 'basic' scenario. If any of the unilateral improvements are guaranteed improvements, the basic scenario is not stable: otherwise return to C and check for another player.

G. If having checked for all players, none of them has a guaranteed improvement, the basic scenario is **stable**.

To see the connection between the algorithm A–G and metagame analysis, let us concentrate on a two-person game G and recall the conditions (6.16)–(6.23) which guaranteed $(\bar{x}_1,\bar{x}_2)$ is metarational in the games $1G$, $21G$, $12G$ and $2G$. In (6.16), $(\bar{x}_1,\bar{x}_2)$ is metarational for 1 in $1G$ if $e_1(\bar{x}_1,\bar{x}_2) \geqslant \max_{x_1} e_1(x_1,\bar{x})$, which says there is no unilateral improvement for 1 from $(\bar{x}_1,\bar{x}_2)$. (6.18) says $(\bar{x}_1,\bar{x}_2)$ is metarational for 1 in $21G$ if the x_2 which minimises the L.H.S. is a sanction. In the metagames $12G$ and $2G$, the metarationality conditions (6.20) and (6.22) say there is no guaranteed improvement for 1. So the analysis of options recognises the basic scenario as metarational for 1 in $2G$ at D, metarational for 1 in $21G$ at E, and metarational for 1 in $12G$ and $2G$ at F. If it passes all these steps, and so has a guaranteed improvement, it is not stable.

The object of doing the analysis this way is to try and minimise the number of times step F is attempted. Finding unilateral improvements and sanctions is relatively easy since only the player or the remaining players' strategies are changed. However, in finding guaranteed improvements you have to consider two outcomes where no one has a common strategy, and this is much more difficult to envisage than the other comparisons.

6.7 ANALYSIS OF OPTIONS APPLIED TO MARKET STRATEGY

To see analysis of options at work consider the following business problem. A U.K. computer manufacturer must decide whether to build a personal computer and, if so, whether to aim it exclusively at the U.K. or to try and get sales also world-wide (which is mainly in the U.S.). An American company already is marketing its personal computer in the U.S. and must decide whether to sell it in the U.K. and if so at a high or low price. Following the steps set down in the previous section we get the following analysis of whether the 'status quo', i.e. no sales by either company in the U.K., is stable when we are assisting the U.K. company.

A: List of Players

Player 1: U.K. firm   Player 2: U.S. firm

Options open to each player are given in terms of columns of 0's and

1's which represent whether that action is taken in that option (1) or not (0) (Table 6.1).

Table 6.1

U.K. firm	Option 1	Option 2	Status quo	Infeasible
Make for U.K. market	1	1	0	0
Make for U.S.	0	1	0	1
U.S. firm	**Option 1**	**Option 2**	**Status quo**	**Infeasible**
Enter U.K. market (high price)	1	0	0	1
Enter U.K. market (low price)	0	1	0	1

Thus, both forms have two feasible options apart from the status quo and one infeasible option (the U.K. firm can't sustain exports without a home base).

B: We concentrate on the status quo scenario where neither firm does anything new.

C_1: Consider the U.K. firm first

D_1: To find if player 1 has any unilateral strategies, we ask the U.K. company to order its feasible options given no change by the U.S. company. Suppose its preferences are given by Table 6.2.

Table 6.2

		Preferred to status quo	Status quo	Not preferred to status quo
1	Make for U.K. market	1	0	1
	Make for U.S. market	1	0	0
2	Enter U.K. (high)	0	0	0
	Enter U.K. (low)	0	0	0

E_1: There is a unilateral improvement (make for both markets) so are there any sanctions? Here the U.K. firm must order its preferences if the U.S. firm acts. These are given by Table 6.3 where—means either action can be taken and the preference remains.

Table 6.3

		Preferred to status quo			Status quo	Not preferred to status quo
1	Make for U.K.	1	0	1	0	—
	Make for U.S.	0	0	1	0	—
2	Enter U.K. (high)	1	1	1	0	0
	Enter U.K. (low)	0	0	0	0	1

Firm 2 entering the U.K. market with a low price is a sanction since no matter what he does the U.K. firm doesn't like the outcome as much as the status quo. So we can now turn to firm 2.

C: Look at the scenarios in terms of U.S. firm's preferences. This will be obtained by a mixture of gleaning information and using the U.K. firm's judgement.

D_2: Player 2's preferences if the U.K. firm does nothing is given by Table 6.4.

Table 6.4

		Preferred to status quo		Status quo	Not preferred to status quo
1	Make for U.K.	0	0	0	
	Make for U.S.	0	0	0	
2	Enter U.K. (high)	1	0	0	
	Enter U.K. (low)	0	1	0	

So both strategies are unilateral improvements for the U.S. firm and we must check for sanctions.

E_2: Considering possible sanctions for the U.K. firm, we ask for each is it true the U.S. firm, no matter what it does, will prefer the status quo? The result could be as shown in Table 6.5 which gives the U.S. firm's preference and includes the preferences in Table 6.4.

Table 6.5

		Preferred to status quo	Status quo	Not preferred to status quo
1	Make for U.K.	0 0 1 1	0	1 1 1 1
	Make for U.S.	0 0 0 1	0	0 1 0 1
2	Enter U.K. (high)	0 1 1 0	0	0 1 0 0
	Enter U.K. (low)	1 0 0 1	0	1 0 0 0

Since for each strategy by the U.K. firm (0,0), (1,0) and (1,1) there is some counter by the U.S. firm which the U.S. firm prefers, there is no sanction.

F_2: So we must check the unilateral improvements for the U.S. firm to see if they are guaranteed improvements. Since strategy (1,1,1,0) is not preferred by the U.S. firm to status quo in 6.5, entering the U.K. with a high price is not a guaranteed improvement. Similarly for the option of entering U.K. with a low price, the strategy (1,0,0,1) is not preferred to the status quo by firm 2. So it is not a guaranteed improvement. Thus there are no guaranteed improvements and the status quo is stable.

Obviously by taking other preference relations for the two firms the analysis will proceed in different ways. The procedure appears rather pedantic, but it is obviously easy to put on a microcomputer and throughout the decision-maker is only asked for his preference between the outcome under discussion and an alternative.

6.8 FURTHER READING

Metagame theory was started by Howard in a series of papers in *General Systems* (1966a,1966b,1970) and described in his book *Paradoxes of Rationality* (1971). The latter has a great deal on the philosophy behind the metagame approach, and the papers are probably a quicker way to get into the ideas. There have been one or two papers which describe experiments to check that stable outcomes are the only ones that occur, see Thomas (1974).

The transformation of metagames into analysis of options, though begun by Howard, was continued by Radford both in his books *Managerial Decision Making* (1975) and *Complex Decision Problems* (1977) and in some cases studies such as the proposed merger between Simpsons and Simpsons–Seers in Canada (1980). Frazer and Hipel wrote up a computer code, CAP, to implement the analysis and have applied it in a real conflict about water resources (1980) and a hypothetical trade union negotiation problem (1981) among others. Considering the amount of material written about metagames, the actual number of written-up case studies is disappointing. However, the technique is used by international firms on both sides of the Atlantic and, as is often the case, it may well be that the most successful implementations will not and cannot be published.

An even more recent development along the lines of metagame analysis is the 'methodique' of hypergame analysis introduced by Bennet (Bennet, 1977; Bennet and Huxham, 1982). A methodique is supposed to be a half-way house between a methodology and a technique. This looks at the interesting and important question of how in game theory you allow for the fact that you often don't know your opponent's strategies, or his preferences or even correctly identify his strategies when he plays them. This problem of misperception in a game has been looked at in different ways by several authors from as far back as Luce and Adams (1956). In the hypergame approach, for an n-person game there are n first-order perception games—one for each player. Each of these games describes the strategies that a player believes he and his opponents have, and his belief of their preferences over the outcomes. Thus, not only can the outcomes be differently ordered in the different games, but they may have a different number of strategies for the same player. We also have a mapping from the strategies of one player in his own game to his set of strategies in another player's perception of the game. The formal theory has not yet been published completely, but the connection with metagames is now obvious. We start with a player's perception of the game, then move on to his perception of other people's view of the game. At the third level we would have A's perception of B's perception of C's view of the game, and so on.

Hypergame theory has been used to describe errors in conflict situations that have occurred, like the fall of France in 1940 (Bennett and Dando, 1979) or soccer holliganism (Bennet, Dando and Sharp, 1980), and it is good fun to hypothesise why decision-makers took certain options which proved disastrous. It is thus useful as a descriptive tool, but, in my view, it has yet to be applied satisfactorily to aid in current or future decisions. As a prescriptive tool, it has the disadvantage that any misperception that you perceive cannot then be a misperception.

PROBLEMS FOR CHAPTER 6

6.1. Write out the 4×16 payoff matrix of the metagame $12G$, where G is Battle of the Sexes (Example 3.2 in Chapter 3). Find the outcomes which are rational for player 1, and those which are rational for player 2. Hence find the metaequilibrium in $\hat{E}(12G)$.

6.2. Consider the two-person game G:

$$\begin{array}{ccc} & & 2 \\ & & \begin{array}{cc} \text{c} & \text{d} \end{array} \\ 1 & \begin{array}{c} \text{a} \\ \text{b} \end{array} & \begin{pmatrix} (1,4) & (3,1) \\ (2,0) & (0,3) \end{pmatrix} \end{array} .$$

Use the metarationality theorem to find the metarational outcomes and the metaequilibria for the games $1G$, $2G$, $21G$ and $12G$. Find all the symmetric metaequilibria.

6.3. Consider the game G with payoff matrix:

$$\begin{array}{ccc} & & 2 \\ & & \begin{array}{cc} \text{a} & \text{b} \end{array} \\ 1 & \begin{array}{c} \text{a} \\ \text{b} \end{array} & \begin{pmatrix} (-1,-1) & (6,0) \\ (0,0) & (3,3) \end{pmatrix} \end{array} .$$

Write down the metagames $1G$, $2G$ and find the meta-equilibrium for them. Use the theorem on metarationality to check this result and to find the metaequilibria in $12G$ and $21G$.

6.4. During the Vietnam War there were several negotiation sessions to try and end the conflict. The North Vietnamese had two strategies to continue the conflict (c) or to withdraw their forces (w), while the U.S. government had the same options. If the preferences of the two governments are given by the matrix:

$$\begin{array}{ccc} & & \text{U.S.} \\ & & \begin{array}{cc} \text{w} & \text{c} \end{array} \\ \begin{array}{l} \text{North} \\ \text{Vietnam} \end{array} & \begin{array}{c} \text{w} \\ \text{c} \end{array} & \begin{pmatrix} (2,4) & (1,3) \\ (4,2) & (3,1) \end{pmatrix} \end{array} ,$$

where in each pair the first component is the North Vietnam preference of that outcome, and the second the U.S. preference (4 is preferred to 3 to 2 to 1). Use a metagame analysis to find all metaequilibria in G, $1G$, $2G$, $12G$ and $21G$, and explain the bargaining stances that might result in each metaequilibria.

6.5. Question 6.4 is in fact a simplification of the negotiations because the South Vietnamese government was also involved. They also have two options: to continue to fight (c) or to stop fighting (s). The preferences among the eight options are given below where 8 is most preferred option, 1 is least preferred.

N.V.	U.S.	S.V.	N.V. Preference	U.S. Preference	S.V. Preference
s	w	c	2	8	6
s	c	c	1	6	8
c	w	c	7	4	3
c	c	c	4	3	4
s	w	s	6	7	5
s	c	s	3	5	7
c	w	s	8	2	1
c	c	s	5	1	2

By finding the metarational outcomes in 123*G*, 213*G*, 312*G*, 132*G*, 231*G* and 321*G*, find all the metaequilibria.

6.6. In an analysis of the U.S.–U.S.S.R. conflict over Europe, it is assumed the Soviet Union has three strategies—a conventional attack on Europe (C), a limited nuclear strike (L) on Europe, and a full nuclear attack on Europe and the U.S. (F). In response, the U.S. has a choice of three similar strategies. The preferences among the outcomes are given in the following table where 9 means most preferred, 1 least preferred and in each pair the U.S. preference is first:

$$\text{U.S.}\quad \begin{array}{c} \\ C \\ L \\ F \end{array} \begin{array}{c} \text{U.S.S.R.} \\ \begin{array}{ccc} C & L & F \end{array} \\ \begin{pmatrix} (5,8) & (4,9) & (1,7) \\ (9,4) & (6,6) & (2,5) \\ (8,1) & (7,2) & (3,3) \end{pmatrix} \end{array}$$

Analyse this game and for each outcome determine whether it is rational in the basic game, or metarational for each player. Hence, find which outcomes are metaequilibria.

6.7. In 1967, Syria and Israel almost went to war. The U.S.S.R. was supplying Syria with weaponry. Use analysis of options to consider if the status quo is stable, where no one is firing or advancing and Syria is accepting the Soviet aid. The feasible options are

Syria:	Fire	0 1 1 1 1 0
	Attempt to advance	0 0 0 1 1 0
	Accept Soviet aid	0 0 1 0 1 1
U.S.S.R:	Give aid to Syria	1 0
Israel:	Fire	0 1 1
	Attempt to advance	0 0 1

Using your own choice of preferences find the unilateral improvements, the sanctions if any, and the guaranteed improvements for each country. Explain your analysis.

6.8. The union in a firm has presented the management with a list of wage and benefit demands, threatening a strike if they are not met. The management can do nothing and so encourage the strike, meet only the wage demands or meet all the demands. There is also an arbitration procedure, where the matter is taken before an outside arbitrator, whose fee depends on the speed with which matters are settled. Set this up as an analysis of options problem and by choosing your own preferences, analyse whether the scenario where only the wage demands are met is stable.

CHAPTER 7

Multi-stage games

'Those who cannot remember the past are condemned to repeat it'—George Santayana

7.1 MULTI-STAGE GAMES

A multi-stage game is one in which the outcome can be a real payoff and a requirement to play the same or another game again. So far the games we have looked at can all be considered as 'one-off' games, in the sense that there is no obligation to play another game after them. However, when we discussed mixed strategies we mentioned explicitly the idea of repeating the game and the concept of repeatedly seeking prices for the product is implicit in the ideas of oligopoly theory.

Before looking at multi-stage games in detail, we will look at a 'one-off' type of game which we have already solved, and show that we can think of this also as a multi-stage game. The reason for doing this is that the way of solving this game as a multi-stage one carries through to more complicated cases.

Example 7.1—Noisy Duel (see Example 2.4). Two duellists start $2N$ steps apart and move towards each other. At each step, they can decide whether or not to fire their one bullet. When they are $2i$ steps apart, the chance of killing an opponent is $(i + 1)^{-1}$. The duel is noisy and the payoff to player I is 1 if he is alive and 0 if not. We can think of this as N, 2×2 games, $\Gamma_0, \Gamma_1, \Gamma_2, \ldots, \Gamma_N$. Γ_i is the game when they are $2i$ steps apart and both players have the choice of whether to shoot I_s, II_s or not $I_{n.s.}$, $II_{n.s.}$. The payoff matrix is:

$$\begin{array}{cc} & \begin{array}{cc} II_s & II_{n.s.} \end{array} \\ \begin{array}{c} I_s \\ I_{n.s.} \end{array} & \begin{pmatrix} i/i+1 & 1/i+1 \\ i/i+1 & \Gamma_{i-1} \end{pmatrix}, \end{array} \tag{7.1}$$

since if II shoots now, I's chance of remaining alive is $i/i + 1$, no matter what. If I shoots but II does not, then I had better kill II, otherwise since it is a noisy duel II will know I has missed and will wait until they are 0 steps apart ($i = 0$) when he can kill I with certainty. So I's chance of survival is $(i + 1)^{-1}$. If neither shoot they move into the game where they are $2(i - 1)$ steps apart.

In order to solve this game we need to be able to calculate $v_i = \text{val}\,\Gamma_i$ for $i = 0,1,2,\ldots,N$, where $\text{val}\,\Gamma_i$ is the value of Γ_i, see Definition 2.1, and the corresponding optimal strategies. Γ_0 is easy to solve because there is no such game as Γ_{-1}, and so if both don't shoot at this stage, I will survive. The payoff matrix (7.1) is then:

$$\begin{array}{ccc} \Gamma_0 & \text{II}_\text{s} & \text{II}_\text{n.s.} \\ \text{I}_\text{s} & \multicolumn{2}{c}{\multirow{2}{*}{$\begin{pmatrix} 0 & 1 \\ 0 & 1 \end{pmatrix}$}} \\ \text{I}_\text{n.s.} & & \end{array} . \tag{7.2}$$

II_s dominates $\text{II}_\text{n.s.}$ not surprisingly and so $v_0 = \text{val}\,\Gamma_0 = 0$. Now we can solve Γ_1 because we can replace the payoff 'play Γ_0' in it with the value, v_0, of the game Γ_0. This is the vital idea in solving multi-stage games. You replace the commitment of having to play the game by a payoff equal to the value of the game. Thus, the game Γ_1 can now be thought of as having the payoff matrix

$$\begin{array}{c|cc} \Gamma_1 & \text{II}_\text{s} & \text{II}_\text{n.s.} \\ \text{I}_\text{s} & ½ & ½ \\ \text{I}_\text{n.s.} & ½ & 0 \end{array} . \tag{7.3}$$

By domination of $\text{I}_\text{n.s.}$ by I_s we get $v_1 = \text{val}\,\Gamma_1 = ½$. Substituting this for Γ_1 in the payoff matrix of Γ_2 gives:

$$\begin{array}{c|cc} \Gamma_2 & \text{II}_\text{s} & \text{II}_\text{n.s.} \\ \text{I}_\text{s} & ⅔ & ⅓ \\ \text{I}_\text{n.s.} & ⅔ & ½ \end{array} . \tag{7.4}$$

Domination of I_s by $\text{I}_\text{n.s.}$ and II_s by $\text{II}_\text{n.s.}$ leads to $v_2 = \text{val}\,\Gamma_2 = ½$. In fact we can now prove recursively that for all other $i > 2$ $v_i = ½$. Assuming $v_{i-1} = ½$, the payoff matrix for Γ_i is:

$$\begin{array}{c|cc} & \text{II}_\text{s} & \text{II}_\text{n.s.} \\ \text{I}_\text{s} & i/i + 1 & 1/i + 1 \\ \text{I}_\text{n.s.} & i/i + 1 & ½ \end{array} , \tag{7.5}$$

and it is trivial that $\text{I}_\text{n.s.}$ dominates I_s and $\text{II}_\text{n.s.}$ dominates II_s. This gives the solution $v_i = ½$ and the induction is proved. The value of the game is ½ and it is best for I to shoot when they are two steps apart.

We could have set up this game in normal form with each player having $N + 1$ pure strategies, and obtained the same answer. The object of doing it this way was to show that replacing a game by its value is a valid operation which aids solution of the game.

Throughout this chapter we will concentrate mainly on multi-stage games which have only two players and are zero-sum. Obviously you can define non-zero-sum and n-player versions of these games, but one complication at a time is enough.

7.2 STOCHASTIC GAMES, RECURSIVE GAMES AND SUPER-GAMES

Example 7.1 was easy to solve because each game Γ_i depended only on a 'smaller' game Γ_{i-1} and we could work back from Γ_0 to solve all the rest. In general, the connections between the games are more complicated and the payoff from any game can be partly a numerical payoff and partly probabilities of playing other games.

Definition 7.1. A two-person zero-sum **multi-stage game**, Γ, is a set of N games $\Gamma_1, \Gamma_2, \ldots, \Gamma_N$, hereafter called **subgames**. The normal form of subgame Γ_k is an $n_k \times m_k$ payoff matrix whose entries are of the form:

$$e_{ij}^k = a_{ij}^k \quad \text{and} \quad \sum_{l=1}^{N} p_{ij}^{kl} \Gamma_l, \tag{7.6}$$

where e_{ij}^k is the payoff if player I plays I_i and II plays II_j. It consists of a numerical reward a_{ij}^k where we assume $|a_{ij}^k| \leqslant M$, and a probability p_{ij}^{kl} of playing Γ_l for $l = 1,2,\ldots,N$, where $p_{ij}^{kl} \geqslant 0$ and $\sum_{l=1}^{n} p_{ij}^{kl} \leqslant 1$. Each time we play one of the subgames constitutes a **stage** of the game Γ. If $\sum_{l=1}^{N} p_{ij}^{kl} < 1$, then there is a positive probability $1 - \sum_{l=1}^{N} p_{ij}^{kl}$ that the game ends at this stage.

The difficulty with such general multi-stage games is that they could lead to unbounded payoffs, and we cannot compare two strategies that both have such payoffs. There are two ways of thinking about the rewards from the game which overcome this, namely:

(a) Discounting the payoffs, so that the payoff from the game played at the rth stage is discounted by β^{r-1}, where $\beta \leqslant 1$. This incorporates the idea of inflation, so a reward of 1 next stage is really only equal to a reward of β now, and a reward of 1 two stages hence is equal to one of β^2 now.

(b) Taking the average reward per stage. So if E_s is the reward from the first s stages of the game for a strategy, we look at $\lim_{s\to\infty} E_s/s$ as the average reward for that strategy.

We will concentrate on the discounted case, because the conditions under which there is a solution to the average cost case are only now being discovered (Mertens and Neyman, 1981) and there are no good solution algorithms for it. It is usual to call these general multi-stage games, **stochastic** games or **Markov** games, following Shapley (1953).

There are two variants of such games which have also been studied.

Definition 7.2. A **recursive** game is one where the payoffs e_{ij}^k consist of a numerical payoff a_{ij}^k or of probabilities p_{ij}^{kl} of playing other subgames Γ_l. Thus, in this game the players receive a reward only at the last stage, when they don't have to play any other subgames, and before this they just move from subgame to subgame without receiving any reward. These were introduced by Everett (1957).

Definition 7.3. A **supergame** is one where there is only one subgame Γ_1 and $p_{ij}^{11} = 1$ for all i and j. This is a Markov game where you play the same game over and over again, and is used in somewhat the same vein as metagames to analyse the stability of strategies in the original subgame.

In stochastic, recursive and supergames, we assume that both players know all the payoffs and that each game is zero sum. We can generalise this to n-person non-zero-sum games, but one of the more interesting generalisations is to assume that the players do not know the payoff matrix. They learn something each time they play the game, and use this information the next time they play it. These are called **repeatable games with partial information.**

7.3 DISCOUNTED STOCHASTIC GAMES

In a **discounted stochastic** game, Γ, a typical strategy for a player, is very complicated. For player I it consists of a collection of strategies $\{x^k(t,h_1,h_2,\ldots,h_{t-1}),\ k = 1,2,\ldots,N,\ t = 1,2,\ldots\}$, where x^k is a (possibly mixed) strategy in the subgame Γ_k and depends on which stage, t, you are playing and on the history $h_1,h_2,\ldots,h_{t-1}$ of what happened in the first $t-1$ stages. h_i says which game was played at the ith stage and which strategy each player played in it.

If $x^k(t,h_1,h_2,\ldots,h_{t-1}) = x^k$, for all t, k, and all possible histories $h_1,h_2,\ldots,h_{t-1}$, the strategy is called a **stationary Markov** one, and we will show that the optimal strategies are always of this form.

Although each subgame Γ_k of the stochastic game Γ has a finite number of pure strategies, it is obvious from the above description of strategies that the overall game Γ has an infinite number of such strategies, since each such strategy must choose a strategy for each subgame at each of an infinite number of stages. So we can not use the von Neumann minimax theorem (Theorem 2.1) to guarantee that the stochastic game Γ has a solution, since that theorem needs a finite number of pure strategies. The next section will furnish a proof that every discounted stochastic game has a solution.

The way we prove it is to use the same trick of replacing a game by its value that we used in Example 7.1. Obviously, if the game Γ has a value it will depend on which subgame we started playing at the first stage. Thus, the value v^* of Γ (if it exists) will be vector $\mathbf{v}^* = (v_1^*, v_2^*, \ldots, v_N^*)$, where $v_i^* =$ val ($\Gamma|$starting in Γ_i at stage one). If we substitute the value of Γ starting in Γ_l instead of Γ_l in (7.6) we might think that this new game has payoff entries $a_{ij}^k + \sum_{l=1}^N p_{ij}^{kl} v_l^*$. However, if we start Γ in Γ_i, then if we go to Γ_l at the second stage, the value of the game from there on is not v_l^* but βv_l^*. We are not playing Γ starting at Γ_l at the first stage but playing Γ with Γ_l at the second stage and so all the payoffs will be one stage later than under Γ starting in Γ_l. The discounting criterion means the resultant value is βv_l^*.

The trick of replacing the game by its value then suggests that $\mathbf{v}^*$, if it exists, should satisfy the equation:

$$v_k = \text{val}(\Gamma_k(\mathbf{v})), \qquad k = 1,2,\ldots,N, \tag{7.7}$$

where $\Gamma_k(v)$ is an $n_k \times m_k$ game with payoff matrix:

$$e_{ij}^k = a_{ij}^k + \beta \sum_{l=1}^N p_{ij}^{kl} v_l. \tag{7.8}$$

In general we write $\Gamma_k(\mathbf{w})$ for the $n_k \times m_k$ game with payoff matrix

$$e_{ij}^k = a_{ij}^k + \beta \sum_{l=1}^N p_{ij}^{kl} w_l. \tag{7.9}$$

In the next section we use this idea that the value of the game, if it exists, should satisfy (7.7) to prove its existence.

7.4 EXISTENCE OF VALUE

The object of this section is to prove the following result.

Theorem 7.1.
(a) There is always a solution to (7.7).
(b) There is only one solution to (7.7).
(c) This unique solution of (7.7) is the value of the game Γ.

Before we start the proof we need one result that helps along the way.

Lemma 7.1. If E is an $n \times m$ game with payoff entries e_{ij} and F is an $n \times m$ games with payoff entries f_{ij}, then if val E and val F are the values of games E and F:

$$\left|\text{val } E - \text{val } F\right| \leq \max_{ij} \left|e_{ij} - f_{ij}\right|. \tag{7.10}$$

Proof. For any of II's strategies $\mathbf{y}$:

$$\sum_{j=1}^m f_{ij}y_j - \sum_{j=1}^m e_{ij}y_j \leq \max_{ij} \left|f_{ij} - e_{ij}\right| \sum_{j=1}^m y_j = \max_{ij} \left|e_{ij} - f_{ij}\right|. \tag{7.11}$$

Then, if y is II's optimal strategy for game E,, (7.11) gives us:

$$\sum_{j=1}^{m} f_{ij}y_j \leqslant \sum_{j=1}^{m} e_{ij}y_j + \max_{ij} |e_{ij} - f_{ij}| \leqslant \text{val } E + \max_{ij} |e_{ij} - f_{ij}|, \tag{7.12}$$

since any strategy for I against II's optimal gives a payoff no greater than val E. Now as (7.12) holds no matter which pure and (hence mixed) strategy I plays, suppose he plays his optimal strategy x in game F. We get:

$$\text{val } F \leqslant \sum_{j=1}^{m} \sum_{i=1}^{n} x_i f_{ij} y_j \leqslant \text{val } E + \max_{ij} |e_{ij} - f_{ij}|, \tag{7.13}$$

where the first inequality follows because if I plays his optimal strategy in F, the payoff must be at least val F, and the second inequality follows from (7.12). Similarly, using I's optimal strategy in E and II's optimal in F we can show that:

$$\text{val } F \geqslant \text{val } E - \max_{ij} |e_{ij} - f_{ij}|. \tag{7.14}$$

(7.13) and (7.14) combined give the desired result (7.10).

We are now in a position to prove the existence theorem, Theorem 7.1. One of the reasons for going through the proof is that it is constructive—that is, it gives us an algorithm for actually finding the value.

Proof of Theorem 7.1. Consider a game where you start in subgame Γ_k, play Γ for n stages and then stop. Depending on where the game is when it is stopped, I receives an amount $\mathbf{w}^0 = (w_1^0, w_2^0, \ldots, w_N^0)$. It is obvious that if $n = 0$, the value of this game is $\mathbf{w}^0$, while if $n = 1$, a moment's thought tell us that the value of the game $\mathbf{w}^1$ must satisfy $w_k^1 = \text{val } \Gamma_k(\mathbf{w}^0)$, where $\Gamma_k(\mathbf{w})$ was defined in (7.9). Since in the game with n stages, we get the reward from the first stage plus the reward, discounted by β, of the game over the remaining $n - 1$ stages, then:

$$w_k^n = \text{val } \Gamma_k(\mathbf{w}^{n-1}) \tag{7.15}$$

is the value of the game described above. We will show that as n tends to infinity, w_k^n converges and its limit satisfies (a), (b) and (c) of the theorem.

If $u_k = \text{val } \Gamma_k(\mathbf{w})$, for $k = 1,2,\ldots,N$, we write this as $\mathbf{u} = T\mathbf{w}$ and for each vector $\mathbf{w}$ let $\|\mathbf{w}\| = \max_k |w_k|$. $\|\mathbf{w}\|$ is a norm, which is a measure of how large is w. Now:

$$\begin{aligned}
\|T\mathbf{w} - T\mathbf{u}\| &= \max_k |\text{val } \Gamma_k(w) - \text{val } \Gamma_k(u)| \\
&\leqslant \max_k \max_{i,j} \Big| \Big(a_{ij}^k + \beta \sum_{l=1}^{N} p_{ij}^{kl} w_l\Big) \\
&\quad - \Big(a_{ij}^k + \beta \sum_{l=1}^{N} p_{ij}^{kl} u_l\Big)\Big| \\
&= \max_k \max_{i,j} \Big|\beta \sum_{l=1}^{N} p_{ij}^{kl}(w_l - u_l)\Big| \\
&\leqslant \max \beta \Big|\sum p_{ij}^{kl}\Big| \max_l |w_l - u_l| \leqslant \beta \|\mathbf{w} - \mathbf{u}\|,
\end{aligned} \tag{7.16}$$

where the first inequality follows from Lemma 7.1. Since $\mathbf{w}^n = T\mathbf{w}^{n-1}$, by substituting into (7.16) over and over we get:

$$\|\mathbf{w}^{n+1} - \mathbf{w}^n\| \leqslant \beta \|\mathbf{w}^n - \mathbf{w}^{n-1}\| \leqslant \beta^n \|\mathbf{w}^1 - \mathbf{w}^0\|. \qquad (7.17)$$

It is a well-known property of norms that they satisfy the triangle inequality $\|a + b\| \leqslant \|a\| + \|b\|$, for all a, b. (If you do not believe it, just realise for our norm that there is one maximisation in $\|a + b\|$, but two allowed in $\|a\| + \|b\|$.) Using this several times we get:

$$\begin{aligned}\|\mathbf{w}^{n+k} - \mathbf{w}^n\| &\leqslant \|\mathbf{w}^{n+k} - \mathbf{w}^{n+k-1}\| + \|\mathbf{w}^{n+k-1} - \mathbf{w}^{n+k-2}\| + \cdots + \|\mathbf{w}^{n+1} \\ &\quad - \mathbf{w}^n\| \\ &\leqslant (\beta^{n+k-1} + \beta^{n+k-2} + \ldots + \beta^n) \|\mathbf{w}^1 - \mathbf{w}^0\| \\ &\leqslant \beta^n \|\mathbf{w}^1 - \mathbf{w}^0\| / (1 - \beta). \end{aligned} \qquad (7.18)$$

If we let n be very large, the R.H.S. of (7.18) will be as small as we like and this proves that all subsequent $\mathbf{w}^{n+k}$ are close to $\mathbf{w}^n$ and so the sequence converges. Let $\mathbf{v}^* = \lim_{n\to\infty} \mathbf{w}^n$, then trivially:

$$T\mathbf{v}^* = T \lim_{n\to\infty} \mathbf{w}^n = \lim_{n\to\infty} T\mathbf{w}^n = \lim_{n\to\infty} \mathbf{w}^{n+1} = \mathbf{v}^*, \qquad (7.19)$$

and (7.19) is exactly equation (7.7) when we recall that $T\mathbf{v}^*_k = \text{val}\,\Gamma_k(\mathbf{v}^*)$. Thus, $\mathbf{v}^*$ is a solution to (7.7) and part (a) of the theorem is proved.

To prove (b) suppose there was another solution $\mathbf{v}'$ of (7.7) so $\mathbf{v}^* = T\mathbf{v}^*$ and $\mathbf{v}' = T\mathbf{v}'$. Then by (7.16):

$$\|\mathbf{v}^* - \mathbf{v}'\| = \|T\mathbf{v}^* - T\mathbf{v}'\| \leqslant \beta \|\mathbf{v}^* - \mathbf{v}'\| \qquad (7.20)$$

As $\beta < 1$, this can only hold if $\|\mathbf{v}^* - \mathbf{v}'\| = 0$, and so $\mathbf{v}^* = \mathbf{v}'$.

Lastly, we have to prove (c) — that $\mathbf{v}^*$ is the value of the game Γ. Think again of the game we introduced earlier where we start in Γ_k, play Γ for n goes, and stop, getting w^0, which we set equal to 0 now. This has a finite number of pure strategies and so by von Neumann's theorem has a value, which is w_n^k and optimal strategies. Suppose in Γ proper, starting in Γ_k, I plays the optimal strategy for this finite horizon version for the first n stages and anything thereafter, what is his payoff? (It is as if he thought the game would last only n stages, played optimally for these and then went to pieces when he found it lasted longer.) I gets w_k^n from the first n stages and because, from Definition 7.1, all payoffs lie between $-M$, and M, the worst that can happen to I thereafter is $-\beta^n M - \beta^{n+1} M - \beta^{n+2} M \ldots$. Since I has a strategy which guarantees $w_k^n - \beta^n M - \beta^{n+1} M \ldots$, the value of the game Γ starting in Γ_k if it exists must satisfy:

$$v_k \geqslant w_k^n - \beta^n M - \beta^{n+1} M - \ldots \geqslant w_k^n - \beta^n M/(1 - \beta). \qquad (7.21)$$

If II plays the same sort of strategy, in Γ an analogous argument shows that

v_k if it exists must satisfy:

$$v_k \leqslant w_k^n + \beta^n M/(1-\beta). \tag{7.22}$$

Letting n go to infinity in (7.21) and (7.22) gives:

$$v_k^* = \lim_{n\to\infty} w_k^n \geqslant v_k \geqslant \lim_{n\to\infty} w_k^n = v_k^*, \tag{7.23}$$

and so $\mathbf{v}^*$ is the value of the game Γ. We have described, at least in the limit, strategies for both players that guarantee $\mathbf{v}^*$.

7.5 ADVERTISING EXAMPLE

One of the advantages of the proof of the previous theorem (yes, it was worth working through it), is that it tells us an algorithm for finding the value of stochastic games. If we calculate the $\mathbf{w}^n$ corresponding to the game stopped after n stages via (7.15), we will converge to $\mathbf{v}^*$ eventually. Let us see this in action.

Example 7.2. Two companies make the same line of products, and each year they can decide whether or not to advertise. The advertising affects their profits, the future image of their company, and the total advertising also affects whether the demand for the product will continue. Think of them as toy manufacturers who are trying to ensure that each Christmas (when 70% of toys are sold) children still want their product. If company I has the better image, the game is Γ_1, whereas if II has the better image it is Γ_2. We take as numerical payoffs the difference between I and II's profit each year. Profits are discounted by a factor $\beta = 0.75$, and a set of hypothetical though reasonable payoffs are given below, where I_a is when I advertises, and I_n when I does not, and similarly for II.

Γ_1:

$$\begin{array}{cc} & \begin{array}{cc} II_a & II_n \end{array} \\ \begin{array}{c} I_a \\ I_n \end{array} & \begin{pmatrix} 1 \;\&\; \tfrac{2}{3}\Gamma_1 + \tfrac{1}{3}\Gamma_2 & 2 \;\&\; \tfrac{2}{3}\Gamma_1 \\ 2 \;\&\; \tfrac{2}{3}\Gamma_2 & 2 \;\&\; \tfrac{1}{3}\Gamma_1 \end{pmatrix} \end{array}. \tag{7.24}$$

Γ_2:

$$\begin{array}{cc} & \begin{array}{cc} II_a & II_n \end{array} \\ \begin{array}{c} I_a \\ I_n \end{array} & \begin{pmatrix} -1 \;\&\; \tfrac{1}{3}\Gamma_1 + \tfrac{2}{3}\Gamma_2 & 1 + \tfrac{2}{3}\Gamma_1 \\ -2 \;\&\; \tfrac{2}{3}\Gamma_2 & 0 \;\&\; \tfrac{1}{3}\Gamma_2 \end{pmatrix} \end{array}. \tag{7.25}$$

Notice $\sum_{l=1}^{2} P_{ij}^{kl} < 1$ in some cases and so the game and the demand for the product range might not last for ever. If the game lasts zero period, $\mathbf{w}^0 = (0,0)$ by definition. Replacing Γ_1 by βw_1^0 and Γ_2 by βw_2^0 in (7.24) and (7.25), we can solve for $\mathbf{w}^1$. (All right, so βw_1^0 and βw_2^0 are both zero, but it is the method that is important.) So:

$$w_1^1 = \text{val}\begin{pmatrix} 1+0+0, & 2+0 \\ 2+0 \quad , & 2-0 \end{pmatrix} = 2. \tag{7.26}$$

$$w_2^1 = \text{val}\begin{pmatrix} -1+0+0, & 1+0 \\ -2+0 \quad , & 0+0 \end{pmatrix} = -1. \tag{7.27}$$

Substituting $\beta w_1^1 = \frac{3}{4} \times 2$ for Γ_1 and $\beta w_2^1 = -\frac{3}{4}$ for Γ_2 gives $\mathbf{w}^2$. Hence,

$$w_1^2 = \text{val}\begin{pmatrix} 1+\frac{2}{3}\cdot\frac{3}{2}+\frac{1}{3}\cdot-\frac{3}{4}, & 2+\frac{2}{3}\cdot\frac{3}{2} \\ 2+\frac{2}{3}\cdot-\frac{3}{4}, & 2+\frac{1}{3}\cdot\frac{3}{2} \end{pmatrix} = \text{val}\begin{pmatrix} 1\frac{3}{4} & 3 \\ 1\frac{1}{2} & 3\frac{1}{2} \end{pmatrix}$$
$$= 1\tfrac{3}{4}. \tag{7.28}$$

$$w_1^2 = \text{val}\begin{pmatrix} -1+\frac{1}{3}\cdot\frac{3}{2}+\frac{2}{3}\cdot-\frac{3}{4}, & 1+\frac{2}{3}\cdot\frac{3}{3} \\ -2+\frac{2}{3}\cdot-\frac{3}{4}, & 0+\frac{1}{3}\cdot-\frac{3}{4} \end{pmatrix} = \text{val}\begin{pmatrix} -1 & 2 \\ -2\frac{1}{2} & -\frac{1}{4} \end{pmatrix}$$
$$= -1. \tag{7.29}$$

So $\mathbf{w}^2 = (1\frac{3}{4}, -1)$ and the next calculation gives $\mathbf{w}^3 = (1\frac{5}{8}, -1\frac{1}{16})$. In the limit we get $\mathbf{w}^n$ converges to $(1\frac{1}{3}, -1\frac{1}{3})$, which is the value of the game.

7.6 BOUNDS ON VALUE ITERATION

The algorithm described above is called value iteration, and one of its disadvantages as a solution technique is that it never arrives at the actual value of the game in a finite number of iterations. If, as in Example 7.2, the iteration values keep being obtained from the same combination of optimal pure strategies, you could guess these would be optimal, for the non-truncated game. Find the value of the game assuming these strategies were optimal and then substitute this value back into the game to check it is the solution. Thus, in Example 7.2 for both $\mathbf{w}^1$ and $\mathbf{w}^2$ the optimal strategies were $(\mathrm{I}_1,\mathrm{II}_1)$ in Γ_1 and Γ_2. Assuming this was the case for the non-truncated game, (7.24) and (7.25) tell us that the solution (v_1,v_2) would satisfy:

$$v_1 = 1 + \tfrac{2}{3}(\tfrac{3}{4}v_1) + \tfrac{1}{3}(\tfrac{3}{4}v_2), \qquad v_2 = -1 + \tfrac{1}{3}(\tfrac{3}{4}v_1) + \tfrac{2}{3}(\tfrac{3}{4}v_2), \tag{7.30}$$

which gives $v_1 = 4/3$, $v_2 = -4/3$. Substituting this back into (7.24) and (7.25) we have to check these are the values of the games Γ_1 and Γ_2:

Γ_1:

$$\begin{array}{c} \\ \mathrm{I_a} \\ \mathrm{I_n} \end{array}\begin{array}{c} \mathrm{II_a} \qquad\qquad\qquad\qquad\qquad\qquad \mathrm{II_n} \\ \begin{pmatrix} 1 + 2/3 \times 3/4 \times 4/3 + 1/3 \times 3/4 \times -4/3, & 2 + 2/3 \times 3/4 \times 4/3 \\ 2 + 2/3 \times 3/4 \times -4/3, & 2 + 1/3 \times 3/4 \times 4/3 \end{pmatrix} = \begin{pmatrix} 1\tfrac{1}{3} & 2\tfrac{2}{3} \\ 1\tfrac{1}{3} & 2\tfrac{1}{3} \end{pmatrix} \end{array} \tag{7.31}$$

and

Γ_2:

$$\begin{array}{c} \\ \mathrm{I_a} \\ \mathrm{I_n} \end{array}\begin{array}{c} \mathrm{II_a} \qquad\qquad\qquad\qquad\qquad\qquad \mathrm{II_n} \\ \begin{pmatrix} -1 + 1/3 \times 3/4 \times 4/3 + 2/3 \times 3/4 \times -4/3, & 1 + 2/3 \times 3/4 \times 4/3 \\ -2 + 2/3 \times 3/4 \times -4/3 & 0 + 1/3 \times 3/4 \times -4/3 \end{pmatrix} = \begin{pmatrix} -1\tfrac{1}{3} & 1\tfrac{2}{3} \\ -2\tfrac{2}{3} & -1/3 \end{pmatrix} \end{array} \tag{7.32}$$

They are. This confirms $(4/3, -4/3)$ as the solution. It is an example of the bare hands method, where you guess the solution and then confirm your guess.

Most of the time, however, the optimal strategy at each iteration changes from one iterate to the next. In that case, suppose we decide to stop after calculating w^n, the nth iteration, can we say how close this is to the value of the game v? The following lemma gives the answer.

Lemma 7.2. Let $\mathbf{w}^n = (w_1^n, w_2^n, \ldots, w_N^n)$ be the value of the nth iteration of the stochastic game with discount factor β. Then if $\mathbf{v} = (v_1, v_2, \ldots, v_N)$ is the value of the game.

$$w_i^n - (\beta\|\mathbf{w}^n - \mathbf{w}^{n-1}\|/(1-\beta)) \leqslant v_i \leqslant w_i^n + (\beta\|\mathbf{w}^n - \mathbf{w}^{n-1}\|/(1-\beta)). \tag{7.33}$$

(Recall $\|z\| = \max_i |z_i|$.)

Proof. The result follows immediately from the inequality (7.17) in the proof of Theorem 7.1. This said, $\|\mathbf{w}^{n+1} - \mathbf{w}^n\| \leqslant \beta\|\mathbf{w}^n - \mathbf{w}^{n-1}\|$ and using the triangle inequality for norms again (see (7.18)) we have:

$$\begin{aligned} \|\mathbf{w}^{n+k} - \mathbf{w}^n\| &\leqslant \|\mathbf{w}^{n+k} - \mathbf{w}^{n+k-1}\| + \|\mathbf{w}^{n+k-1} - \mathbf{w}^{n+k-2}\| \\ &\quad + \ldots + \|\mathbf{w}^{n+1} - \mathbf{w}^n\| \\ &\leqslant (\beta^k + \beta^{k-1} + \ldots + \beta\|\mathbf{w}^n - \mathbf{w}^{n-1}\| \\ &\leqslant \beta\,\|\mathbf{w}^n - \mathbf{w}^{n-1}\|/(1-\beta). \end{aligned} \tag{7.34}$$

Let k tend to infinity in (7.34) so the L.H.S. becomes $\|\mathbf{v} - \mathbf{w}^n\|$. For all i, $i = 1, \ldots, N$:

$$\begin{aligned} -\beta\|\mathbf{w}^n - \mathbf{w}^{n-1}\|/(1-\beta) \leqslant -\|\mathbf{v} - \mathbf{w}^n\| \leqslant v_i - w_i^n \leqslant \|\mathbf{v} - \mathbf{w}^n\| \\ \leqslant \beta\|\mathbf{w}^n - \mathbf{w}^{n-1}\|/(1-\beta), \end{aligned} \tag{7.35}$$

and the lemma is proved.

Going back to Example 7.2, suppose we stopped after the third iteration. The values of the iterations were $\mathbf{w}^1 = (2,-1)$, $\mathbf{w}^2 = (1\frac{3}{4},-1)$, $\mathbf{w}^3 = (1\frac{5}{8},-1\frac{1}{16})$, so $\|\mathbf{w}^3 - \mathbf{w}^2\| = \max\{|1\frac{5}{8} - 1\frac{3}{4}|, |-1\frac{1}{16} - (-1)|\} = \frac{1}{8}$. $\beta = \frac{3}{4}$, so substituting into (7.33) for $\mathbf{w}^3$ gives:

$$1\tfrac{5}{8} - \tfrac{3}{4}\cdot\tfrac{1}{8}/(1 - \tfrac{3}{4}) \leqslant v_1 \leqslant 1\tfrac{5}{8} + \tfrac{3}{4}\cdot\tfrac{1}{8}/(1 - \tfrac{3}{4}), \tag{7.36}$$

or $1\frac{1}{4} \leqslant v_1 \leqslant 2$. A similar calculation gives $-1\frac{7}{16} \leqslant v_2 \leqslant \frac{11}{16}$, and so even if we did not know the value of the game was $(1\frac{1}{3},-1\frac{1}{3})$, we would have these estimates for it. Obviously the more iterations you do, the more accurate the bounds become.

7.7 RECURSIVE GAMES

Another variant of the multi-stage game is the **recursive** game where the players only receive a real payoff at the last subgame they play. Thus, the payoff to each subgame is either to play another game or this real payoff. These recursive games are not as well behaved as stochastic games as the following examples illustrate. (N.B. They do not fall under Theorem 7.1 because the reward you are interested in is the total payoff, not the discounted one.)

Example 7.3—Game that does not end. Suppose there is only one subgame Γ_1 which is a 2×1 game with payoff matrix Γ_1, $\Gamma_1 = \binom{\Gamma_1}{-1}$. It is natural to assume that if a game does not reach a conclusion, the payoff will be 0 to both players. Thus, I would always choose his first strategy to ensure the game does not end.

Example 7.4—Game with no optimal strategies. There is only one subgame Γ_1, which has a 2×2 payoff matrix:

$$\begin{array}{ccc} \Gamma_1 & \mathrm{II}_1 & \mathrm{II}_2 \\ \mathrm{I}_1 & \Gamma_1 & 1 \\ \mathrm{I}_2 & 1 & 0 \end{array} \qquad \begin{pmatrix} \Gamma_1 & 1 \\ 1 & 0 \end{pmatrix}. \tag{7.37}$$

If I plays the mixed strategy $(1 - x,x)$, he ensures a payoff of 1 against II_1 and $1 - x$ against II_2. So as x tends to zero, he can guarantee a payoff as near 1 as he likes, but he cannot ensure himself 1, since if $x = 0$, the payoff against II_1 is zero. The value of the game must be 1 since I can ensure himself as near 1 as he likes and II_2 ensures he does not get more than 1. Yet I does not have a strategy that guarantees himself 1.

Example 7.5—Value of game is not limit of value of truncated game. Γ has two subgames Γ_1 and Γ_2 where the payoff matrices are:

$$\Gamma_1: \quad \begin{array}{c} \\ I_1 \\ I_2 \\ I_3 \end{array} \begin{array}{c} \begin{array}{cc} II_1 & II_2 \end{array} \\ \begin{pmatrix} \Gamma_1 & \Gamma_1 \\ \Gamma_2 & 10 \\ 10 & \Gamma_2 \end{pmatrix} \end{array} \qquad \Gamma_2: \quad \begin{array}{cc} & II_1 \\ I_1 & (-2)\,. \end{array} \tag{7.38}$$

Since going to Γ_2 pays -2, then it is obvious that we could replace Γ_2 by -2 and if I plays $(0,½,½)$ in Γ_1 and II plays $(½,½)$, these ensure that the value of the game is 4 starting in Γ_1 and -2 starting in Γ_2. However, if the game stops after n stages, then I can guarantee himself 5 if he starts in Γ_1 by playing I_1 for the first $n-1$ stages and $(0,½,½)$ at the nth. Thus, for any value of n, the truncated game has value $(5,-2)$ which does not converge to $(4,-2)$ the value of Γ.

7.8 SOLUTION OF RECURSIVE GAMES

Everett (1957) showed that if we change slightly the idea of a solution to a game, then every recursive game can be said to have a solution, provided the subgames have a finite number of pure strategies. We have to replace the idea of optimal strategies by ε-optimal strategies, which as in Example 7.4 ensures a payoff within ε of the value of the game, where ε is taken to be very small.

Definition 7.1. A recursive game possesses a **solution** if there is a vector $\mathbf{v} = (v_1,v_2,\ldots,v_N)$ so that for all $\varepsilon > 0$ there exist strategies $\mathbf{x}^\varepsilon,\mathbf{y}^\varepsilon$ such that if $e_i(\mathbf{x},\mathbf{y})$ is the payoff of Γ starting in Γ_i when I plays $\mathbf{x}$ and II plays $\mathbf{y}$:

$$e_i(\mathbf{x}^\varepsilon,\mathbf{y}) \geqslant v_i - \varepsilon, \quad \text{for all } \mathbf{y}, \quad i = 1,2,\ldots,N, \tag{7.39}$$

$$e_i(\mathbf{x},\mathbf{y}^\varepsilon) \leqslant v_i + \varepsilon, \quad \text{for all } \mathbf{x}, \quad i = 1,2,\ldots,N. \tag{7.40}$$

Everett proved the existence of a solution for such games essentially using the same idea of value iteration introduced in Theorem 7.1. One has to be a little more careful, however, since the value of the game is only one of several possible limit points of value iteration. We won't give the proof but refer the readers to the papers of Everett (1957) or Orkin (1972).

Theorem 7.2. Every recursive game with subgames which have finite numbers of pure strategies has a solution.

7.9 EXAMPLES OF RECURSIVE GAMES

In this section we describe two types of recursive games that have received some attention.

Example 7.6—Colonel Blotto games. 'Colonel Blotto' is the generic title for a whole class of war games where the two sides split their forces to fight in two or more areas. The side with the greatest force in a particular area wins the battle in that area. We are interested in the game where there are two areas—Blotto's camp and that of the enemy. Blotto has m divisions, and must decide how many defend his own camp and how many attack the enemy's camp. The enemy must make a similar choice with its n divisions. To capture a camp the attackers must have more divisions than the defenders, otherwise they retire and try again later. The payoff is +1 if Blotto captures the enemy camp, without losing his own, and −1 for losing his own, no matter what happens to the enemy camp. If neither side wins their opponents' camp the game Γ is played again. The game with $m = 2$, $n = 1$ gives Blotto three strategies I_i, $i = 0$, 1 and 2, corresponding to defending with i divisions, and two for the opponents—II_0 defends with none and II_1 defends with one division. The payoff matrix is:

$$\begin{array}{c} \\ I_0 \\ I_1 \\ I_2 \end{array} \begin{array}{c} \begin{array}{cc} II_0 & II_1 \end{array} \\ \begin{pmatrix} -1 & 1 \\ 1 & \Gamma \\ \Gamma & \Gamma \end{pmatrix} \end{array} . \qquad (7.41)$$

I_2 never produces a result, so it is obvious that I should play I_1 most of the time, occasionally using I_0. If I plays $(\frac{1}{2}\varepsilon, 1 - \frac{1}{2}\varepsilon, 0)$ he can guarantee himself $1 - \varepsilon$ while any strategy for II ensures that I does not get more than 1. So the value of the game is 1, $\mathbf{x}^\varepsilon = (\frac{1}{2}\varepsilon, 1 - \frac{1}{2}\varepsilon, 0)$ and $\mathbf{y}^\varepsilon = (y, 1 - y)$ for any y, $0 \leq y \leq 1$.

Example 7.7—Gambling games. These try to reflect the fact that in gambling against people, as opposed to roulette, your strategies change depending on how much either of you have left to lose. Suppose we have a two-person zero-sum 2 × 2 game with payoff matrix:

$$\begin{array}{cc} & \begin{array}{cc} \mathrm{II}_1 & \mathrm{II}_2 \end{array} \\ \begin{array}{c} \mathrm{I}_1 \\ \mathrm{I}_2 \end{array} & \begin{pmatrix} 1 & -1 \\ -2 & 2 \end{pmatrix}, \end{array} \tag{7.42}$$

where the payoffs represent the actual money that I receives from II. initially I starts with i units of money and II with $N - i$, and they keep playing this game until one of them runs out of money. The payoff in the overall game Γ is 1 if I wins all the money and 0 if he loses all his money. The subgames are $\Gamma_0,\Gamma_1,\Gamma_2,\ldots\Gamma_N$, where Γ_i is the game when I has i units and II has $N - i$. Trivially, $\Gamma_0 = (0)$ and $\Gamma_N = (1)$ and for $1 \leqslant i \leqslant N - 1$, Γ_i has payoff matrix:

$$\begin{pmatrix} \Gamma_{i+1} & \Gamma_{i-1} \\ \Gamma_{i-2} & \Gamma_{i+2} \end{pmatrix}, \tag{7.43}$$

where Γ_{N+1} is identified with Γ_N and Γ_{-1} with Γ_0. Let us solve this for the case $N = 4$, so the solution is a vector $\mathbf{v} = (v_0,v_1,v_2,v_3,v_4)$. To solve it we can use value iteration. Starting with $\mathbf{w}^0 = (0,0,0,0,0)$ if we follow the value iteration scheme used in Example 7.2 with $\beta = 1$ since there is no discounting we get,

$$\mathbf{w}^1 = (0,0,0,0,1), \quad \mathbf{w}^2 = (0,0,0,\tfrac{1}{2},1), \quad \mathbf{w}^3 = (0,0,\tfrac{1}{3},\tfrac{1}{2},1)$$
$$\mathbf{w}^4 = (0,\tfrac{1}{5},\tfrac{1}{3},\tfrac{3}{5},1), \quad \mathbf{w}^5 = (0,\tfrac{3}{14},\tfrac{3}{7},\tfrac{7}{11},1). \tag{7.44}$$

These iterates eventually converge to $\mathbf{v} = (0,1 - 1/\sqrt{2},\tfrac{1}{2},1/\sqrt{2},1)$.

An alternative way of finding the optimal values is to guess that the optimal solutions in each subgame will be mixed, since all the solutions of the iterates of value iteration were obtained by mixed strategies. If this is the case, then we can use the formula (2.42) for the value of 2 × 2 games with no domination, to connect v_0,v_1,v_2,v_3,v_4. This gives:

$$v_0 = 0, \quad v_4 = 1, \quad v_1 = v_2v_3/(v_2 + v_3), \quad v_2 = v_3/(1 + v_3 - v_1),$$
$$v_3 = (1 - v_1v_2)/(2 - v_1 - v_2), \tag{7.45}$$

which has solution $v_1 = 1 - 1/\sqrt{2}$, $v_2 = \tfrac{1}{2}$, $v_3 = 1/\sqrt{2}$.

7.10 SUPERGAMES

Supergames are the special sorts of stochastic games where there is only one subgame Γ_1, which is played an infinite number of times, i.e. $p_{ij}^{11} = 1$, for all i and j. Since the structure is so much simpler than general stochastic games, we will deal with non-zero-sum games and n-person games as well as zero-sum ones. However, we still have the difficulty that the total payoff might be infinite, and so we again must use the discounted criterion or the average reward per stage. We will concentrate on the discounted criterion.

The term **supergame** was introduced by Luce and Raiffa (1957), when discussing Prisoners' Dilemma played repeatedly. The object of their analysis was to find out what the payoffs of the equilibrium pairs of strategies were, in order to get a feel for the 'stability' properties of the payoffs. The word 'stability' might strike a chord in your memory (even if it is only the lost chord), and indeed the idea of metagames was introduced for the same purpose. There is a close connection between metagames and supergames, as we shall try to show by an example in the next section.

A supergame is really the extensive form of playing a game repeatedly, where the dependence of what is being played at this stage on what happened previously is made plain. Suppose the game Γ_1 is an n-person one where the payoff to player i is $e_i(x_1,\ldots,x_n)$, $i = 1,2,\ldots,n$, when they each play x_i, respectively. We have to define the infinite horizon discounted payoff for such games. In such games, a typical strategy for player i is:

$$X_i = \{x_i(1),x_i(2,h_1),x_i(3,h_{1,2}),\ldots,x_i(n,h_1,h_2,\ldots,h_{n-1})\ldots\}, \tag{7.46}$$

where $x_i(n,h_1,h_2,\ldots.,h_{n-1})$ is the strategy played by i at the nth stage of the game, where h_1 to h_{n-1} describe what occurred at stages 1 to $n-1$. $E_i(X_1,X_2,\ldots,X_n)$ is the infinite horizon discounted reward to player i, when players 1 to n play X_1 to X_n, respectively. It is defined by:

$$E_i(X_1,X_2,\ldots,X_n) = \sum_{k=1}^{\infty} \beta^{k-1} e_i(x_1(k,H_{k-1}),x_2(k,H_{k-1}),\ldots,x_n(k,H_{k-1})), \tag{7.47}$$

where $H_{k-1} = (h_1,h_2,\ldots,h_{k-1})$.

We are interested in the equilibrium n-tuples of the supergame (recall Definition 4.1 of equilibrium n-tuple) and in particular the ones that give the same strategy at every stage of the game.

Definition 7.2. An n-tuple of strategies $\mathbf{x} = (x_1,\ldots,x_n)$ in the subgame Γ_1 is called **supergame stable** if there is an equilibrium, n-tuple $(X_1,X_2,\ldots,X_n)$ in the supergame, which results in $\mathbf{x} = (x_1,x_2,\ldots,x_n)$ being played at each stage of the supergame.

Obviously, Theorem 7.1, which guarantees a solution of general stochastic two-person zero-sum discounted games, implies there is at least one equilibrium pair of a two-person zero-sum supergame. However, it is easy to show for all n-person supergames, that provided Γ_1 has an equilibrium n-tuple so does the supergame.

Theorem 7.3. If an n-tuple of strategies $\mathbf{x} = (x_1,\ldots,x_n)$ is an equilibrium n-tuple in the n-person game Γ_1, it is supergame stable in the discounted supergame based on Γ_1.

Proof. Consider the n-tuple of strategies $(X_1,X_2,\ldots,X_n)$ in the supergame Γ_1 where, from (7.46), X_i is defined by:

$$X_i(n,h_{n-1}) = x_i, \quad \text{for all } n = 0,1,2,\ldots \text{ and all } H_{n-1}. \tag{7.48}$$

Then

$$E_i(X_1,\ldots,X_n) = \sum_{l=1}^{\infty} \beta^{k-1}\, e_i(x_1,x_2,\ldots,x_n). \tag{7.49}$$

Taking any other strategy $\tilde{X}_i$ for player i in the supergame could lead to strategies $\tilde{x}_i(1),\tilde{x}_i(2),\ldots,\tilde{x}_i(k),\ldots$ being played by i at each stage in response to $X_1,X_2,\ldots,X_{i-1},X_{i+1},\ldots,X_n$ in the supergame. However, since $x_1,\ldots,x_n$ is an equilibrium n-tuple, $e_i(x_1,x_2,\ldots,x_i,\ldots,x_n) \geqslant e_i(x_1,x_2,\ldots,\tilde{x}_i(k)\ldots,x_n)$ for all stages k. Hence, $E_i(X_1,\ldots,X_i,\ldots,X_n) \geqslant E_i(X_1,\ldots\tilde{X}_i,\ldots,X_n)$ and so $X_1,\ldots,X_n$ is an equilibrium n-tuple in the supergame, which results in $(x_1,\ldots,x_n)$ being played at each stage.

What is of more interest to us is which other strategies are supergame stable. We will not go into this in great detail for abstract games, but instead investigate this problem for an oligopoly game in the next section. Aubin (1979) in his book looks at the problem in a more general setting.

7.11 RELATIONSHIP BETWEEN SUPERGAMES AND META-GAMES: OLIGOPOLY EXAMPLE

This section connects some of the ideas introduced in the previous three chapters. In the duopoly and oligopoly examples of Chapter 5, we find their metagame equilibria (Chapter 6) and their supergame stable payoffs (Chapter 7), and compare the two answers. We do this for a numerical duopoly example, and then sketch how a similar result will occur in the general oligopoly case.

Example 7.8—Duopoly game (see Example 5.1). Two firms, 1 and 2, make the same commodity which they sell at p_1 and p_2, respectively. The demands for the firms' product are $f_1(p_1,p_2)$ and $f_2(p_1,p_2)$, respectively, where

$$\begin{aligned} f_1(p_1,p_2) &= \max\{1 + \tfrac{1}{3}p_2 - \tfrac{1}{2}p_1,0\}, \\ f_2(p_1,p_2) &= \max\{1 + \tfrac{1}{4}p_1 - \tfrac{1}{2}p_2,0\}. \end{aligned} \tag{7.50}$$

There is no production cost so the respective profits are

$$e_1(p_1,p_2) = \max\{p_1 + \tfrac{1}{3}p_1p_2 - \tfrac{1}{2}p_1^2,0\}$$

and

$$f_2(p_1,p_2) = \max\{p_2 + \tfrac{1}{4}p_1p_2 - \tfrac{1}{2}p_2^2,0\},$$

provided $p_1 < 2 + \tfrac{2}{3}p_2$ and $p_2 < 2 + \tfrac{1}{2}p_1$, respectively. In Example 5.1 we showed that the Cournot equilibrium was $p_1 = \tfrac{16}{11}$, $p_2 = \tfrac{15}{11}$ and so by Theorem 7.3 we know that this pair of prices is supergame stable.

In that example we also showed that the way to minimise an opponent's payoff was to set your price to be zero. Thus, the maximin values were $e_1(1,0) = 0.5$ and $e_2(0,1) = 0.5$ (see Fig. 5.5). We will show that any pair of prices $(\bar{p}_1,\bar{p}_2)$, where $e_1(\bar{p}_1,\bar{p}_2) > 0.5$ and $e_2(\bar{p}_1,\bar{p}_2) > 0.5$, is also supergame stable. Consider strategies P_1, P_2 in the supergame where $P_i = (p_i(1),p_i(2),\ldots,p_i(k),\ldots)$ and

$$p_1(k) = \begin{cases} \bar{p}_1, & \text{if } p_2(s) = \bar{p}_2, \quad \text{for } s = 1,2,\ldots,k-1, \\ 0, \text{ otherwise;} \end{cases} \tag{7.51}$$

$$p_2(k) = \begin{cases} \bar{p}_2, & \text{if } p_1(s) = \bar{p}_1, \quad \text{for } s = 1,2,\ldots,k-1, \\ 0, & \text{otherwise.} \end{cases} \tag{7.52}$$

It is easy to see that $E_i(P_1,P_2) = \sum_{k=0}^{\infty} \beta^{k-1}\, e_i(\bar{p}_1,\bar{p}_2)$, since $(\bar{p}_1,\bar{p}_2)$ occurs at each stage. Suppose player 1 plays some other strategy $\tilde{P}_1$ in the supergame. Playing this against P_2 means that $(\bar{p}_1,\bar{p}_2)$ will occur for the first K games (where K could be zero), then $(p_1',\bar{p}_2)$ at the $K + 1$st game where $p_1' \neq \bar{p}_1$. Thereafter player 2 will set his price to zero no matter what player 1 does. So the difference between (P_1,P_2) and $(\tilde{P}_1,P_2)$ in the supergame only occurs at the $K + 1$st stage and thereafter, and

$$\begin{aligned} E_1(P_1,P_2) - E_1(\tilde{P}_1,P_2) = {} & \beta^K(e_1(\bar{p}_1,\bar{p}_2) - e_1(p_1',\bar{p}_2)) \\ & + \sum_{k=K+2}^{\infty} \beta^{k-1}(e_1(\bar{p}_1,\bar{p}_2) - e_1(p_1(k),0)). \end{aligned} \tag{7.53}$$

No matter what $p_1(k)$ is, $e_1(p_1(k),0) \leqslant \tfrac{1}{2} < e_1(\bar{p}_1,\bar{p}_2)$, since the maximin value for player 1 is 0.5 when player 2 puts his price at zero. Thus (7.53) gives:

$$E_1(P_1,P_2) - E_1(\tilde{P}_1,P_2) \geqslant \beta^k[e_1(\bar{p}_1,\bar{p}_2) - e_1(p_1',\bar{p}_2) + (e_1(\bar{p}_1,\bar{p}_2) - \tfrac{1}{2})/(1-\beta)^{-1}]. \quad (7.54)$$

As β tends to 1, the last term of the R.H.S. of (7.54) gets very large, since $(e_1(\bar{p}_1,\bar{p}_2) - \frac{1}{2})$ is positive and so the whole R.H.S. of (7.54) eventually becomes positive. Thus, for all β's greater than some value β^*, (P_1,P_2) is an equilibrium pair and so $(\bar{p}_1,\bar{p}_2)$ is supergame stable.

In the metagame analysis introduced in Chapter 6, we want to find the metaequilibria of this game. By the results of Section 6.3, it is enough to look at the games $12G$ and $21G$. The metarational outcomes in these games are given by (6.18) to (6.23). In the game $12G$ $(\bar{p}_1,\bar{p}_2)$ is metarational for 1 if:

$$e_1(\bar{p}_1,\bar{p}_2) \geqslant \max_{p_1} \min_{p_2} e_1(p_1,p_2) = 0.5, \quad (7.55)$$

and for 2 if:

$$e_2(\bar{p}_1,\bar{p}_2) \geqslant \min_{p_1} \max_{p_2} e_2(p_1,p_2) = 0.5. \quad (7.56)$$

Similarly, in the game $21G$, $(\bar{p}_1,\bar{p}_2)$ is metarational for 1 (2), respectively, if:

$$e_1(\bar{p}_1,\bar{p}_2) \geqslant \min_{p_2} \max_{p_1} e_1(p_1,p_2) = 0.5, \quad (7.57)$$

$$e_2(\bar{p}_1,\bar{p}_2) \geqslant \max_{p_2} \min_{p_1} e_2(p_1,p_2) = 0.5, \quad (7.58)$$

where in (7.55)–(7.58) it turns out the minimiser should always choose $p_i = 0$. Thus, the metaequilibria, which are all symmetric for this game, are the set $(\bar{p}_1,\bar{p}_2)$, where

$$e_1(\bar{p}_1,\bar{p}_2) \geqslant 0.5 \quad \text{and} \quad e_2(\bar{p}_1,\bar{p}_2) \geqslant 0.5. \quad (7.59)$$

The supergame stable pairs of prices are the set of prices where there are strict inequalities in (7.59). In fact if we were to take the average cost per stage version of the supergame, which approximately corresponds to the case when $\beta = 1$ in the discounted case, the supergame stable pairs would be identical with the metaequilibria.

This connection between supergames and metagames is not a fluke of the numbers. If we look at the general n-person oligopoly, where the payoff to the ith player is $e_i(p_1,p_2,\ldots,p_n)$ when the prices are $p_1,p_2,\ldots,p_n$, respectively, then we can put conditions on the e_i to guarantee a similar result. Suppose

$$e_i(p_1,p_2,\ldots,p_n) = p_i f_i(p_1,p_2,\ldots,p_n), \quad (7.60)$$

where f_i is the demand for the ith firm's product. Then if the f_i are monotone increasing in p_j, $j \neq i$, but monotone decreasing in p_i, so that the worst a firm can do to any competitor is set its price equal to zero, a similar result will occur. Any set of prices $(\bar{p}_1,\bar{p}_2,\ldots,\bar{p}_n)$ is supergame stable for some suitable discount factor $\beta < 1$ if:

$$e_i(\bar{p}_1,\bar{p}_2,\ldots,\bar{p}_n) > \min_{p_1,p_2,p_{i-1},p_{i+1},\ldots,p_n} \max_{p_i} e_i(p_1,p_2,\ldots,p_n), \tag{7.61}$$

for all $i = 1,2,\ldots,n$,

whereas the metagame analysis says that the 'stable' outcomes—the metaequilibria—are the sets of prices $(\bar{p}_1,\bar{p}_2,\ldots,\bar{p}_n)$ which satisfy (7.61) with a 'greater than or equal to' inequality replacing the 'greater than' one. Those who wish to check this result should try Problem 7.10.

7.12 FURTHER READING

Although the idea of a dynamic game was used as a proof technique by von Neumann and Morgenstern (1947), the first paper to isolate the idea of multi-stage games was Shapley's work (1953) on stochastic games. It introduced discounted stochastic games and gave a proof of the existence of a value for such a game, which is essentially the one we use here. If ever there was a paper ahead of its time, this was a prime example because it was not until seven years later that people realised how important was the simpler one-person version of this game. This is where there is only one decision-maker who can make decisions at each stage. The decisions affect his immediate payoff and his chances of where he will be next stage, and the whole theory goes under the title of Markov Decision Processes. Since Shapley's paper there have been new solution algorithms suggested, like the variants on value iteration given by van der Wal (1977), or ones called policy iteration which try to improve the strategies used at each iteration—see Rao, Chandrasekaran and Nair (1973). For an easy introduction to the subject see van der Wal and Wessels' (1976) review paper or that of Parthasarathy and Stern (1977).

The actual applications of stochastic games have been far less widespread than one would have expected. Kirman and Sobel (1974) used them to model oligopoly problems which concentrated on the inventories held by the firms and Sobel (1982) uses them to model the effect of advertising on oligopoly. Deshmukh and Winston (1978) also model duopoly problems in terms of a zero-sum stochastic game. Another area where stochastic games make good models is in searching for a hider, who does not want to be found. Charnes and Schroeder (1967) looked at this problem in anti-submarine warfare terms.

The theory behind 'average reward per stage' stochastic games is only now being understood. For many years it was an open question whether or not such games had a value or not. The difficulties that arise can be illustrated by the game called the 'Big Match' which has three subgames Γ_1, Γ_2 and Γ_3 where the payoffs are:

$$\begin{matrix} \Gamma_1 & \Gamma_2 & \Gamma_3 \\ \begin{pmatrix} 1 \,\&\, \Gamma_1, & \Gamma_1 \\ \Gamma_2, & 1 \,\&\, \Gamma_3 \end{pmatrix} & (\Gamma_2) & (1 \,\&\, \Gamma_3). \end{matrix} \tag{7.62}$$

Blackwell and Ferguson (1968) constructed a history-remembering strategy for this game which is better than every Markov, or history-forgetting strategy, and so the clean analysis of discounted games where you concentrate only on the 'nice' stationary Markov strategies must fail. Bewley and Kohlberg (1976) made a large step on the road to solving the problem of existence of a value by showing that if we write the value of a discounted game with discount factor β as an expansion in powers of β, then the expansion has a limit as β tends to 1. Building on this, Mertens and Neyman (1981) proved the existence of a value for average reward stochastic games, provided this limit exists. However, the best strategies might again be ε-optimal rather than optimal. Algorithms for solving such games are discussed in van der Wal (1980).

Returning to the discounted case, Rogers (1969) generalised Shapley's existence result to show the existence of an equilibria in two-person non-zero-sum stochastic games. Later Sobel (1971) proved a similar result in the n-person case.

Recursive games, on the other hand, have a much less drawn-out history. The idea of such games and all the basic results were introduced by Everett (1957) and all the i's were dotted and t's crossed in a paper by Orkin (1972). Luce and Raiffa (1957) give a good discussion of what we called gambling games, and the Colonel Blotto game we discuss are a multi-stage version of the ones introduced by Blackett (1954).

Supergames were first discussed by Luce and Raiffa (1957) in trying to understand what happens when people play Prisoners' Dilemma over and over again. Friedman used the non-cooperative version of such games as models for oligopoly problems and his book (Friedman, 1977) gives a detailed account of the results. Meanwhile Aumann (1959, 1960) had used the cooperative version of supergames to look for suitable solutions concepts in n-person games.

Another area of multi-stage games, which we have not dealt with here, was also introduced by Robert Aumann. This deals with repeated games with incomplete information, where the players repeatedly play a game in which they do not have all the information about their payoffs or their opponents. At each stage of the game they may learn some extra information, which relates to the strategy they chose then, and the object is to prove such games have a value and to find the optimal strategies. In the original paper by Aumann and Maschler (1967a, 1967b), player I knows all about the game and player II only knows it is one of a finite set of games. At each stage he is told which strategy his opponent played, but nothing else, not even his own payoff. These papers are difficult to acquire and a good review of the material is to be found in Kohlberg (1975). Since then there have been many generalisations of the original problem, most based on the work of Mertens and Zamir (1971) who assume both players only have partial knowledge of the game. There have been other approaches to this problem, see for example the paper of Sweat (1968) where the players are told the rewards they obtain at each stage.

PROBLEMS FOR CHAPTER 7

7.1. A hider can hide in either Area 1 or Area 2. At each move of the game he decides which of the two areas to hide in for that move, and the seeker decides which area to look at for that move in order to find him. If the hider is in area 1, and the searcher looks in area 1, his chance of finding him is ¼, while if the hider is in area 2 and the searcher looks in area 2, the chance of finding him is ½. Obviously, if the searcher looks in the wrong area he will have no chance of finding the hider. The payoff is 1 to the searcher if he finds the hider within three moves, 0 otherwise. Find the value of the game setting it up as a multi-stage game with three subgames.

7.2. In a simplified game of Space Invaders there is only one monster who moves down through a 3 × 3 grid at each step moving to the next lowest row either vertically or diagonally (see Fig. 7.1). (Thus, from square (3,1) it can move to (2,1) or (2,2) and from (3,2) to

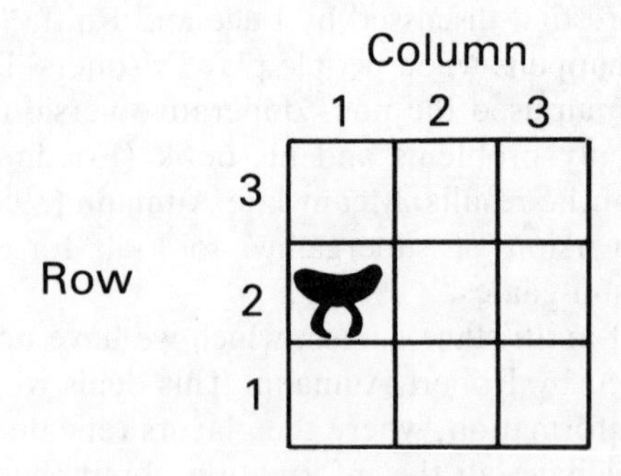

Fig. 7.1—Space invader game.

(2,1), (2,2) or (2,3).) Its opponent can fire vertically up one column at each step and if that is the column that the monster moves to at that step, the monster is eliminated (payoff +1 to I, who is the firer). If the monster reaches row 1 without being eliminated, the game ends (payoff 0 to I). This can be considered as a multi-stage game where $\Gamma_{i,j}$ is the game when the monster is in square i,j. Show that the payoff matrices are:

$\Gamma_{ij},\ i = 2,3,\ j = 1,3$ $\qquad\qquad$ $\Gamma_{i2},\ i = 2,3$

$$\begin{pmatrix} 1 & \Gamma_{i-1\,2} \\ \Gamma_{i-1\,j} & 1 \\ \Gamma_{i-1\,j} & \Gamma_{i-1\,2} \end{pmatrix} \quad \begin{pmatrix} 1 & \Gamma_{i-1\,2} & \Gamma_{i-1\,3} \\ \Gamma_{i-1\,1} & 1 & \Gamma_{i-1\,3} \\ \Gamma_{i-1\,1} & \Gamma_{i-1\,2} & 1 \end{pmatrix}$$

$\Gamma_{1j} = (0),\ j = 1,2,3.$

Using the fact that the solution of the game with payoff matrix

$$\begin{pmatrix} c & a & b \\ b & c & b \\ b & a & c \end{pmatrix}$$

is $(c^2 - 2ab + bc)/(3c - 2a - b)$, show the value of the game is $5/7$ if the monster starts in (3,1) or (3,3) and $7/11$ if it starts in (3.2).

If a monster is eliminated, then another one immediately reappears in square (3,2). If the payoff is the number of monsters you can eliminate before one reaches the first row, set this up as a stochastic game and show that if it starts with a monster in square (3,2) its value is $7/4$.

7.3. In a discounted stochastic game, Γ, with discount fctor $\beta = 0.8$ there are two subgames, Γ_1 and Γ_2, which have payoffs and transition probabilities given by:

$$\Gamma_1: \quad \begin{matrix} & \mathrm{II}_1 & \mathrm{II}_2 \\ \mathrm{I}_1 & \\ \mathrm{I}_2 & \end{matrix} \begin{pmatrix} 16 \,\&\, (\tfrac{3}{4}\Gamma_1 + \tfrac{1}{4}\Gamma_2) & 16 \,\&\, (\tfrac{1}{4}\Gamma_1 + \tfrac{3}{4}\Gamma_2) \\ 22 \,\&\, (\tfrac{1}{8}\Gamma_1 + \tfrac{7}{8}\Gamma_2) & 10 \,\&\, (\tfrac{7}{8}\Gamma_1 + \tfrac{1}{8}\Gamma_2) \end{pmatrix}.$$

$$\Gamma_2: \quad \begin{matrix} & \mathrm{II}_1 & \mathrm{II}_2 \\ \mathrm{I}_1 & \end{matrix} \begin{pmatrix} -8 \,\&\, (\tfrac{1}{2}\Gamma_1 + \tfrac{1}{2}\Gamma_2) & 0 \,\&\, (\tfrac{1}{4}\Gamma_1 + \tfrac{3}{4}\Gamma_2) \end{pmatrix}.$$

Starting with $\mathbf{w}^0 = (0,0)$, apply two iterations of the value iteration algorithm to this problem. What limits on the value of the optimal solution can you obtain by using the bounds given by the first two iterations?

7.4. In the following discounted stochastic game apply value iteration three times, and use the bounds to find limits on the value of the game, where $\beta = 0.5$.

Γ_1 has payoff matrix:

$$\begin{array}{c} \\ I_1 \\ I_2 \end{array} \begin{array}{c} \begin{array}{cc} II_1 & II_2 \end{array} \\ \begin{pmatrix} (0.8\Gamma_1 + 0.2\Gamma_2) & 4 \ \& \ (0.4\Gamma_1 + 0.2\Gamma_2) \\ -2 \ \& \ (0.4\Gamma_2) & 2 \ \& \ (0.2\Gamma_) \end{pmatrix} \end{array}$$

Γ_2 has payoff matrix:

$$\begin{array}{c} \\ I_1 \\ I_2 \end{array} \begin{array}{c} \begin{array}{cc} II_1 & II_2 \end{array} \\ \begin{pmatrix} -1 & 2 \ \& \ (0.2\Gamma_1 + 0.2\Gamma_2) \\ 2 \ \& \ (0.6\Gamma_1 + 0.4\Gamma_2) & 4 \ \& \ (0.2\Gamma_1 + 0.6\Gamma_2) \end{pmatrix} \end{array}$$

Can you solve the game exactly?

7.5. Two broadcasting companies, Enu Broadcasting Corporation (E.B.C.) and Muppet Broadcasting Authority (M.B.A.) compete for advertising revenue. Each year each company can decide either to increase advertising time or to concentrate on improving the standard of programmes, and each year the advertisers have to decide whether they feel E.B.C. had the better programmes (i.e. highest audience), which we denote by state Γ_1, or M.B.A. had the better programmes, state Γ_2, or that the standard of programmes is so low it is no longer worth using TV advertising. The extra advertising revenue that E.B.C. gets compared with an even division of revenue between the companies is given for the various states and possible actions of the companies in the tables below.

State Γ_1:	M.B.A. increases advertising	M.B.A. improves programmes
E.B.C. Increases advertising	2	5
E.B.C. Improves programmes	−1	0

State Γ_2	M.B.A. increases advertising	M.B.A. improves programmes
E.B.C. Increases advertising	−2	0
E.B.C. Improves programmes	−2	−2

The probability that next year we will be in state Γ_1 is given by the following tables.

If we are in state Γ_1 this year:

	M.B.A. increases advertising	M.B.A. improves programmes
E.B.C. Increases advertising	¼	¼
E.B.C. Improves programmes	¾	½

If we are in state Γ_2 this year:

	M.B.A. increases advertising	M.B.A. improves programmes
E.B.C. Increases advertising	0	0
E.B.C. Improves programmes	½	¼

Similarly, the probabilities of being in state Γ_2 next year are given by:

In state Γ_1 this year:

	M.B.A. increases advertising	M.B.A. improves programmes
E.B.C. Increases advertising	¼	½
E.B.C. Improves programmes	0	½

In state Γ_2 this year:

	M.B.A. increases advertising	M.B.A. improves programmes
E.B.C. Increases advertising	0	½
E.B.C. Improves programmes	0	½

If the revenue each year is discounted by ⅔, set this up as a discounted stochastic game. Apply three iterations of the value iteration algorithm starting with $\mathbf{w}^0 = (0,0)$. Use the bounds to get limits on the solution of the game.

Find the value of the game if the companies use the same type of strategies that were optimal in the third iteration of value iteration

(I_1, II_1 in Γ_1 mixed strategies in Γ_2). By showing these values solve the optimality equation, or otherwise, find the value of the game.

7.6. An alternative algorithm for solving stochastic games was suggested by Pollatschek and Avi-itzhak (1969). Consider again Problem 7.5, and start with $\mathbf{w}^0 = (0,0)$. Do one value iteration and note what are II's strategies for this iteration. Check it is $\mathbf{y}_1 = (1,0)$ in Γ_1 and $\mathbf{y}_2 = (1,0)$ in Γ_2. Now find the value of the game if II always plays these strategies. Check that this value $\mathbf{w}^1 = (w_1^1, w_2^1)$ satisfies:

$$w_1^1 = \max\{2 + \tfrac{2}{3}(\tfrac{1}{4}w_1^1 + \tfrac{1}{4}w_2^1),\ -1 + \tfrac{2}{3}\cdot\tfrac{3}{4}w_1^1\}.$$

$$w_1^2 = \max\{-2, -2 + \tfrac{2}{3}\cdot\tfrac{1}{2}w_1^1\}.$$

Now do one value iteration starting with $\mathbf{w}^1$ and repeat the procedure. What are the iterates $\mathbf{w}^2$, $\mathbf{w}^3$ of this algorithm?

7.7. Blotto has three divisions, and must capture an enemy position held by two divisions. However, he must also watch the enemy does not capture his camp while he is attacking. An attacker needs one more unit than a defender to be successful; if it is not it retreats to its own camp and tries the next day. Payoff is +1 if Blotto takes the enemy position without losing his own, −1 if he loses his own camp, no matter what else happens. Set this up as a recursive game and prove value is +1. Find the ε-optimal strategies.

7.8. Two players, I and II, play the simplified Poker game of Example 2.1 over and over again until one of the players loses all his money and the payoff to I is 1 if he wins all II's money and 0 if he loses his own. A player may raise the pot even though he has not got the extra £1 during one play of the game, but if he then loses that game, he will have lost overall. Suppose the sum of I's and II's money is £4, set this up as a recursive game with five subgames Γ_i, $i = 0,1,2,3,4$. If v_i is the payoff of the game to I given that he starts with £i, it is obvious that $0 = v_0 \leqslant v_1 \leqslant v_2 \leqslant v_3 \leqslant v_4 = 1$. If we assume $v_2 \leqslant \tfrac{1}{2}$, show by looking at the inequalities between the entries in the payoff matrices that:

(i) $v_1 = \tfrac{1}{2}v_2$,

(ii) $v_2 = (v_1v_3 - \tfrac{1}{2}v_3 - \tfrac{1}{2}v_1)/(v_1 - 1)$,

(iii) $v_3 = \tfrac{1}{2} + \tfrac{1}{2}v_1$

Solve for v_1, v_2 and v_3 and describe the optimal strategies in Γ_1, Γ_2 and Γ_3.

7.9. What pure strategies are supergame stable in the discounted supergame based on Prisoners' Dilemma (Example 3.3), which has payoff matrix:

$$
\begin{array}{ccc}
 & \mathrm{II}_c & \mathrm{II}_n \\
\mathrm{I}_c & \left(\begin{array}{c} (-9,-9) \\ -10,0) \end{array}\right. & \left.\begin{array}{c} (0,-10) \\ (-1,-1) \end{array}\right) . \\
\mathrm{I}_n & &
\end{array}
$$

7.10. Consider the oligopoly problem with n firms, where the payoff to firm i is $e_i(p_1,p_2,\ldots,p_n)$ given by (7.60) if the firms' prices are $p_1,p_2,\ldots,p_n$ respectively. Prove that a set of prices $(\bar{p}_1,\bar{p}_2,\ldots,\bar{p}_n)$ is supergame stable for some discount factor $\beta < 1$ if:

$$e_i(\bar{p}_1,\bar{p}_2,\ldots,\bar{p}_n) > \min_{p_1,p_2,\ldots,p_{i-1},p_{i+1},\ldots,p_n} \ \max_{p_i} \ e_i(p_1,p_2,\ldots,p_n),$$

$i = 1,2,\ldots,n.$

Prove that $(\bar{p}_1,\bar{p}_2,\ldots,\bar{p}_n)$ is a symmetric metaequilibria if:

$$e_i(\bar{p}_1,\bar{p}_2,\ldots,\bar{p}n) \geqslant \min_{p_1,p_2,\ldots,i1,p_{i+1},\ldots p_n} \ \max_{p_i} \ e_i(p_1,p_2,\ldots,p_n),$$

$i = 1,2,\ldots,n.$

7.11. Look at the stochastic game which has three subgames Γ_1, Γ_2, Γ_3, where the payoff matrices are:

$$
\Gamma_1: \quad
\begin{array}{c|cc}
 & \mathrm{II}_1 & \mathrm{II}_2 \\
\mathrm{I}_1 & 1 \,\&\, \Gamma_1 & \Gamma_2 \\
\mathrm{I}_2 & \Gamma_1 & 1 \,\&\, \Gamma_3
\end{array}
\qquad
\Gamma_2: \quad
\begin{array}{c|c}
 & \mathrm{II}_1 \\
\mathrm{I}_1 & (\Gamma_2)
\end{array}
\qquad
\Gamma_3: \quad
\begin{array}{c|c}
 & \mathrm{II}_1 \\
\mathrm{I}_1 & (1 \,\&\, \Gamma_3)
\end{array}
$$

Find the value of the game if the discount factor is $\beta = 0.9$ and if it is $\beta = 0.999$. You can solve it directly or use value iteration. Explain why, if instead of discounting we take the average payoff per position then the analysis breaks down. (For more details of the average cost game, see Gillette, 1957.)

CHAPTER 8

Evolutionary games

The survival of the fittest'—H. Spencer

8.1 INTRODUCTION

One of the most unusual applications of game theory in recent years has been to model how animal behaviour evolves from generation to generation. The way that most genetic characteristics evolve has long been modelled satisfactorily using Mendel's principles, e.g. eye colouring in humans. However, there are some characteristics whose evolution cannot be analysed so simply, because these characteristics affect the reproducing capacity of the individual. The evolution of these characteristics depends on how individuals with such characteristics interact with individuals with other characteristics.

These characteristics deal with the mating behaviour of animals, and the way they look after their offsprings. Before mating, males of a species frequently engage in contests to establish who is 'top man'. Stags have pushing matches with their antlers, crabs have fights, and snakes have wrestling matches. Yet in most cases, the contest is not as fierce as one would expect. The stags do not attack each other in the side where it might hurt, and the snakes do not bite one another. In many contests, it is a matter of waiting to see who gives up first. This seems quite sensible for the species as a whole, since not many males are badly injured or killed. However, this idea that animals behave in a way that is best for the group is not considered usual in biology. It is more often assumed that the individual does what is best for itself. So why don't we have stags that are very aggressive and 'fight dirty'?

To try and answer these questions Maynard-Smith and Price (1973) modelled these conflicts as games.

They assume that each individual uses a strategy, $\mathbf{x}$ say, from a set of possible strategies and when he meets an opponent who uses a strategy $\mathbf{y}$, there is a payoff $e(\mathbf{x},\mathbf{y})$ to the first individual. This payoff is related to the fitness of the individual—which is the biologists' version of utility—and in some sense measures his likely reproductive success. However, this is not a game in the way we have thought of it previously. The animals are not assumed to be 'thinking' and making rational choices of their strategies. It is assumed that an animal whose genes make him use strategy $\mathbf{x}$, will have offspring with the same genes. If the payoffs to $\mathbf{x}$, $e(\mathbf{x},.)$, are high, then it will produce more offspring using strategy $\mathbf{x}$ than a less successful gene type and will predominate in the population.

The set of strategies will as usual consist of a set of pure strategies together with the mixed ones, which consist of playing each pure one with a certain probability. In the stag contest you could think of two pure-strategies: 'Hawk' which always attacks and 'Dove' which always runs away. A mixed strategy could be interpreted by saying either that each individual has a random mechanism which triggers off each of the appropriate pure strategies, or that the whole population is divided into animals using the different pure strategies, but in proportions given by the mixed strategy. Either interpretation is satisfactory for what follows.

8.2 EVOLUTIONARY STABLE STRATEGIES (ESS)

In ordinary game theory the equilibrium point is one of the main solution concepts. At an equilibrium, it does not pay any player to change from his strategy to another one, if the rest of the players keep to their strategy, as the change will not improve his payoff. In evolutionary games we are also interested in strategies that are in equilibrium in some sense. If all the population is using a certain strategy, then it does not pay some to change to a different strategy since their expected fitness will be worse than the rest of the population. We do not assume that the animals decide to change strategies, only that genetic mutation produces a mutant strategy. The strategies we are interested in are those whose fitness will be greater than that of any mutant strategy that should arise. In that case, the next generation will have a smaller proportion of the mutant strategy, and eventually the mutation will die out.

The following definition encapsulates this idea, where only a small proportion ε of the population have a mutant strategy.

Definition 8.1. Let X be the set of strategies in an evolutionary game. A strategy $\mathbf{x}^* \in X$ is called an **ESS (evolutionary stable strategy)** if for every $\mathbf{y} \in X$, $\mathbf{y} \neq \mathbf{x}^*$, with $\bar{\mathbf{x}} = (1 - \varepsilon)\mathbf{x}^* + \varepsilon\mathbf{y}$, then:

$$e(\mathbf{x}^*,\bar{\mathbf{x}}) > e(\mathbf{y},\bar{\mathbf{x}}), \tag{8.1}$$

for sufficiently small $\varepsilon > 0$.

In this definition $\mathbf{y}$ is the mutation which affects ε of the population.

In fact we can change this definition into a form that is more useful when actually calculating ESS. When $\varepsilon = 0$ $\bar{\mathbf{x}}$ becomes $\mathbf{x}^*$ and in order for (8.1) to hold for small ε, we must have that when ε is zero $e(\mathbf{x}^*,\mathbf{x}^*) \geqslant e(\mathbf{y},\mathbf{x}^*)$. If $e(\mathbf{x}^*,\mathbf{x}^*) > e(\mathbf{y},\mathbf{x}^*)$, then it certainly will be true that for small ε, (8.1) still holds. However, if $e(\mathbf{x}^*,\mathbf{x}^*) = e(\mathbf{y},\mathbf{x}^*)$ we must look at (8.1) more closely. Expanding both sides gives:

$$(1-\varepsilon)e(\mathbf{x}^*,\mathbf{x}^*) + \varepsilon e(\mathbf{x}^*,\mathbf{y}) > (1-\varepsilon)e(\mathbf{y},\mathbf{x}^*) + \varepsilon e(\mathbf{y},\mathbf{y}). \tag{8.2}$$

So if $e(\mathbf{x}^*,\mathbf{x}^*) = e(\mathbf{y},\mathbf{x}^*)$ we must have $e(\mathbf{x}^*,\mathbf{y}) > e(\mathbf{y},\mathbf{y})$. This leads to an alternative definition.

Definition 8.2. A strategy $\mathbf{x}^* \in X$ is ESS if for every $\mathbf{y} \in X$, $\mathbf{y} \neq \mathbf{x}^*$:

$$e(\mathbf{x}^*,\mathbf{x}^*) \geqslant e(\mathbf{y},\mathbf{x}^*); \tag{8.3}$$

and if $e(\mathbf{x}^*,\mathbf{x}^*) = e(\mathbf{y},\mathbf{x}^*)$, then:

$$e(\mathbf{x}^*,\mathbf{y}) > e(\mathbf{y},\mathbf{y}). \tag{8.4}$$

This was the form originally given by Maynard-Smith (1974).

We can think of evolutionary games as two-player non-zero-sum games where the payoffs, if the first player plays $\mathbf{x}$ and the second $\mathbf{y}$, are $e_1(\mathbf{x},\mathbf{y}) = e(\mathbf{x},\mathbf{y})$ to the first and $e_2(\mathbf{x},\mathbf{y}) = e(\mathbf{y},\mathbf{x})$ to the second. It follows straightaway from (8.3) that, if $\mathbf{x}^*$ is an ESS, $(\mathbf{x}^*,\mathbf{x}^*)$ will be an equilibrium pair for the corresponding non-zero-sum game, but not all such equilibrium pairs are ESS because of equation (8.4).

Let us look at some examples of evolutionary games to see which strategies are ESS.

8.3 EXAMPLES OF EVOLUTIONARY GAMES

As pointed out above, we can think of evolutionary games as two-person non-zero-games, but because of the symmetry it is enough to give the payoff (fitness) matrix for the first (row) player.

Example 8.1—Hawk *v*. Dove. For the stag contest, assume that there are only two pure strategies: 'Hawk' which means keep fighting until you or your opponent is injured and 'Dove' which means run away. If the winner gets a prize of V, and injury means your fitness goes down by D, we get the fitness matrix:

$$\begin{array}{c} \\ \text{Hawk} \\ \text{Dove} \end{array} \begin{array}{c} \begin{array}{cc} \text{Hawk} & \text{Dove} \end{array} \\ \begin{pmatrix} \tfrac{1}{2}(V-D) & V \\ 0 & \tfrac{1}{2}V \end{pmatrix} \end{array}. \tag{8.5}$$

The halves occur since there are equal chances of success and injury when two similar strategies meet. Look at two different cases.

Example 8.2—$V \geqslant D$. Put $V = 4$, $D = 2$ say, so that the advantage of winning outweighs any disadvantage because of injury. Thus we get:

$$\begin{array}{ccc} & \text{Hawk} & \text{Dove} \\ \text{Hawk} & \left(\begin{array}{c} 1 \\ 0 \end{array}\right. & \left.\begin{array}{c} 4 \\ 2 \end{array}\right) \\ \text{Dove} & & \end{array}.$$

In this case if you look at the pure Hawk strategy, $\mathbf{x} = (1,0)$ and take any other strategy $\mathbf{y} = (y,1 - y)$, $y \neq 1$:

$$e(\mathbf{x},\mathbf{x}) = 1 > y = e(\mathbf{y},\mathbf{x}), \tag{8.6}$$

and so Hawk is ESS.

Example 8.3—$V < D$. Put $V = 2$, $D = 4$, so that injury outweighs the advantage of winning. We get the following matrix:

$$\begin{array}{lcc} & \text{Hawk} & \text{Dove} \\ \text{Hawk} & \multicolumn{2}{c}{\begin{pmatrix} -1 & 2 \\ 0 & 1 \end{pmatrix}} \\ \text{Dove} & & \end{array}. \tag{8.7}$$

For this game the strategy $\mathbf{x} = (½,½)$ is ESS, as the following calculations show. Let $\mathbf{y} = (y,1 - y)$ $y \neq ½$, then:

$$e(\mathbf{x},\mathbf{x}) = ½ = e(\mathbf{y},\mathbf{x}). \tag{8.8}$$

Thus to show $\mathbf{x}$ is ESS, we must look at condition (8.4):

$$e(\mathbf{y},\mathbf{y}) = 1 - 2y^2, \qquad e(\mathbf{x},\mathbf{y}) = 3/2 - 3y. \tag{8.9}$$

Since $(3/2 - 2y) - (1 - 2y^2) = 2(y - ½)^2 > 0$, if $y \neq ½$, we have

$e(\mathbf{x},\mathbf{y}) - e(\mathbf{y},\mathbf{y}) > 0$, which shows $\mathbf{x}$ is ESS.

Example 8.4—General Hawk *v.* Dove. Using the same ideas as in the numerical examples above, we can find the ESS in Example 8.1 for all $V,D \geqslant 0$.

If $V > D$, then the Hawk strategy $\mathbf{x}^* = (1,0)$ is always ESS, since for any other strategy $\mathbf{y} = (y,1 - y)$ $y \neq 1$:

$$e(\mathbf{x},\mathbf{x}) = ½(V - D) > ½(V - D)y = e(\mathbf{y},\mathbf{x}). \tag{8.10}$$

If $V = D$, then the Hawk strategy $\mathbf{x}^* = (1,0)$ is still ESS, but this time we will need to use condition (8.4) to show it. For any other strategy $\mathbf{y} = (y, 1-y)$, we have:

$$e(\mathbf{x},\mathbf{x}) = 0 = e(\mathbf{y},\mathbf{x}), \tag{8.11}$$

but

$$e(\mathbf{x},\mathbf{y}) = V(1-y) > Vy(1-y) + \tfrac{1}{2}V(1-y)^2 = e(\mathbf{y},\mathbf{y}). \tag{8.12}$$

Finally, if $V < D$, then the ESS strategy is the mixed strategy $\mathbf{x} = (V/D, D - V/D)$. This follows because, if we take any other strategy $\mathbf{y} = (y, 1-y)$, $y \neq V/D$, then:

$$e(\mathbf{x},\mathbf{x}) = (D-V)/2D = e(\mathbf{y},\mathbf{x}), \quad \text{for all } \mathbf{y}; \tag{8.13}$$

but

$$e(\mathbf{x},\mathbf{y}) = (V-D)Vy/2D + V^2(1-y)/D + V(D-V)(1-y)/2D,$$

and

$$e(\mathbf{y},\mathbf{y}) = \tfrac{1}{2}(V-D)y^2 + Vy(1-y) + \tfrac{1}{2}V(1-y)^2,$$

so

$$e(\mathbf{x},\mathbf{y}) - e(\mathbf{y},\mathbf{y}) = \tfrac{1}{2}D(y - V/D)^2. \tag{8.14}$$

Thus, for any $\mathbf{y}$ with $y \neq V/D$, $e(\mathbf{x},\mathbf{y}) > e(\mathbf{y},\mathbf{y})$.

Notice as the injuries become serious, so that D is large, the ESS strategy tends to the 'Dove' strategy. This might explain why snakes wrestle rather than bite!

Example 8.5—'Scratch–Bite–Trample'. Let us suppose that an animal had three forms of attack in a contest, namely to scratch with its claws, bite with its teeth, or trample with its legs. Make the payoffs in this hypothetical context like those of 'Scissor, Stone and Paper', so that scratching will beat biting, which beats trampling, which beats scratching. If both animals use the same strategy it is a draw, so the fitness matrix is:

$$\begin{array}{l} \\ \text{Scratch} \\ \text{Bite} \\ \text{Trample} \end{array} \begin{array}{c} \begin{array}{ccc} \text{Scratch} & \text{Bite} & \text{Trample} \end{array} \\ \begin{pmatrix} 0 & 1 & -1 \\ -1 & 0 & 1 \\ 1 & -1 & 0 \end{pmatrix} \end{array}. \tag{8.15}$$

There is no ESS strategy for this game, as we now show. Suppose strategy $\mathbf{p} = (p_1, p_2, 1 - p_1 - p_2)$ plays strategy $\mathbf{q} = (q_1, q_2, 1 - q_1 - q_2)$, then:

$$e(\mathbf{p},\mathbf{q}) = (q_1 - q_2) - (p_1 - p_2) + 3p_1q_2 - 3p_2q_1. \tag{8.16}$$

This means for any strategy $\mathbf{p}$, $e(\mathbf{p},\mathbf{p}) = 0$, and if we look at the three pure strategies $\mathbf{p}_1 = (1,0,0)$, $\mathbf{p}_2 = (0,1,0)$ and $\mathbf{p}_3 = (0,0,1)$, then the only q for which $e(\mathbf{p},\mathbf{q}) \leqslant 0$ for all three is $\mathbf{q} = (\frac{1}{3}, \frac{1}{3}, \frac{1}{3})$. Thus, the only strategy $\mathbf{p}$ which has $e(\mathbf{p},\mathbf{p}) \geqslant e(\mathbf{q},\mathbf{p})$ for all $\mathbf{q}$ is $\mathbf{p} = (\frac{1}{3}, \frac{1}{3}, \frac{1}{3})$. However, this is not ESS because, against any strategy $\mathbf{q}$, $\mathbf{p} = (\frac{1}{3}, \frac{1}{3}, \frac{1}{3})$ gives $e(\mathbf{p},\mathbf{q})$ equal to zero which is the same as $e(\mathbf{q},\mathbf{q})$. So no ESS.

8.4 PROPERTIES OF EVOLUTIONARY STABLE STRATEGIES

(a) Some games have no ESS. This was shown by the Scratch–Bite–Trample example.

(b) Some games have more than one ESS. Look at the game with payoff matrix:

$$\begin{pmatrix} 2 & 0 \\ 0 & 1 \end{pmatrix}.$$

Both pure strategies $p_1 = (1,0)$ and $p_2 = (0.1)$ are ESS. Haigh (1975) gives an example where there are two mixed strategies which are ESS, namely:

$$\begin{pmatrix} 5 & 7 & 2 \\ 8 & 6 & 5 \\ 1 & 8 & 4 \end{pmatrix}.$$

Here both $\mathbf{p}_1 = (\frac{1}{4}, \frac{3}{4}, 0)$ and $\mathbf{p}_2 = (0, \frac{1}{3}, \frac{2}{3})$ are ESS.

If $\mathbf{x}$ is an ESS, then the pure strategies which make up $\mathbf{x}$ are called the **worthwhile** strategies of $\mathbf{x}$ (just as in equilibrium pairs). We define $W(\mathbf{x}) = \{I_i | x_i > 0$, where $\mathbf{x} = (x_1,x_2,\ldots,x_n)\}$ to be the set of worthwhile strategies for $\mathbf{x}$.

We will now show you that you cannot have two ESS where one's worthwhile strategies include the other's worthwhile strategies, but first we prove the following lemma.

Lemma 8.1. For any ESS strategy $\mathbf{x}$, the payoff of playing any of its worthwhile strategy against $\mathbf{x}$, is the same as the payoff of playing $\mathbf{x}$ against $\mathbf{x}$.

Proof. If $\mathbf{x} = (x_1,x_2,\ldots,x_n)$, then:

$$e(\mathbf{x},\mathbf{x}) = \sum_{i=1}^{n} x_i e(I_i,\mathbf{x}). \tag{8.16}$$

Since $\mathbf{x}$ is ESS, then:

$$e(\mathbf{x},\mathbf{x}) \geqslant e(I_i,\mathbf{x}), \quad \text{for all } i. \tag{8.17}$$

Since $\sum_{i=1}^{n} x_i = 1$, we must have equality in (8.17) for all i, where $x_i > 0$ for (8.16) to hold. So $e(I_i,\mathbf{x}) = e(\mathbf{x},\mathbf{x})$ for all worthwhile strategies I_i. This is the same proof as Lemma 2.2 which proves the same result for equilibrium pairs.

Lemma 8.2. If $\mathbf{x},\mathbf{y}$ are two ESS, then $W(\mathbf{x}) \not\subset W(\mathbf{y})$, so that one ESS cannot contain all the worthwhile strategies of another ESS.

Proof. Now suppose $W(\mathbf{x}) \subset W(\mathbf{y})$, then Lemma 8.1 means $e(\mathbf{y},\mathbf{y}) = e(I_j,\mathbf{y})$ for all I_j in $W(\mathbf{y})$, and hence for all I_j in $W(\mathbf{x})$. Then:

$$e(\mathbf{x},\mathbf{y}) = \sum_j x_j\, e(I_j,\mathbf{y}) = e(\mathbf{y},\mathbf{y}). \tag{8.18}$$

Thus, as $\mathbf{y}$ is ESS we must have $e(\mathbf{y},\mathbf{x}) > e(\mathbf{x},\mathbf{x})$, but this contradicts $\mathbf{x}$ being an ESS.

Notice that as in Haigh's example we can have a non-empty intersection of $W(\mathbf{x})$ and $W(\mathbf{y})$.

(c) We can add a constant to all entries of the payoff matrix without changing the ESS strategies. This follows immediately since if $e'_{ij} = e_{ij} + c$:

$$e'(\mathbf{x},\mathbf{y}) = e(\mathbf{x},\mathbf{y}) + c, \quad \text{for all strategies } \mathbf{x} \text{ and } \mathbf{y}.$$

8.5 HOW TO FIND EVOLUTIONARY STABLE STRATEGIES

8.5.1 Pure ESS Strategies

If I_i is to be a pure ESS strategy it is enough if the payoff matrix e_{ij} has $e_{ii} > e_{ji}$ for all $j \neq i$. This means that the diagonal element is the largest in that column. I_i is also ESS if $e_{ii} \geqslant e_{ji}$ and for those j where $e_{ii} = e_{ji}$ we must have $e_{ij} > e_{jj}$. Thus, it is easy to check if there are any pure strategy ESSs.

Note that a pure ESS strategy need not dominate all other strategies, where domination was defined in Chapter 2, Section 2.5. Again, a pure strategy that dominates all others need not be a pure ESS. Look at

$$\begin{pmatrix} 3 & 2 \\ 1 & 4 \end{pmatrix} \quad \text{and} \quad \begin{pmatrix} 2 & 0 & 3 \\ 2 & 0 & 2 \\ 1 & 0 & 1 \end{pmatrix}. \tag{8.19}$$

In the first, both $\mathbf{p}_1 = (1,0)$ and $\mathbf{p}_2 = (0,1)$ are ESS but neither dominates the other, while in the second example $\mathbf{p}_1 = (1,0,0)$ dominates all other strategies, but taking $\mathbf{p}_2 = (0,1,0)$ we have $e(\mathbf{p}_1,\mathbf{p}_1) = e(\mathbf{p}_2,\mathbf{p}_1)$ and $e(\mathbf{p}_1,\mathbf{p}_2) = e(\mathbf{p}_2,\mathbf{p}_2)$. So $\mathbf{p}_1$ is not ESS.

8.5.2 2×2 Games

Let us concentrate first on solving 2×2 games. First check if there are any pure ESS strategies. If there are, then by Lemma 8.2 there cannot be any mixed ESS strategies, since a mixed ESS has both pure strategies as worthwhile ones.

Suppose $\mathbf{x} = (x,1 - x)$ is a mixed ESS strategy for a 2×2 game with non-trivial payoff matrix:

$$\begin{pmatrix} a & b \\ c & d \end{pmatrix}. \tag{8.20}$$

A trivial payoff matrix has $a = c$ and $b = d$ so the strategies duplicate each other. Since both I_1 and I_2 are worthwhile for $\mathbf{x}$, we must have $e(I_1,\mathbf{x}) = e(I_2,\mathbf{x}) = e(\mathbf{x},\mathbf{x})$. This gives:

$$ax + b(1 - x) = cx + d(1 - x), \tag{8.21}$$

so

$$x = (b - d)/(b + c - a - d).$$

If there is no pure ESS strategy, then $c \geqslant a$ and $b \geqslant d$ (but not both equalities holding) so $b + c - a - d$ is positive. For this $\mathbf{x}$ we know $e(\mathbf{x},\mathbf{x}) = e(\mathbf{p},\mathbf{x})$ for all $\mathbf{p}$. Hence, in order that $\mathbf{x}$ is ESS we want $e(\mathbf{x},\mathbf{q}) > e(\mathbf{q},\mathbf{q})$ for all $\mathbf{q}$. In order to show this, first notice that if $\mathbf{p} = (p,1 - p)$, $\mathbf{q} = (q,1 - q)$:

$$e(\mathbf{p},\mathbf{q}) = d + (c - d)q + p[b - d + (a + d - b - c)q]. \tag{8.22}$$

So, for a fixed $\mathbf{q}$, $e(\mathbf{p},\mathbf{q})$ increases as p increases if $b - d + (a + d - b - c)q$ is positive, and $e(\mathbf{p},\mathbf{q})$ increases as p decreases if $b - d + (a + d - b - c)q$ is negative. Thus, if $q > x = (b - d)/(b + c - a - d)$, then $b - d + (a + d - b - c)q$ is negative and so $e(\mathbf{x},\mathbf{q}) > e(\mathbf{q},\mathbf{q})$; whereas if $q < x = (b - d)/(b + c - a - d)$, then $b - d + (a + d - b - c)q$ is positive and so $e(\mathbf{x},\mathbf{q}) > e(\mathbf{q},\mathbf{q})$. Thus, $\mathbf{x}$ is an ESS strategy.

This means that every 2×2 game has at least one ESS strategy unless it is the trivial case $a = c$, $b = d$.

8.5.3 General $m \times m$ games

Suppose we want to check if $\mathbf{p}$ is an ESS for the evolutionary game with fitness matrix $E = (e_{ij})$. By renumbering the pure strategies we can always assume $\mathbf{p} = (p_1,p_2,\ldots,p_r,0,0,\ldots,0)$, where $\sum_{i=1}^{r} p_i = 1$. We must have:

$$e(\mathrm{I}_1,p) = e(\mathrm{I}_2,p) = \ldots = e(\mathrm{I}_r,\mathbf{p}) = \max_{1 \leqslant j \leqslant m} e(I_j,\mathbf{p}) \tag{8.23}$$

Otherwise for some I_j we will have $e(I_j,\mathbf{p}) > e(I_i,\mathbf{p})$, for all $i = 1, \ldots, r$, and hence:

$$e(I_j,\mathbf{p}) > \sum_{i=1}^{r} p_i e(I_i,p) = e(\mathbf{p},\mathbf{p}), \tag{8.24}$$

which contradicts condition (8.3) on ESS. Suppose, furthermore, we have numbered the pure strategies so that $e(I_i,\mathbf{p}) = \max_j e(I_j,\mathbf{p})$, for $i = 1,2,\ldots,k$, where $k \geqslant r$. Then it is obvious that the only strategies $\mathbf{q}$ for which $e(\mathbf{q},\mathbf{p}) = e(\mathbf{p},\mathbf{p})$ are $\mathbf{q} = (q_1,q_2,\ldots,q_k,0,0,\ldots,0)$.

Let B be the $k \times k$ submatrix of E consisting of the first k rows and columns, then for the $\mathbf{q}$ with $e(\mathbf{q},\mathbf{p}) = e(\mathbf{p},\mathbf{p})$, we have:

$$\begin{aligned} e(\mathbf{p},\mathbf{q}) - e(\mathbf{q},\mathbf{q}) &= \mathbf{p}^{\mathrm{T}}B\mathbf{q} - \mathbf{q}^{\mathrm{T}}B\mathbf{q} = (\mathbf{p}^{\mathrm{T}} - \mathbf{q}^{\mathrm{T}})B(\mathbf{q} - \mathbf{p} + \mathbf{p}) \\ &= -(\mathbf{p} - \mathbf{q})^{\mathrm{T}}B(\mathbf{p} - \mathbf{q}) + e(\mathbf{p},\mathbf{p}) - e(\mathbf{q},\mathbf{p}) = -(\mathbf{p} - \mathbf{q})^{\mathrm{T}}B(\mathbf{p} - \mathbf{q}). \end{aligned} \tag{8.25}$$

So, for condition (8.4) to hold for $\mathbf{p}$ we want all

$$(\mathbf{p} - \mathbf{q})^{\mathrm{T}}B(\mathbf{p} - \mathbf{q}) < 0, \tag{8.26}$$

for all such $\mathbf{q}$.

So to find all the ESS in a game we have the following algorithm.

1. For each one of the $2^m - 1$ possible subsets of $\{I_1,I_2,\ldots,I_m\}$ think of the strategies $\mathbf{p}$ which have that subset as worthwhile strategies (say $I_{i_1}, I_{i_2},\ldots,I_{i_r}$)
2. Find the $\mathbf{p}$ which satisfies:
$$e(I_{i_1},\mathbf{p}) = e(I_{i_2},\mathbf{p}) = \ldots = e(I_{i_r},\mathbf{p}). \tag{8.27}$$
3. Check that the solution of (8.27) has $p_i \geqslant 0$, $\sum_i p_i = 1$ and satisfies (8.23).
4. Let
$$M(p) = \{j | e(I_j,\mathbf{p}) = \max_i e(I_i,\mathbf{p})\}. \tag{8.28}$$
Then for all strategies $\mathbf{q}$ which are mixtures of the pure strategies in $M(p)$, show that (8.26) holds for $\mathbf{p}$.

For details of how to proceed with Step 4, look at Problems 8.3 and 8.4.

8.6 WAR OF ATTRITION

Many mating characteristics are not as violent as the fights discussed in Hawk versus Dove. They consists of 'displays' by the animals, in which victory goes to the animal which displays longest. The prize consists of control of a particular territory or female, and the disadvantage in such displays is the time wasted by both animals, which could have been used more profitably. We model this (see Maynard-Smith, 1974, and Bishop and Canning, 1978) by saying that the victor receives a payoff of value v and both receive a penalty equal to the length of the contest.

The set of pure strategies for this contest is $X = [0,\infty)$, where $m \in X$ means the contestant will display for at most m and then stop. If 1 chooses m_1 and 2 chooses m_2 the payoff is:

$$e(\mathbf{m}_1,\mathbf{m}_2) = \begin{cases} v - m_2, & \text{if } m_1 > m_2, \\ \tfrac{1}{2}v - m_1, & \text{if } m_1 = m_2, \\ -m_1 & \text{if } m_2 > m_1. \end{cases} \tag{8.29}$$

It is easy to see that no pure strategy can be ESS in such a contest since:

$$e(\mathbf{m}_1,\mathbf{m}_1) = \tfrac{1}{2}v - m_1, \quad e(\mathbf{m}_1 + h,\mathbf{m}_1) = v - m_1, \quad \text{if } h > 0, \tag{8.30}$$

and so $e(\mathbf{m}_1 + h,\mathbf{m}_1) > e(\mathbf{m}_1,\mathbf{m}_1)$.

A mixed strategy in such a contest is given by the **distribution function** $F(x)$, which is the probability that the strategy will stop no later than x. If this is differentiable, the derivative of $F(x)$ is called the **probability density function,** $f(x)$, and the chance of stopping in a time between x and $x + h$, where h is small, is $f(x).h$. Suppose we look at a mixed strategy with density function f, where $f(x) > 0$ for all $x \geqslant 0$, then the payoff of playing the pure strategy m against it, is:

$$e(\mathbf{m},\mathbf{f}) = \int_0^m (v - x)f(x)\mathrm{d}x + \int_m^\infty (-m)f(x)\mathrm{d}x. \tag{8.31}$$

All the pure strategies are worthwhile for **f**. In order for **f** to be ESS recall that the payoff of playing any worthwhile strategy against **f** is the same as playing **f** against itself, and so the same as the payoff of any other worthwhile strategy against **f**. So for **f** to be ESS we certainly need $e(\mathbf{m},\mathbf{f}) = C$, for all **m**. In particular, if $e(\mathbf{m} + \mathrm{h},\mathbf{f}) - e(\mathbf{m},\mathbf{f}) = 0$ we get:

$$\int_0^{m+h} (v - x)f(x)\mathrm{d}x - (m + h) \int_{m+h}^\infty f(x)\mathrm{d}x - \int_0^m (v - x)f(x)\mathrm{d}x +$$
$$m \int_m^\infty f(x)\mathrm{d}x = \int_m^{m+h} (v - x + m)f(x)\mathrm{d}x - h \int_{m+h}^\infty f(x)\mathrm{d}x = 0. \tag{8.32}$$

If h is very small, this becomes:

$$vf(m) = \int_m^\infty f(x)\mathrm{d}x = 1 - F(m). \tag{8.33}$$

Thus, $f(m)/(1 - F(m)) = 1/v$, and if we use

$$-\frac{\mathrm{d}}{\mathrm{d}m}(\log(1 - F(m)) = \frac{f(m)}{1 - F(m)} = \frac{1}{v}, \tag{8.34}$$

we get:

$$F(m) = 1 - e^{-m/v} \qquad f(m) = \frac{1}{v} e^{-m/v}, \tag{8.35}$$

i.e. the **negative exponential distribution**.

In order to show that **f** is an ESS we must prove that $e(\mathbf{f},\mathbf{g}) > e(\mathbf{g},\mathbf{g})$ for any other **g**. What we will show is that

$$T(\mathbf{f},\mathbf{g}) = e(\mathbf{f},\mathbf{f}) - e(\mathbf{f},\mathbf{g}) - e(\mathbf{g},\mathbf{f}) + e(\mathbf{g},\mathbf{g}) \leqslant 0, \tag{8.36}$$

for all **f** and **g**—with strict inequality if $\mathbf{f} \neq \mathbf{g}$. Since we know $e(\mathbf{f},\mathbf{f}) = e(\mathbf{g},\mathbf{f})$, this implies $e(\mathbf{f},\mathbf{g}) > e(\mathbf{g},\mathbf{g})$, which is what is required.

$e(\mathbf{f},\mathbf{g}) + e(\mathbf{g},\mathbf{f})$ is the sum of the payoffs to the two players if one plays **f** and the other **g**. Thus, if the game lasts x, this sum must be $v - 2x$, and the chance the game lasts until exactly x is $f(x)(1 - G(x)) + g(x)(1 - F(x))$, since one player will stop at x and the other must not have stopped before x.

Thus,

$$e(\mathbf{f},\mathbf{g}) + e(\mathbf{g},\mathbf{f}) = v - 2\int_0^\infty x(f(x)(1 - G(x) + g(x)(1 - F(x))\mathrm{d}x, \tag{8.37}$$

and so

$$e(\mathbf{f},\mathbf{f}) = \tfrac{1}{2}v - 2\int_0^\infty x(f(x)(1 - F(x))\mathrm{d}x. \tag{8.38}$$

From (8.37) and (8.38) we get:

$$T(\mathbf{f},\mathbf{g}) = 2\int_0^\infty x(f(x) - g(x))(F(x) - G(x))\mathrm{d}x.$$

Integrating by parts gives:

$$2\int_0^\infty x(F(x) - G(x))(f(x) - g(x))\mathrm{d}x = 2[x(F(x) - G(x)]^2]_0^\infty$$

$$- 2\int_0^\infty x(f(x) - g(x))(F(x) - G(x))\mathrm{d}x - 2\int_0^\infty (F(x) - G(x))^2\mathrm{d}x. \tag{8.39}$$

Since $x(F(x) - G(x))^2$ tends to 0 as x goes to 0 and ∞, we get:

$$T(\mathbf{f},\mathbf{g}) = -\int_0^\infty (F(x) - G(x))^2\mathrm{d}x \leqslant 0, \tag{8.40}$$

and the result holds. Thus, we have shown the negative exponential distribution is an ESS strategy, and since all the pure strategies are worthwhile for it, an infinite dimensional analogue of Lemma 8.2 says there cannot be any other ESS.

One example of a mixed strategy which fits the negative exponential distribution predicted for such contests is given by Parker (1970). Male dung flies stay on cow pats waiting for females to arrive so they can mate with them. A fresh cow pat attracts more females than a stale one, so after a time the male flies go in search of a fresher pat. Although this is an n-fly game rather than a 2-fly one, the problem is the same, namely that the 'fitness' of the fly depends on the others' strategies. If most flies leave a cow pat early, then the fly that stays will mate all the females who arrive afterwards, whereas if most males stay on a cow pat for a long time, it is worth leaving to search for a fresh cow pat. Parker measured the length of time flies stayed on a cow pat and showed it fitted the negative exponential distribution very well. It is amazing where game theory leads you!

8.7 DYNAMIC EVOLUTIONARY GAMES

Although we have set up evolutionary games as one-off games, there is obviously an underlying dynamic game. The important property you require of an ESS strategy is that it will persist from generation to generation. Thus, we can set up an evolutionary game as a multi-stage game, where stage k represents the kth generation. To do this, we take the interpretation that if the population has a strategy $\mathbf{p} = (p_1,p_2,\ldots,p_n)$, then a fraction p_i of the population are of a genotype that uses pure strategy i.

Suppose $\mathbf{p}^k = (p_1^k,p_2^k,\ldots,p_n^k)$ describes the distribution of strategies in the kth generation; and suppose r_i is the average number of offspring of those using the ith pure strategy. Then the next generation we would expect:

$$p_i^{k+1} = r_i p_i^k \Big/ \sum_{j=1}^{n} r_j p_j^k . \tag{8.41}$$

Obviously, r_i is related to the fitness of the ith strategy which is given in the kth generation by $e(\mathrm{I}_i,\mathbf{p}^k)$. If we take $r_i = ce(\mathrm{I}_i,\mathbf{p}^k) = c\sum_j e_{ij}p_j^k$, we get the following relation between $\mathbf{p}^{k+1}$ and $\mathbf{p}^k$:

$$p_i^{k+1} = \frac{ce(\mathrm{I}_i,\,\mathbf{p}^k)p_i^k}{c\sum_{j=1}^{n} e(I_j,\mathbf{p}^k)p_j^k)} = \sum_{j=1}^{n} p_i^k\, e_{ij}p_j^k \Big/ \sum_{i,j=1}^{n} p_i^k e_{ij} p_j^k . \tag{8.42}$$

One would expect all the payoffs $e(\mathbf{p},\mathbf{q})$ to be non-negative, in this interpretation. If we add a constant to all the elements in the fitness matrix, we don't change the ESS. This seems the obvious way to make all $e_{ij} \geqslant 0$. However, as we shall see, this does make some differences in the dynamic version of the game. Nevertheless, from now on we assume $e_{ij} \geqslant 0$, for all i and j. Let us look at three examples.

Example 8.6—Hawk *v*. Dove. Look at the first example in Section 8.3 with payoff matrix:

	Hawk	Dove
Hawk	1	4
Dove	0	2

,

and suppose initially Hawk and Dove are equally likely, so $\mathbf{p}^0$ = (0.5,0.5). Then as $e(\mathbf{p}^0,\mathbf{p}^0) = \frac{1}{4}.1 + \frac{1}{4}.4 + \frac{1}{4}.2 = \frac{7}{4}$, $p_1^1 = e(\mathbf{I}_2,\mathbf{p}^0).p_1^0/e(\mathbf{p}^0,\mathbf{p}^0) = \frac{5}{2}.\frac{1}{2}/\frac{7}{4} = \frac{5}{7}$. $p_2^1 = e(1_i,\mathbf{p}^0)p_2^0/e(\mathbf{p}^0,\mathbf{p}^0) = 1.\frac{1}{2}/\frac{7}{4} = \frac{2}{7}$. So $\mathbf{p}^1 = (\frac{5}{7},\frac{2}{7})$ are repeated calculations given $\mathbf{p}^2 = (\frac{65}{73},\frac{8}{73})$, $\mathbf{p}^3$ = (0.98, 0.02), $\mathbf{p}^4$ = (0.9993, 0.0007). Thus, very rapidly the population becomes all Hawks (which was the only ESS). If the population starts all Hawks ($\mathbf{p}^0$ = (1,0)) it will remain so. However, notice that if it starts all Doves, i.e. $\mathbf{p}^0$ = (0,1), then $\mathbf{p}^k$ = (0,1) for all k, so this is also in equilibrium.

Example 8.7—Game with two ESS strategies. The payoff matrix $\left(\begin{smallmatrix}2 & 0\\ 0 & 1\end{smallmatrix}\right)$ gave us an evolutionary game with two ESS strategies, so what happens in the dynamic evolutionary game built on this? It all depends on the starting distribution. If $\mathbf{p}^0 = (\frac{1}{2},\frac{1}{2})$, $\mathbf{p}^1 = (\frac{2}{3},\frac{1}{3})$, $\mathbf{p}^2 = (\frac{8}{9},\frac{1}{9})$, $\mathbf{p}^3 = (\frac{128}{129},\frac{1}{129})$ and the population is rapidly all using the first strategy. If, however, $\mathbf{p}^0 = (\frac{1}{4},\frac{3}{4})$, $\mathbf{p}^1 = (\frac{2}{11},\frac{9}{11})$, $\mathbf{p}^3 = (\frac{8}{89},\frac{81}{89},)$, $\mathbf{p}^4$ = (0.02, 0.98) and the population soon all use the second strategy. Finally, if $\mathbf{p}^0 = (\frac{1}{3},\frac{2}{3})$, $\mathbf{p}^1 = \mathbf{p}^k = (\frac{1}{3},\frac{2}{3})$, for all k.

Example 8.8—Game with no ESS strategy. If we add 1 to the example of Scratch–Bite–Trample, we get a payoff matrix:

$$\begin{pmatrix} 1 & 2 & 0 \\ 0 & 1 & 2 \\ 2 & 0 & 1 \end{pmatrix}, \tag{8.43}$$

and since the latter matrix has no ESS neither can this one. If we start $\mathbf{p}^0$ = (0.4,0.4,0.2) we get the sequence:

$$\begin{aligned}
\mathbf{p}^1 &= (0.48,0.32,0.20), & \mathbf{p}^2 &= (0.5376,0.2304,0.2320),\\
\mathbf{p}^3 &= (0.5367,0.1600,0.3033), & \mathbf{p}^4 &= (0.4598,0.1227,0.4175),\\
\mathbf{p}^5 &= (0.3242,0.1175,0.5583), & \mathbf{p}^6 &= (0.1813,0.1450,0.6737),\\
\mathbf{p}^7 &= (0.0853,0.2163,0.6984), & \mathbf{p}^8 &= (0.0442,0.3489,0.6069),\\
\mathbf{p}^9 &= (0.0328,0.5452,0.4220), & \mathbf{p}^{10} &= (0.0368,0.7574,0.2058).
\end{aligned}$$

This should convince most people that there is no convergence.

Although we refer to these examples as **dynamic** or **multi-stage** games, they are much simpler than the multi-stage games of the previous chapter. Once you have specified the initial distribution of the genotypes there are no decisions made, and the game model is just a useful way of doing the dynamical calculation. It is as if you are interested in only one stationary, state-independent strategy played against itself, and want to know its evolution in time.

One could also construct a continuous time version of the dynamic evolutionary game. Thus, if $p_i(t)$ is the proportion of genotypes using pure strategy i at time t, one would describe the dynamics of the game by:

$$\frac{\mathrm{d}p_i}{\mathrm{d}t}(t) = r_i p_i(t),$$

where r_i is the relative rate of fitness of the ith strategy. Thus, we would expect it to depend on $e(\mathrm{I}_i,\mathbf{p}(t))$, the fitness of this strategy at time t. However, since $\sum_{i=1}^{n} p_i(t)$ must be 1 at all time t, we want:

$$\frac{\mathrm{d}}{\mathrm{d}t}\sum_{i=1}^{n} p_i(t) = \sum_{i=1}^{n} r_i p_i(t) = 0. \tag{8.44}$$

In order for (8.44) to hold, the obvious choice is $r_i = e(\mathrm{I}_i,p(t)) - e(\mathbf{p}(t),\mathbf{p}(t))$. Thus, we have:

$$\frac{d}{dt}p_i(t) = [e(\mathrm{I}_i,\mathbf{p}(t)) - e(\mathbf{p}(t),\mathbf{p}(t))]p_i(t). \tag{8.45}$$

We now return to the discrete version and look more formally at the stability of ESS strategies.

8.8 STABILITY IN DYNAMIC EVOLUTIONARY GAMES

If a strategy (or probability distribution of strategies) persists from generation to generation, then we say it is a **dynamic equilibrium** where

Definition 8.3. A strategy **x** is a dynamic equilibrium for the above games if, when $\mathbf{p}^k = \mathbf{x}$, then $\mathbf{p}^{k+1} = \mathbf{x}$ for any $k \geqslant 0$.

Lemma 8.3. All ESS strategies are dynamic equilibria.

Proof If when $p^k = \mathbf{x}$, we require $p^{k+1} = \mathbf{x}$, then (8.42) becomes:

$$x_i = e(\mathrm{I}_i,\mathbf{x}).x_i/e(\mathbf{x},\mathbf{x}), \text{ for all } i \tag{8.46}$$

So it is sufficient that $e(\mathrm{I}_i,\mathbf{x}) = e(\mathbf{x},\mathbf{x})$, for all $x_i > 0$. But this is immediately true for ESS strategies by Lemma 8.1.

There are a lot of strategies which are dynamic equilibria, but which are not ESS strategies, though. In the previous section, the pure strategy Dove in the first example, $\mathbf{x} = (⅓,⅔)$ in the second, and $\mathbf{x} = (⅓,⅓,⅓)$ in the third, are all dynamic equilibria, but are not ESS. However, they are **unstable** because, if they are slightly perturbed, successive generations have mixtures which differ widely from them. This leads us to the idea of the stability of an equilbrium, which is a concept that appears in many different areas of differential equations and dynamical systems (see Hirsch and Smale, 1974, for examples).

An equilibrium is **stable** if strategies which start close to it, stay close. It is **asymptotically stable** if all strategies close by, not only stay close, but actually tend to the equilibrium strategy. We measure closeness by the norm:

$$\|p - q\|_2 = \left(\sum_{i=1}^{m} (p_i - q_i)^2\right)^{1/2}$$

Definition 8.4. An equilibrium strategy $\mathbf{x}$ is stable if, for any $\varepsilon > 0$, there is a $\delta > 0$ so that if $\|\mathbf{x} - \mathbf{p}^0\|_2 < \delta$, then $\|\mathbf{x} - \mathbf{p}^k\|_2 < \varepsilon$,

for all $k = 1,2....$

Definition 8.5. An equilibrium strategy $\mathbf{x}$ is asymptotically stable if it is stable as above and

$$\lim_{k \to \infty} \mathbf{p}^k = \mathbf{x}. \tag{8.47}$$

What we would like to happen is that every ESS strategy is asymptotically stable in the corresponding dynamic game since this justifies the concept of evolutionary stable strategies. The next result shows that is true for 2×2 games.

Theorem 8.1. In a 2×2 evolutionary game, every ESS strategy is asymptotically stable.

Proof. Suppose $\mathbf{x}$ is an ESS strategy where $\mathbf{x} = (x_1,x_2)$, then either $x_1 = 0$ or $x_1 = 1$ or $e(\mathrm{I}_1, \mathbf{x}) = e(\mathrm{I}_2, \mathbf{x}) = e(\mathbf{x},\mathbf{x})$, if it is a mixed strategy. Any strategy close to $\mathbf{x}$ can be written as either $\tilde{\mathbf{x}} = (1 - h)\mathbf{x} + h(1,0)$ or $\tilde{\tilde{\mathbf{x}}} = (1 - h)\mathbf{x} + h(0,1)$. We concentrate on $\tilde{\mathbf{x}}$ which also means $x_1 < 1$, for there to be possible increase in $\mathbf{x}$'s first component.

Since $\mathbf{x}$ is an ESS, by Definition 8.2 we have:

$$e(\mathbf{x},\tilde{\mathbf{x}}) > e(\tilde{\mathbf{x}},\tilde{\mathbf{x}}) > e(\mathrm{I}_1,\tilde{\mathbf{x}}), \tag{8.48}$$

where we think of the mutants as playing the first pure strategy. What we want to show is that if $\mathbf{p}^k = \tilde{\mathbf{x}}$, then p^{k+1} lies between $\tilde{\mathbf{x}}$ and $\mathbf{x}$ since this is enough to ensure asymptotic stability. Since

$$p^{k+1} = \left(\frac{e(\mathrm{I}_1, \tilde{\mathbf{x}})}{e(\tilde{\mathbf{x}}, \tilde{\mathbf{x}})}\,\tilde{x}_1,\ \frac{e(\mathrm{I}_2, \tilde{\mathbf{x}})}{e(\tilde{\mathbf{x}}, \tilde{\mathbf{x}})}\,\tilde{x}_2\right),$$

we will show that the first component of $\mathbf{p}^{k+1}$ is less than that of $\tilde{\mathbf{x}}$, i.e.

$$\frac{e(\mathrm{I}_1, \tilde{\mathbf{x}})}{e(\tilde{\mathbf{x}}, \tilde{\mathbf{x}})}\,\tilde{x}_1 < \tilde{x}_1, \tag{8.49}$$

and the second component of $\mathbf{p}^{k+1}$ is less than that of $\mathbf{x}$, i.e.

$$\frac{e(\mathrm{I}_2, \tilde{\mathbf{x}})}{e(\tilde{\mathbf{x}}, \tilde{\mathbf{x}})} \tilde{x}_2 < x_2. \tag{8.50}$$

(8.49) follows immediately from (8.48), but (8.50) is harder. We can rewrite it as $e(\mathrm{I}_2,\tilde{\mathbf{x}})(1 - h)x_2 < e(\tilde{\mathbf{x}},\tilde{\mathbf{x}})x_2$, and so we need to show that $e(\tilde{\mathbf{x}},\tilde{\mathbf{x}}) > (1 - h)e(\mathrm{I}_2,\tilde{\mathbf{x}})$. Now,

$$e(\tilde{\mathbf{x}},\tilde{\mathbf{x}}) = (1 - h)^2 e(\mathbf{x},\mathbf{x}) + h(1 - h)e(\mathrm{I}_1,\mathbf{x}) + h(1 - h)e(\mathbf{x},\mathrm{II}_1) + h^2 e(\mathrm{I}_1,\mathrm{II}_1). \tag{8.51}$$

Since $e(\mathrm{I}_1,\mathbf{x}) = e(\mathrm{I}_2,\mathbf{x}) = e(\mathbf{x},\mathbf{x})$ or $x_2 = 0$ (which makes (8.50) trivial), we get:

$$\begin{aligned} e(\tilde{\mathbf{x}},\tilde{\mathbf{x}}) &= (1 - h)e(\mathrm{I}_2,\mathbf{x}) + h(1 - h)e(\mathbf{x},\mathrm{II}_1) + h^2 e(\mathrm{I}_1,\mathrm{II}_1) \\ &= (1 - h)e(\mathrm{I}_2,\tilde{\mathbf{x}}) + h(1 - h)\,[e(\mathrm{I}_2,\mathbf{x}) + e(\mathbf{x},\mathrm{II}_1) - e(\mathrm{I}_2,\mathrm{II}_1)] + h^2 e(\mathrm{I}_1,\mathrm{II}_1). \end{aligned} \tag{8.52}$$

Since $e(\mathrm{I}_1,\mathrm{II}_1) > 0$ and

$$e(\mathrm{I}_2,\mathbf{x}) + e(\mathbf{x},\mathrm{II}_1) > x_1 e(\mathrm{I}_2,\mathrm{II}_1) + x_2 e(\mathrm{I}_2,\mathrm{II}_1) = e(\mathrm{I}_2,\mathrm{II}_1), \tag{8.53}$$

we have $e(\tilde{\mathbf{x}},\tilde{\mathbf{x}}) > (1 - h)e(\mathrm{I}_2,\tilde{\mathbf{x}})$ and so (8.50) holds.

A similar argument works starting with $\tilde{\tilde{\mathbf{x}}} = (1 - h)\mathbf{x} + h(0,1)$ and the theorem is proved.

Unfortunately, the same result does not hold in general. It is not true that for $n \times n$ evolutionary games all ESS are asymptotically stable. Taylor and Jonker (1978) give a slight modification of the Scratch–Bite–Trample example where this happens. Take as the matrix:

$$\begin{pmatrix} 1 - h & 2 & 0 \\ 0 & 1 - h & 2 \\ 2 & 0 & 1 - h \end{pmatrix} \tag{8.54}$$

(compare with Example 8.8) and for any positive h it is easy to show that $\mathbf{x} = (\frac{1}{3},\frac{1}{3},\frac{1}{3})$ is ESS. However, for small enough h, no matter how close you start to this ESS, you will eventually start oscillating wildly, just as in Example 8.8.

This is not such a body-blow to the justification of ESS as you would imagine. The dynamic game which describes what happens from generation to generation is really an approximation of the more accurate continuous time model given by the differential equation (8.45). In the continuous time model it is true that every ESS is asymptotically stable.

Theorem 8.2. In any continuous time evolutionary game, every ESS strategy is asymptotically stable.

The proof of this theorem is beyond the scope of this book because it depends on decompositions of matrices (Taylor and Jonker, 1978) or the Liapunov function (Hofbauer, Schuster and Sigmund, 1979).

So, what goes wrong in the discrete version of the game? Essentially the strategies in each generation of this game are an approximation of the strategies at equally spaced times of the continuous version. Provided the latter converge in an orderly way, then the discrete approximation should also converge. However, if you look at how the strategies change in time in the continuous time version of the counter example (8.54) (see Fig. 8.1) they spiral in towards the ESS strategy. In the discrete version the errors are sufficient to just push the strategies out beyond a circular orbit and they spiral outward.

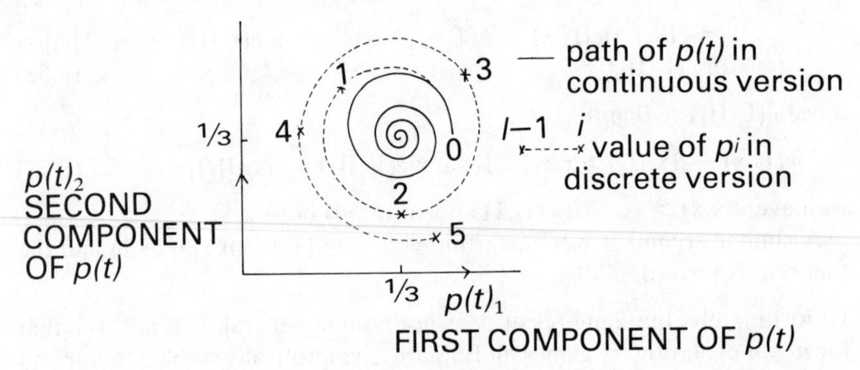

Fig. 8.1—Paths of continuous and discrete versions of (8.54).

However, it is not true in either the discrete or continuous dynamic games that every asymptotically stable strategy is an ESS one. Counter-examples arise when one has a mutant subpopulation which is fitter than the stable strategies, but as this subpopulation develops in time the different genotypes in it develop at different rates and thus swing the resulting overall strategy around in a spiral towards the stable strategy.

8.9 EVOLUTIONARY STABLE STRATEGIES IN MULTI-TYPE GAMES

So far we have discussed evolutionary games where there is only one type of player, but in many interesting cases there are several types of players. For example, in rearing behaviour one would expect different behaviour in the two sexes, and evolutionary models of predator–prey obviously involve two types of players.

We concentrate on games with two types of players but the definition of ESS for these obviously extends to games with more players. Suppose we have a game with two types of players and suppose there are m possible pure strategies for type 1 players and n for type 2 players. The overall strategy for the population is then given by mixed strategies $(\mathbf{x},\mathbf{y})$, where $\mathbf{x} = (x_1,x_2,\ldots,x_m)$, $\mathbf{y} = (y_1,\ldots,y_n)$, so x_i is the proportion of type 1 who play the ith strategy and y_j is the proportion of type 2 who play the jth strategy. The fitness of type 1's playing $\mathbf{r}$ in such a population is $e_1(\mathbf{r},(\mathbf{x},\mathbf{y}))$ and the fitness of type 2's playing $\mathbf{s}$ is $e_2(\mathbf{s},(\mathbf{x},\mathbf{y}))$.

Definition 8.5. A pair of strategies $(\mathbf{x}^*,\mathbf{y}^*)$ are **evolutionary stable strategies** (ESS) if for any $\mathbf{r} \neq \mathbf{x}^*$ or $\mathbf{s} \neq \mathbf{y}^*$ we let $\bar{\mathbf{x}} = (1-\varepsilon)\mathbf{x}^* + \varepsilon\mathbf{r}$, $\bar{\mathbf{y}} = (1 - \varepsilon\mathbf{y}^* + \varepsilon\mathbf{s}$, then:

$$e_1(\mathbf{r},(\bar{\mathbf{x}},\bar{\mathbf{y}})) + e_2(\mathbf{s},(\bar{\mathbf{x}},\bar{\mathbf{y}})) < e_1(\mathbf{x}^*,(\bar{\mathbf{x}},\bar{\mathbf{y}})) + e_2(\mathbf{y}^*,(\bar{\mathbf{x}},\bar{\mathbf{y}}))\,, \tag{8.55}$$

for sufficiently small $\varepsilon > 0$.

As in the one-type case there is an equivalent definition in the finite fitness matrices case when

$$e_1(\mathbf{r},(\mathbf{x},\mathbf{y})) = \mathbf{r}(A\mathbf{x} + B\mathbf{y}) \quad \text{and} \quad e_2(\mathbf{s},(\mathbf{x},\mathbf{y})) = \mathbf{s}(C\mathbf{x} + D\mathbf{y}), \tag{8.56}$$

which is more useful for calculating ESS. Here A, B, C *and* D are the $m \times m$, $m \times n$, $n \times m$ *and* $n \times n$ matrices, respectively, corresponding to the payoffs of type 1 against type 1, type 1 against type 2, type 2 against type 1 and type 2 against type 2.

Definition 8.6. A pair of strategies $(\mathbf{x}^*,\mathbf{y}^*)$ is ESS if, for any $\mathbf{r} \neq \mathbf{x}^*$, or $\mathbf{s} \neq \mathbf{y}^*$,

$$e_1(\mathbf{x}^*,(\mathbf{x}^*,\mathbf{y}^*)) + e_2(\mathbf{y}^*,(\mathbf{x}^*,\mathbf{y}^*)) \geqslant e_1(\mathbf{r},(\mathbf{x}^*,\mathbf{y}^*)) + e_2(\mathbf{s},(\mathbf{x}^*,\mathbf{y}^*)), \tag{8.57}$$

and if there is equality in (8.57):

$$e_1(\mathbf{x}^*,(\mathbf{r},\mathbf{s})) + e_2(\mathbf{y}^*,(\mathbf{r},\mathbf{s})) > e_1(\mathbf{r},(\mathbf{r},\mathbf{s})) + e_2(\mathbf{s},(\mathbf{r},\mathbf{s})). \tag{8.58}$$

By putting $\mathbf{r} = \mathbf{x}^*$ and $\mathbf{s} = \mathbf{y}^*$ in turn in (8.57) we get:

$$e_1(\mathbf{x}^*,(\mathbf{x}^*,\mathbf{y}^*)) \geqslant e_1(\mathbf{r},(\mathbf{x}^*,\mathbf{y}^*)); \qquad e_2(\mathbf{y}^*,(\mathbf{x}^*,\mathbf{y}^*)) \geqslant e_2(\mathbf{s},(\mathbf{x}^*,\mathbf{y}^*)), \tag{8.59}$$

so in some sense $(\mathbf{x}^*,\mathbf{y}^*)$ is an equilibrium pair.

Let us look at a very simple model of rearing offspring. Assume males have two possible strategies—to be faithful or to philander—while females can be coy by demanding a long 'engagement' period or they can be 'fast'. Philandering males made no headway with coy females, so if there are lots of coy females it is worth being faithful. However, if there are lots of faithful males, a fast female who skips the engagement period will do better than her coy counterpart. Yet in a population of fast females, a philanderer will have easy pickings. So what mixtures of strategies will persist.

Suppose the reward to each parent of producing an offspring is v, but there is a cost of $2c$ of rearing it, which is either shared between male and female if the male is faithful, or borne totally by the female otherwise. The cost of courtship is d, and a coy female and philandering male both have payoff 0, when they meet. Then using the notation of (8.56) this gives fitness matrices:

$$A = \begin{pmatrix} 0 & 0 \\ 0 & 0 \end{pmatrix}, \quad B = \begin{matrix} \text{faithful} \\ \text{philanderer} \end{matrix} \begin{pmatrix} v-c-d & v-c \\ 0 & v \end{pmatrix}, \quad (\text{columns: coy, fast})$$

$$C = \begin{matrix} \text{coy} \\ \text{fast} \end{matrix} \begin{pmatrix} v-c-d & 0 \\ v-c & v-2c \end{pmatrix}, \quad D = \begin{pmatrix} 0 & 0 \\ 0 & 0 \end{pmatrix} \tag{8.60}$$

Lets look at some cases.

Example 8.9—$v = 20$, $c = 6$, $d = 2$; so it is worth bringing up offsprings alone. In this case, since $A = D = [0]$, (8.57) implies that any ESS pair of strategies must be an equilibrium pair in the non-zero-sum game, given by B and C. So first find all equilibrium pairs for this game:

$$\begin{matrix} & \text{coy} & \text{fast} \\ \text{faithful} & (12,12) & (14,14) \\ \text{philanderer} & (10,0) & (20,8) \end{matrix} \tag{8.61}$$

Since 'fast' strictly dominates 'coy' for the females, it is obvious that any equilibrium pair must involve the female playing fast. By domination, the male part of the equilibrium pair must be to philander, i.e. $x^* = (0,1)$, $y^* = (0,1)$. This is an ESS since on checking (8.57) with $\mathbf{r} = (r, 1-r)$, $\mathbf{s} = (s, 1-s)$:

$$\begin{aligned} e_1(\mathrm{I}_2,(\mathrm{I}_2,\mathrm{II}_2)) + e_2(\mathrm{II}_2,(\mathrm{I}_2,\mathrm{II}_2)) &= 20 + 8 > 20 - 6r + 8 - 8s \\ &= e_1(\mathbf{r},(\mathrm{I}_2,\mathrm{II}_2)) + e_2(\mathbf{s},(\mathrm{I}_2,\mathrm{II}_2)). \end{aligned} \tag{8.62}$$

Example 8.10—$v = 10$, $c = 6$, $d = 2$. This gives the non-zero-sum game matrix:

$$\begin{matrix} & \text{coy} & \text{fast} \\ \text{faithful} & (2,2) & (4,4) \\ \text{philanderer} & (0,0) & (10,-2) \end{matrix}. \tag{8.63}$$

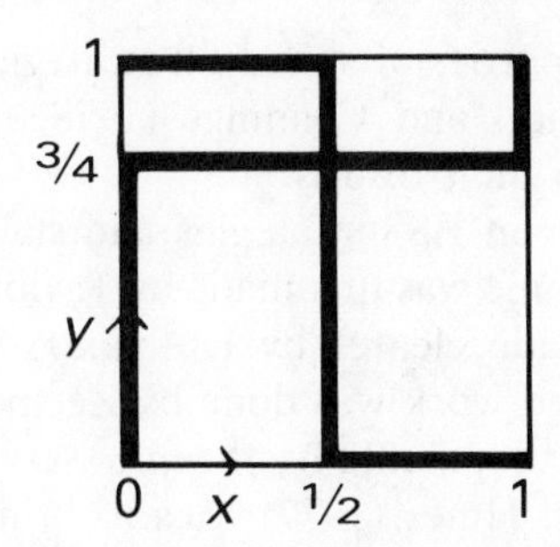

Fig. 8.2—Swastika diagram for Example 8.10.

Again we first find all equilibrium pairs since any ESS pair must be an equilibrium pair. Using the swastika method of Chapter 3, Section 3.5, let $\mathbf{x} = (x,1-x)$, $\mathbf{y} = (y,1-y)$. Then $e_1(\mathbf{x},\mathbf{y}) = x(8y-6) + 10 - 10y$. So, if $y < 3/4$, $x = 0$, if $y = 3/4$. x is anything and if $y > 3/4$, $x = 1$. Since $e_2(\mathbf{x},\mathbf{y}) = y(2-4x) + 6x - 2$, we again have that if $x < 1/2$, $y = 1$; if $x = 1/2$, y is anything and if $x > 1/2$, $y = 0$. From Fig. 8.2 we see that the only equilibrium is $\mathbf{x} = (1/2,1/2)$, $\mathbf{y} = (3/4,1/4)$. However, for any $\mathbf{r},\mathbf{s}$ we have $e_1(\mathbf{r},(\mathbf{x},\mathbf{y})) + e_2(\mathbf{s},(\mathbf{x},\mathbf{y})) = 3\frac{1}{2}$, and so must check condition (8.58). We find:

$$\begin{aligned} e_1((1/2,1/2),\ (\mathrm{I}_2,\mathrm{II}_2)) + e_2((3/4,1/4),),(\mathrm{I}_2,\mathrm{II}_2)) &= 7 + -1/2 < \\ e_1(\mathrm{I}_2,(\mathrm{I}_2,\mathrm{II}_2)) + e_2(\mathrm{II}_2,(\mathrm{I}_2,\mathrm{II}_2)) &= 10 - 2. \end{aligned} \tag{8.64}$$

So there is no ESS for this game.

Example 8.11—$v = 7$, $c = 6$, $d = 2$; little reward in rearing offspring. This leads to the matrix:

$$\begin{array}{lcc} & \text{coy} & \text{fast} \\ \text{faithful} & \multicolumn{2}{c}{\begin{pmatrix} (-1,-1) & (1,1) \\ (0,0) & (7,-5) \end{pmatrix},} \\ \text{philanderer} & & \end{array} \tag{8.65}$$

and a similar argument to Example 8.9 shows that $((0,1),(1,0))$, i.e. philanderer and coy form an ESS pair strategies. In such a situation the population is likely to die out.

8.10 FURTHER READING

Evolutionary games are the most recent application of game theory which we deal with in this book. The idea of an evolutionary stable strategy was introduced in Maynard-Smith and Price (1973) and Maynard-Smith (1974). (See Maynard-Smith, 1982, for an up-to-date account of the progress.) The basic mathematical properties of such ESS strategies were worked out by Haigh (1975) and Bishop and Cannings (1978), with Abakuks (1980) pointing out an error in their results.

The connection between ESS strategies and stability properties in the dynamic evolutionary game was first made by Taylor and Jonker in (1978). The connection was made clearer by Hofbauer, Schuster and Sigmund (1979) and more detailed work was done by Zeeman (1979).

One of the other developments has been to look at more complicated war of attrition models. Hines (1977) tries to bring in factors about the surrounding environment, while Haigh and Rose (1980) allow for 'overshoot', i.e. the winner will be in fact 'bid' more than the loser's final bid. Also, several authors have looked at the case where the contestants put different values on the prize, both for 'war of attrition' games and other types of games; see Maynard-Smith and Parker (1976), Selten (1980) and Hammerstein (1981).

Another variation was introduced by Riley (1979) when he looked at finite populations, rather than the implicitly infinite population assumed by the usual theory. He introduces two alternative concepts of ESS for this case, one allowing no invasion by mutants and the other no overwhelming invasion by mutants.

Maynard-Smith (1978) raised the problem of games played between relatives, where one should allow for the fact that they are more likely to have the same strategy than two animals picked at random from the population. Grafen (1979) describes how this modification should be made by looking at gene frequency.

Lastly, the evolutionary games with two type of players were introduced by Taylor (1979) and the parental strategy game described in Section 8.9 is based on the game of this type considered by Schuster and Sigmund (1981).

PROBLEMS FOR CHAPTER 8

8.1. Find which strategies are ESS in the games with the following payoff matrices:

$$\text{(a)}\begin{pmatrix} 3 & 2 \\ 1 & 4 \end{pmatrix}\text{(b)}\begin{pmatrix} 2 & 5 \\ 3 & 1 \end{pmatrix};\ \text{(c)}\quad \begin{pmatrix} 0 & 1 & 0 \\ 0 & 0 & 1 \\ 1 & 0 & 0 \end{pmatrix}.$$

8.2. Prove that in any 2×2 game a pure strategy that dominates the other one must be ESS.

8.3. In Section 8.5 we saw that in order to find the ESS of $n \times m$ games we took $\mathbf{p} = (p_1, p_2, \ldots, p_r, 0, \ldots, 0)$, where $\sum_{i=1}^{r} p_i = 1$, and assumed $e(I_i, \mathbf{p}) = \max_j e(I_j, \mathbf{p})$, for $e = 1,2,\ldots,k$. If B is the $k \times k$ submatrix of E

consisting of the first k rows and columns, then for $\mathbf{p}$ to be ESS we require $(\mathbf{p}-\mathbf{q})^{\mathrm{T}}B(\mathbf{p}-\mathbf{q})<0$, for all $q=(q_1,q_2,\ldots,q_k,0,0,\ldots,0)$. Let C be the $(k-1)\times(k-1)$ matrix defined by:

$$c_{ij}=b_{ij}-b_{ik}-(b_{kj}-b_{kk}). \tag{8.66}$$

Prove that if:

$$\mathbf{z}^{\mathrm{T}}(C+C^{\mathrm{T}})\mathbf{z}<0, \quad \text{for all } z, \tag{8.67}$$

then p is ESS. (Matrices that satisfy condition (8.67) are called negative definite and there is an easy check using the subdeterminants of the matrix which tells when a matrix is negative definite.)

8.4. Haigh (1975) proved Problem 8.3 and also said that (8.67) was a necessary as well as a sufficient condition for p to be ESS. (A sufficient condition is one which, if it holds, then the p we are considering is ESS. A necessary condition is one which all ESS must satisfy.) This is true if $k<r+2$, but Abakuks (1980) gave the following counter-example for $k\geqslant r+2$.

Let the fitness matrix be:

$$\begin{pmatrix} 0 & 2 & 2 \\ 0 & 1 & 0 \\ 0 & 0 & 1 \end{pmatrix}.$$

Show that $p=(1,0,0)$ is ESS. Calculate the C of (8.66) for this p and show it does not satisfy (8.67).

So if $k<r+2$, we can use (8.67) to check if p is ESS. Otherwise we have to use (8.26) directly.

8.5. (a) The fitness matrix in a Hawk v. Dove game is:

	Hawk	Dove
Hawk	−1	6
Dove	0	3

Find the ESS strategy for this game.

(b) The above evolutionary game can be thought of as a two-person non-zero-sum game G with payoff matrix:

		II Hawk	II Dove
I	Hawk	(−1,−1)	(6,0)
I	Dove	(0,6)	(3,3)

Write down the metagame $1G$, and find the metaequilibrium in it.

(c) I has four strategies in the game $1G$, which can be described as always Hawk, always Dove, Bully (Hawk against Dove, Dove against Hawk) and Retaliator (Hawk against Hawk, Dove against Dove). Think of these as strategies in an evolutionary game where each player first chooses either Hawk or Dove and then chooses again knowing his opponent's first choice. Hawk plays Hawk always, and Dove always plays Dove; Bully first plays Hawk, then the opposite to his opponent's choice; Retaliator first plays Dove, then the same as his opponent did. The payoffs are given by these second choices. Write down the fitness matrix and show that the mixed strategy Hawk with probability ¾ and Bully with probability ¼ is ESS. Why isn't Retaliator a pure ESS strategy?

8.6. For the three games in Problem 8.1 describe what happens in the first six generations of the dynamic evolutionary games based on them. Assume $\mathbf{p}_0 = (\frac{1}{2},\frac{1}{2})$ for (a) and (b) and try both $\mathbf{p}_0 = (\frac{1}{3},\frac{1}{3},\frac{1}{3})$ and $\mathbf{p}_0 = (0.4,0.4,0.2)$ for (c).

8.7. The graduated risk game. Each animal can choose a level to which he is prepared to escalate the contest, and if animal I chooses p_1 and animal II chooses p_2, the chance that a contestant is injured is $\min(p_1,p_2)$. If V is the value of the prize and D is the disadvantage caused by injury, assume $D > V$. The winner is the one who escalates the more if neither is injured, otherwise with equal probability one will be the victor and the other will be injured. Show that the payoff of playing p_1 against p_2, where $p_1 > p_2$ is:

$$e(p_1,p_2) = V(1 - p_2) + \tfrac{1}{2}(V - D)p_2.$$

Hence, show that no pure strategy can be ESS.
Show that the payoff of playing a pure strategy p against a mixed strategy given by density function f is:

$$e(p,f) = \int_0^p (V(1 - x) + \tfrac{1}{2}(V - d)x)f(x)\mathrm{d}x + \int_p^1 \tfrac{1}{2}(V - D)pf(x)\mathrm{d}x.$$

Prove that if $f(x) = \alpha(1 - x)^{\alpha-1}$, where $\alpha = (D - V)/2V$, then $e(p,f)$ is independent of p. What is the only possible ESS? Is it ESS?

8.8. In the two types of animal evolutionary games,

$$e_1(r,(x,y)) = rAx + rBy \quad \text{and} \quad e_2(s(x,y)) = sCx + sDy,$$

where A, B, C and D are matrices. Show that in this case the definition of an ESS pair of strategies (x,y) is equivalent to saying for any $r \neq x$ or $s \neq y$:

$$e_1(x,(x,y)) + e_2(y,(x,y)) \geqslant e_1(r,(x,y)) + e_2(s,(s,(x,y)),$$

and if there is equality:

$$e_1(x,(r,s)) + e_2(y,(r,s)) > e_1(r,(r,s)) + e_2(s,(r,s)).$$

8.9. In a predator–prey model, both types of animals can come out either in the Day (D) or the Night (N) to feed. As in Problem 8.8, the fitness of the prey I is given by e_1, where:

$$e_1(r,(x,y)) = r^{\mathrm{T}}Ax + r^{\mathrm{T}}By.$$

So A describes the feeding success of a strategy for the prey compared with the other animals of its type and B describes its vulnerability to attack from the predator. Take

$$A = \begin{array}{c} \\ \text{Day} \\ \text{Night} \end{array} \begin{array}{c} \text{Day} \quad \text{Night} \\ \begin{pmatrix} 2 & 6 \\ 3 & 1 \end{pmatrix} \end{array}, \qquad B = \begin{array}{c} \\ \text{Day} \\ \text{Night} \end{array} \begin{array}{c} \text{Day} \quad \text{Night} \\ \begin{pmatrix} -4 & 0 \\ 0 & -8 \end{pmatrix} \end{array}$$

The fitness of the predator I is given by e_2, where:

$$e_2(s,(x,y)) = sCx + sDy.$$

So C describes its success on feeding on the prey and D describes the effect of having to deal with others of its own species. Take

$$C = \begin{array}{c} \\ \text{Day} \\ \text{Night} \end{array} \begin{array}{c} \text{Day} \quad \text{Night} \\ \begin{pmatrix} 4 & 0 \\ 0 & 8 \end{pmatrix} \end{array}, \qquad D = \begin{array}{c} \\ \text{Day} \\ \text{Night} \end{array} \begin{array}{c} \text{Day} \quad \text{Night} \\ \begin{pmatrix} -2 & 0 \\ 0 & -1 \end{pmatrix} \end{array}$$

Show that to be part of any ESS pair, x must satisfy:

$$e_1(x,(x,y)) \geqslant e_1(r,(x,y)), \qquad r \neq x,$$

and y must satisfy:

$$e_2(y,(x,y)) \leqslant e_2(s,(x,y)), \qquad s \neq y.$$

Hence, find the ESS pairs for this game, if they exist.

CHAPTER 9

Bidding and Auctions

'Pleasant-faced people are generally the most welcome, but an auctioneer prefers to see a man whose appearance is for bidding'—Anon.

9.1 TYPES OF AUCTIONS

Auctioning is one of the oldest and yet most exciting ways of selling items. There can be very few people who have not been attracted by the chance of getting something at a knock-down price, or been spell-bound by the fast-talking monologue of the auctioneer. However, essentially the same method is used by firms and governments when awarding billion dollar and pound contracts to tendering firms. The problem of how to bid in such contests is an interesting many-person decision problem, which can be turned into game format. In this chapter we will take a look at some simple bidding situations and see what a game theoretic approach uncovers.

There are many ways in which auctions can be conducted but three methods stand out.

9.1.1 The English Auction—Sequential Bidding

This is probably the best known and most widely used auctioning rule. The auctioneer describes the item for sale and then invites bids, usually suggesting an initial bid. Bids are made for the item so that each new bid is higher than all previous bids, and the process continues until no one will bid higher than the present bid. The object is then awarded to that bidder. We ignore the complication that the seller may have set a **reserve price** below which he will not sell, since we can think of this as the seller's bid for the item. Notice if there are n bidders who will bid up to $b_1,b_2,\dots,b_n$, respectively, where without loss of generality we assume $b_1 \geqslant b_2 \geqslant \dots \geqslant b_n$, then the item will go to bidder 1 at a price b_2 or the smallest bidding increment above b_2. So essentially it is sold at the price of the second highest bid.

9.1.2 The Dutch Auction

In this auction, which originated with selling flowers in Holland, the auctioneer sets a price well above what he thinks the item will go for. The auctioneer, or some mechanism, then systematically lowers the price until a bid is made. This first bidder is the winner and gets the item at the price he agreed to. Thus, in the n-bidder case, the item will be sold to bidder 1, but at a price b_1, the highest bid.

9.1.3 Sealed Bid Auctions

This procedure is standard for bidding for contracts. Each of the potential bidders submits a bid, sealed so that the other bidders do not know it. At the appropriate time, the auctioneer looks at all the bids submitted and awards it to the highest bidder. So again the price is that of the highest bid.

Cassady (1967) describes some other interesting auctions such as the candle auction and the hand-shake auction, but essentially there are only two different decision criteria used in auctions—the English auction, where the item goes to the highest bidder at the second highest bid price, and the Dutch auction and sealed bid auction, where it goes to the highest bidder at his bid price. If there are ties, so that more than one firm bids the same highest price, we will assume each one is equally likely to be given the item. This might not be the case if one of the bidders is more likely to catch the auctioner's eye than anyone else. Thus, beautiful blondes have a built-in advantage, though females might also do quite well. However, mathematically we will ignore such distractions.

For a particular bidder in a sealed bid auction, suppose he values the item at v, and puts in a bid b, then his payoff is $v - b$ if his is the highest bid; $1/k(v - b)$ if one of k equally valued highest bids; and 0 otherwise. This describes what occurs in most auctions to purchase items. However, if you are bidding for a contract, then you work out the cost c of fulfilling the contract and bid a price p. If p is the lowest price, you get the contract and so a profit of $p - c$; if p is one of k equally priced lowest bids, your expected profit is $l/k(p - c)$ and otherwise you do not get the contract or any profit. The two problems are equivalent if we think of the value of the contract as $v = -c$ and think of bidding a price p as actually bidding $b = -p$. Then again the highest bid b wins and the profit is $v - b = (-c) - (-p) = p - c$ if it wins, and zero otherwise.

We first look at some simple examples where there are only two bidders who know each other's valuations of the item. We then drop the unrealistic assumption that you know the other bidder's valuation. Then we turn to auctions involving several items to be auctioned and finally look at those with several sellers.

9.2 DUTCH AUCTION—DISCRETE BIDS AND KNOWN VALUATIONS

Consider the simplest example of a Dutch auction, where there are two bidders, I and II, who value the item at v_1 and v_2, respectively, and we assume $v_1 \geqslant v_2$. Assume also that each person knows the other one's valuation. Bids have to be made in integer steps, i.e. 0,1,2,3 and so if player I will bid up to b_1 and player II up to b_2, the payoffs are:

$$e_1(b_1,b_2) = \begin{cases} v_1 - b_1, & b_1 > b_2, \\ \tfrac{1}{2}(v_1 - b_1), & b_1 = b_2, \\ 0, & b_1 < b_2, \end{cases}$$

$$e_2(b_1,b_2) = \begin{cases} 0, & b_1 > b_2, \\ \tfrac{1}{2}(v_2 - b_2), & b_1 = b_2, \\ v_2 - b_2, & b_1 < b_2. \end{cases} \tag{9.1}$$

This is a two-person non-zero-sum game, and so we should think of the solution concepts of Chapter 3. We will always assume that bidding cannot exceed an upper bound B, both for practical reasons of constraints on time and money, and also because it seems unrealistic that the auctioneer will start the Dutch Auction at infinity. This restriction means that each player has $B + 1$ pure strategies—the $B + 1$ possible bids—and so the von Neumann–Nash theorem guarantees at least one equilibrium pair.

An equilibrium pair is a non-cooperative game concept and obviously if we think of an auction in that way, we will want to know all the equilibrium pairs. If $v_1 > v_2$, then one obvious way you would expect the bidding to go is that II bids v_2, and I bids $v_2 + 1$ to get the object. A moment's thought, or three lines of work, shows this is almost always an equilibrium pair, but are there any others?

Before answering this question, let's deal with what happens if we think of the game as a cooperative one. This does occur in practice, when 'buyers' rings' form, which agree on a bidding strategy, or rather a 'no-bidding' strategy before hand. If each player bids up to his valuation of the object, he can guarantee himself 0 no matter what his opponent does, and this is the best he can do, since if player II bids up to v_1 (or player I up to v_2), there is no way his opponent can get more than zero. Thus, the maximin values are $v_{\text{I}} = v_{\text{II}} = 0$. Obviously, the Pareto optimal set of bids will be to bid as little as possible. If both players bid 0, the payoff is $(\tfrac{1}{2}v_1,\tfrac{1}{2}v_2)$, whereas if one bids 1 and the other 0, this leads to $(v_1 - 1,0)$ and $(0,v_2 - 1)$, respectively. Thus, the Pareto optimal boundary is the two

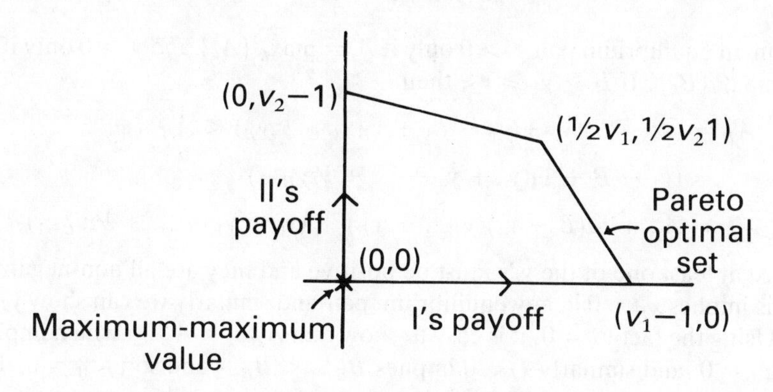

Fig. 9.1—Pareto boundary of Dutch auction.

lines joining $(v_1 - 1,0)$ to $(\frac{1}{2}v_1,\frac{1}{2}v_2)$ and $(\frac{1}{2}v_1,\frac{1}{2}v_2)$ to $(0,v_2 - 1)$ (Fig. 9.1).

We leave it as an exercise (Problem 9.1) to show the maximin bargaining solution is $(\frac{1}{2}v_1,\frac{1}{2}v_2)$, but it is obvious that such cooperative behaviour is unrealistic in general, since the auctioneer would be aware of such reticence and withdraw the item.

So let us look at the non-cooperative case and find out whether there are any other equilibrium pairs beyond the one already described. Let $\mathbf{x} = (x_0,x_1,x_2,\dots,x_B)$ and $\mathbf{y} = (y_0,y_1,\dots,y_B)$ be typical mixed strategies and let $X_i = \sum_{j=0}^{i-1} x_j$, $Y_i = \sum_{j=0}^{i-1} y_j$, then the payoffs (9.1) lead to:

$$e_1(\mathbf{x},\mathbf{y}) = \sum_{i=0}^{B} x_i[(v_1 - i)Y_i + \tfrac{1}{2}(v_1 - i)y_i], \tag{9.2}$$

$$e_2(\mathbf{x},\mathbf{y}) = \sum_{i=0}^{B} y_i[(v_2 - i)X_i + \tfrac{1}{2}(v_2 - i)x_i]. \tag{9.3}$$

To find equilibrium pairs, we use the idea behind the swastika method of Chapter 3, Section 3.5. For a fixed strategy **y** of player II, **x** can be its partner in an equilibrium pair only if it maximises $e_1(\mathbf{x},\mathbf{y})$. Similarly, for a fixed strategy **x** of player I, **y** can be its partner in an equilibrium pair only if it maximises $e_2(\mathbf{x},\mathbf{y})$. We then find which pairs $(\mathbf{x},\mathbf{y})$ satisfy both these conditions. We can rewrite (9.2), (9.3) as:

$$e_1(\mathbf{x},\mathbf{y}) = \sum_{i=0}^{B} x_i A_i, \quad \text{where } A_i = (v_1 - \mathrm{i})(y_0 + y_1 + \dots + y_{i-1} + \tfrac{1}{2}y_i), \tag{9.4}$$

and

$$e_2(\mathbf{x},\mathbf{y}) = \sum_{i=0}^{B} y_j B_j, \quad \text{where } B_j = (v_2 - j)(x_0 + x_1 + \dots + x_{j-1} + \tfrac{1}{2}x_j). \tag{9.5}$$

So in an equilibrium pair $x_i > 0$ only if $A_i = \max_k \{A_k\}$ and $y_j > 0$ only if $B_j = \max_k \{B_k\}$. If $B > v_1 \geqslant v_2$, then:

$$\begin{aligned} A_B &= (v_1 - B)(y_0 + y_1 + \ldots + y_{B-1} + \tfrac{1}{2}y_B) < A_{B-1} \\ &= (v_1 - B + 1)(y_0 + y_1 + \ldots + \tfrac{1}{2}y_{B-1}) \\ &= A_B + \tfrac{1}{2}(B - v_1)(y_{B-1} + y_B) + (y_0 + y_1 + \ldots + \tfrac{1}{2}y_{B-1}), \end{aligned}$$

since at least one of the y_i's must be positive and they are all non-negative. This implies $x_B = 0$ in any equilibrium pair and similarly we can show $y_B = 0$. Using the fact $y_B = 0$, it is easy to show that $A_{B-1} < A_{B-2}$, which implies $x_{B-1} = 0$, and similarly $x_B = 0$ implies $B_{B-1} < B_{B-2}$ and so $y_{B-1} = 0$. The argument repeats itself for all $i > v_1$ and so in an equilibrium pair no one bids more than the highest valuation, and we can set $B = v_1 \geqslant v_2$ without loss.

The precise results thereafter depend on whether $v_1 > v_2 + 2$, $v_1 = v_2 + 2$, $v_1 = v_2 + 1$ or $v_1 = v_2$, and we will give examples of the various cases.

Example 9.1—$v_1 = 4$, $v_2 = 1$ $(v_1 > v_2 + 2)$. The payoff matrix is given below where I_i is the strategy where I bids i, and II_j that where II bids j. It is obtained by substituting the values into (9.1):

$$\begin{array}{c} \\ \mathrm{I}_0 \\ \mathrm{I}_1 \\ \mathrm{I}_2 \\ \mathrm{I}_3 \\ \mathrm{I}_4 \end{array} \begin{array}{c} \begin{array}{ccccc} \mathrm{II}_0 & \mathrm{II}_1 & \mathrm{II}_2 & \mathrm{II}_3 & \mathrm{II}_4 \end{array} \\ \left(\begin{array}{ccccc} (2,\tfrac{1}{2}) & (0,0) & (0,-1) & (0,-2) & (0,-3) \\ (3,0) & (1\tfrac{1}{2},0) & (0,-1) & (0,-2) & (0,-3) \\ (2,0) & (2,0) & (1,-\tfrac{1}{2}) & (0,-2) & (0,-3) \\ (1,0) & (1,0) & (1,0) & (\tfrac{1}{2},-1) & (0,-3) \\ (0,0) & (0,0) & (0,0) & (0,0) & (0,-1\tfrac{1}{2}) \end{array}\right) \end{array}. \tag{9.6}$$

Thus, $A_0 = 2y_0$, $A_1 = 3y_0 + 1\tfrac{1}{2}y_1$, $A_2 = 2y_0 + 2y_1 + y_2$, $A_3 = y_0 + y_1 + y_2 + \tfrac{1}{2}y_3$, $A_4 = 0$, $B_0 = \tfrac{1}{2}x_0$, $B_1 = 0$, $B_2 = -x_0 - x_1 - \tfrac{1}{2}x_2$, $B_3 = -2x_0 - 2x_1 - 2x_2 - x_3$ and $B_4 = -3x_0 - 3x_1 - 3x_2 - 3x_3 - 1\tfrac{1}{2}x_4$ as defined in (9.4) and (9.5). Since some x_i must be positive as they add up to 1, $B_3 > B_4$, which implies $y_4 = 0$. $y_4 = 0$ implies $A_3 > A_4$, and hence $x_4 = 0$ as A_4 cannot maximise A_j, $j = 0,1,2,3,4$. Substituting $x_4 = 0$ into B_j gives $B_2 > B_3$ and so $y_3 = 0$. Notice also that $A_0 < \max(A_1, A_3)$, since if y_0 is positive $A_1 > A_0$, and if $y_0 = 0$, $A_3 > A_0$. This means $x_0 = 0$. Let us draw breath in this bombardment of implications and see what we have proved about equilibrium pairs. So far we have shown that if $(\mathbf{x},\mathbf{y})$ is an equilibrium pair, then $\mathbf{x} = (0,x_1,x_2,x_3,0)$, $\mathbf{y} = (y_0,y_1,y_2,0,0)$, and $A_1 = 3y_0 + 1\tfrac{1}{2}{}_1$, $A_2 = 2y_0 + 2y_1 + y_2$, $A_3 = y_0 + y_1 + y_2$, $A_4 = 0$, $B_0 = B_1 = 0$, and $B_2 = -x_1 - \tfrac{1}{2}x_2$.

Now $A_2 > A_3$ unless $y_2 = 1$ which means $x_3 = 0$ unless $y_2 = 1$. If $y_2 = 1$ is to be part of an equilibrium pair we want $B_2 \geqslant B_0$, $B_2 \geqslant B_1$, i.e. $x_1 = x_2 = 0$. So one possible equilibrium pair is $x_3 = 1$, $y_2 = 1$, i.e. $\mathbf{x} = (0,0,0,1,0)$ $\mathbf{y} = (0,0,1,0,0)$ or I bids 3, II bids 2. If $x_3 = 0$, then $B_2 < 0 = B_1$ and so $y_2 = 0$. We are left with $A_1 = 3y_0 + 1\frac{1}{2}y_1$, $A_2 = 2y_0 + 2y_1$, and $B_0 = B_1 = 0$. This last condition ensures that any combination of y_0 and y_1 maximises $e_2(\mathbf{x},\mathbf{y})$ for any of the possible $\mathbf{x} = (0,x_1,x_2,0,0)$. Thus, the only condition is that $\mathbf{x}$ maximises $e_1(\mathbf{x},\mathbf{y})$. We choose $x_1 = 1$ if $A_1 > A_2$, which occurs if $y_1 < \frac{2}{3}$; x_1 and x_2 arbitrary if $A_1 = A_2$, i.e. $y_1 = \frac{2}{3}$, and $x_2 = 1$ if $A_2 > A_1$, i.e. $y_1 > \frac{2}{3}$. So the equilibrium pairs are:

$$\begin{aligned} &\mathrm{I}_3\ v.\mathrm{II}_2; \qquad \mathrm{I}_1\ v.(1-a,a,0,0,0), \qquad 0 \leqslant a < \tfrac{2}{3}; \\ &(0,x,1-x,0,0)\ v.(\tfrac{1}{3},\tfrac{2}{3},0,0,0), \qquad 0 \leqslant x \leqslant 1, \\ &\mathrm{I}_2\ v.(1-a,a,0,0,0), \qquad \tfrac{2}{3} < a \leqslant 1. \end{aligned} \tag{9.7}$$

Thus, the pure equilibrium pairs are $(v_2 + 2, v_2 + 1)$, $(v_2 + 1, v_2)$ and $(v_2, v_2 - 1)$.

Example 9.2—$v_1 = 4$, $v_2 = 2$. The payoff matrix is:

$$\begin{array}{c} \\ \mathrm{I}_0 \\ \mathrm{I}_1 \\ \mathrm{I}_2 \\ \mathrm{I}_3 \\ \mathrm{I}_4 \end{array} \begin{array}{c} \begin{array}{ccccc} \mathrm{II}_0 & \mathrm{II}_1 & \mathrm{II}_2 & \mathrm{II}_3 & \mathrm{II}_4 \end{array} \\ \left(\begin{array}{ccccc} (2,1) & (0,1) & (0,0) & (0,-1) & (0,-2) \\ (3,0) & (1\frac{1}{2},\frac{1}{2}) & (0,0) & (0,-1) & (0,-2) \\ (2,0) & (2,0) & (1,0) & (0,-1) & (0,-2) \\ (1,0) & (1,0) & (1,0) & (\frac{1}{2},-\frac{1}{2}) & (0,-2) \\ (0,0) & (0,0) & (0,0) & (0,0) & (0,-1) \end{array}\right) \end{array} \tag{9.8}$$

We leave the analysis of this example as a problem (Problem 9.2) but the equilibrium pairs are actually $\mathrm{I}_2 v.\ (y_0,y_1,y_2,0,0)$ where $y_2 + \frac{1}{2}y_1 \geqslant y_0$ and $(0,0,x,1-x,0)\ v.\ \mathrm{II}_2$ for arbitrary x, $0 \leqslant x \leqslant 1$. The pure equilibrium pairs are $(v_2 + 1, v_2)$, (v_2, v_2) and $(v_2, v_2 - 1)$.

Example 9.3—$v_1 = 4$, $v_2 = 3$. This has payoff matrix:

$$\begin{array}{c} \\ \mathrm{I}_0 \\ \mathrm{I}_1 \\ \mathrm{I}_2 \\ \mathrm{I}_3 \\ \mathrm{I}_4 \end{array} \begin{array}{c} \begin{array}{ccccc} \mathrm{II}_0 & \mathrm{II}_1 & \mathrm{II}_2 & \mathrm{II}_3 & \mathrm{II}_4 \end{array} \\ \left(\begin{array}{ccccc} (2,1\frac{1}{2}) & (0,2) & (0,1) & (0,0) & (0,-1) \\ (3,0) & (1\frac{1}{2},1) & (0,1) & (0,0) & (0,-1) \\ (2,0) & (2,0) & (1,\frac{1}{2}) & (0,0) & (0,-1) \\ (1,0) & (1,0) & (1,0) & (\frac{1}{2},0) & ((0,-1) \\ (0,0) & (0,0) & (0,0) & (0,0) & (0,-\frac{1}{2}) \end{array}\right) \end{array}. \tag{9.9}$$

Again we spare us, if not you the reader, the trauma of going through the calculation to find all the equilibrium pairs (see Problem 9.2). This time they are I_3 v. $(y_0,y_1,y_2,y_3,0)$, where $\frac{1}{2}y_3 \geqslant y_0 + y_1$ and $\frac{1}{2}y_3 + y_2 \geqslant 2y_0 + \frac{1}{2}y_1$, $(0,0,x_2,x_3,0)$ v. II_2 for any x_2,x_3. The pure equilibrium pairs are (v_2,v_2), $(v_2,v_2 - 1)$ and $(v_2 - 1,v_2 - 1)$. Notice that $(v_2 + 1,v_2)$ is not an equilibrium pair this time because $v_2 + 1 = v_1$ and so it is better for I to share a profit of 1 half the time than have a sure profit of zero.

Example 9.4—$v_1 = v_2 = 4$.

$$
\begin{array}{c} \\ \text{I}_0 \\ \text{I}_1 \\ \text{I}_2 \\ \text{I}_3 \\ \text{I}_4 \end{array}
\begin{array}{c} \begin{array}{ccccc} \text{II}_0 & \text{II}_1 & \text{II}_2 & \text{II}_3 & \text{II}_4 \end{array} \\
\left(\begin{array}{ccccc}
(2,2) & (0,3) & (0,2) & (0,1) & (0,0) \\
(3,0) & (1\frac{1}{2},1\frac{1}{2}) & (0,2) & (0,1) & (0,0) \\
(2,0) & (2,0) & (1,1) & (0,1) & (0,0) \\
(1,0) & (1,0) & (1,0) & (\frac{1}{2},\frac{1}{2}) & (0,0) \\
(0,0) & (0,0) & (0,0) & (0,0) & (0,0)
\end{array}\right) \end{array} . \qquad (9.10)
$$

We will do the analysis for this symmetrical game where, using the notation of (9.4), and (9.5), $A_0 = 2y_0$, $A_1 = 3y_0 + 1\frac{1}{2}y_1$, $A_2 = 2y_0 + 2y_1 + y_2$, $A_3 = y_0 + y_1 + y_2 + \frac{1}{2}y_3$, $A_4 = B_4 = 0$, $B_3 = x_0 + x_1 + x_2 + \frac{1}{2}x_3$, $B_2 = 2x_0 + 2x_1 + x_2$, $B_1 = 3x_0 + 1\frac{1}{2}x_1$, and $B_0 = 2x_0$. Again, note $A_0 <$ max (A_1,A_3) and so $x_0 = 0$ in any equilibrium pair. Similarly, $B_0 <$ max (B_1,B_3), so $y_0 = 0$. If $A_4 = \max_j \{A_j\}$, this can only occur if $y_4 = 1$ and hence $B_4 = \max_j \{B_j\}$. This in turn requires $x_4 = 1$, which is satisfied by A_4 maximising the A_j. So $x_4 = y_4 = 1$ is an equilibrium pair. If we try $y_4 < 1$, then since some other y is then positive, $A_3 > A_4$ and $x_4 = 0$, which in turn makes $B_3 > B_4$ and $y_4 = 0$. So now assume $y_4 = x_4 = 0$. $A_1 = 1\frac{1}{2}y_1$, $A_2 = 2y_1 + y_2$, so either $y_3 = 1$ and $A_1 = A_2$ or $y_3 < 1$ and $A_2 > A_1$ and $x_1 = 0$. If $y_3 = 1$, $A_3 > A_1$, $A_3 > A_2$ and hence $x_3 = 1$, which makes another equilibrium pair.

So finally suppose $x_1 = 0$. This means $B_3 > B_1$ and so $y_1 = 0$ and we are left with possible strategies $\mathbf{x} = (0,0,x_2,x_3,0)$ and $\mathbf{y} = (0,0,y_2,y_3,0)$, where $A_2 = y_2$, $A_3 = y_2 + \frac{1}{2}y_3$, $B_3 = x_2 + \frac{1}{2}x_3$ and $B_2 = x_2$. Thus, $B_2 = B_3$, i.e. $y_2 > 0$, only if $x_3 = 0$, and similarly $x_2 > 0$, only if $y_3 \neq 0$. This gives two more sets of equilibrium pairs $(0,0,x_2,x_3,0)$ v. II_2 and I_2 v. $(0,0,y_2,y_3,0)$ So the equilibrium pairs are I_4 v. II_4, I_3 v. II_3, $(0,0,x_2,x_3,0)$ v. II_2 and I_2 v. $(0,0,y_2,y_3,0)$ and the pure pairs correspond to bids of (v,v), $(v - 1, v - 1)$ and $(v - 2, v - 2)$.

Even in the simple cases, there are more equilibrium pairs than you would expect, and notice in Examples 9.3 and 9.4 that $(v_2 + 1,v_2)$ is not an equilibrium pair. The results are caused partly by the discrete nature of the bids allowed and partly by the rule for ties. If we allowed bids every c, instead of every 1, we could get the results from the integer bid analysis by taking the rewards as v_1/c, v_2/c, solving the problem with integer bids and then multiplying the final values by c. In the case when $v_1 = v_2$, we can show, see Problem 9.3, that for integer bids the only symmetrical equilibrium pairs (where both players have the same strategy) are I_v v. II_v, I_{v-1} v. II_{v-1} and I_{v-2} v. II_{v-2}. As c tends to zero we get only I_v v. II_v, since $c(v/c - 2) \to v$. In the next section we look to see are there any mixed strategy equilibrium pairs in this continuous bid case.

9.3 DUTCH AUCTION—CONTINUOUS BIDS AND KNOWN VALUATIONS

In this section we assume that in a sealed bid auction the players can bid any value, not just multiples of some unit bid. This is the limit as c tends to zero in the above analysis. We again assume that in the case of ties, each player is equally likely to get the object and that each player knows the other's valuation of the object.

For two bidders with v_1, v_2, we are essentially in the limit of Example 9.1. One would expect the pair of strategies where II bids v_2 and I bids $v_2 + \varepsilon$, where ε is very small to be in equilibrium. This is not really the case as it is even better for I to bid $v_2 + \frac{1}{2}\varepsilon$. In the limit as ε becomes zero there is almost an equilibrium. We denote this as **ε-equilibrium pair.**

Definition 9.1. $(\mathbf{x}^\varepsilon,\mathbf{y}^\varepsilon)$ is an ε-equilibrium pair in a two-person game with strategy sets X and Y if:

$$\begin{aligned} e_1(\mathbf{x}^\varepsilon,\mathbf{y}^\varepsilon) &\geqslant e_1(\mathbf{x},\mathbf{y}^\varepsilon) - \varepsilon, \quad \text{for all } \mathbf{x} \in X, \\ e_2(\mathbf{x}^\varepsilon,\mathbf{y}^\varepsilon) &\geqslant e_2(\mathbf{x}^\varepsilon,\mathbf{y}) - \varepsilon, \quad \text{for all } \mathbf{y} \in Y. \end{aligned} \tag{9.11}$$

This is the same idea as that of ε-optimal solution for recursive games—Definition 7.1 in Chapter 7.

Thus, $I_{v_2+\varepsilon},II_{v_2}$ is an ε-equilibrium pair. In fact, if you take any b, $v_2 \leqslant b < v_1$, $(I_{b+\varepsilon},II_b)$ is an ε-equilibrium pair. These correspond to the 'dog in the manger'attitude by player II, who forces I up way beyond what he, (II), thinks the item is worth. The (I_3,II_2) equilibrium pair is an example of this in Example 9.1. There was another set of equilibrium pairs in that example where player I played v_2 and II played a mixed strategy of bids up to v_2. Does the same thing occur in this continuous version? If I plays v_2, then II's payoff is 0 provided he doesn't bid more than v_2, and so any mixed strategy, F, with bids up to v_2 will maximise $e_2(I_{v_2},F)$. We think of a mixed strategy for these continuous games as given by the distribution function $F(x)$, which is the probability the bid is x or less, i.e. $P(b \leqslant x) = F(x)$. In order for I_{v_2} and II_F to make an equilibrium pair, we must show that I does better bidding v_2 and so getting $v_1 - v_2$ with certainty than bidding any other value b. Obviously, if $b > v_2$, this is worse for I, but if $b < v_2$, I gets $(v_1 - b)$ with probability $F(b)$ and 0 otherwise. Thus, I_{v_2} will be the best response to F if:

$$v_1 - v_2 \geqslant (v_1 - b)F(b), \quad \text{for all } b, \quad 0 \leqslant b \leqslant v_2. \tag{9.12}$$

So, for any distribution function F satisfying $F(b) \leqslant v_1 - v_2/v_1 - b, 0 \leqslant b \leqslant v_2$, we will have an equilibrium pair (I_{v_2}, II_F). We have been rather vague about ties, but this can always be made rigorous by carefully defining $F(b)$ at points where it is discontinuous.

For the case $v_1 = v_2 = v$, obviously the strategies where both bid v is an equilibrium pair. However, it is very much like the Prisoners' Dilemma, because this equilibrium pair gives both players zero payoffs, whereas there are other strategies, where both bid less, which give them both positive payoffs. Notice that in Example 9.4 the equilibrium pair (I_4, II_4) is dominated by (I_0, II_0) which is not an equilibrium pair. Smith and Case (1975) suggested a strategy F given by:

$$F(b) = \begin{cases} (v - c)/(v - b), & 0 \leqslant b \leqslant c, \\ 1, & c < b. \end{cases} \tag{9.13}$$

This consists of a positive probability $v - c/v$ of playing 0 and then a chance of putting in a bid between 0 and c given by the density function $f(x) = F'(x) = (v - c)(v - x)^{-2}$. The advantage of playing this is that if your opponent bids b, his payoff is:

$$e_1(b,F) = \begin{cases} \tfrac{1}{2}(v - c), & b = 0, \\ v - c, & 0 < b \leqslant c, \\ v - b & b > c. \end{cases} \tag{9.14}$$

Thus, it pays him to put in a bid somewhere between 0 and c. If it also was optimal for him to bid 0 as well—it isn't becaue of the 'ties' rule—then the mixed strategy F would be optimal and (F,F) would be an equilibrium pair. If I decides to bid b against any of II's strategies and this bid gives him a payoff $v - c$, then if II changes his strategy to F, defined by (9.13), II's payoff is optimised subject to no change in I's payoff. Also, I's payoff will remain the same if he now bids any value less than b, though such a bid will increase II's payoff. Thus, it gives I the greatest encouragement to change his strategy to help II. However, there are no other equilibrium pairs apart from (I_v, II_v), and the ones we obtained in Example 9.4 all converge to this one as c tends to zero.

9.4 ENGLISH AUCTION—KNOWN VALUATIONS

The analysis of English auctions turns out to be a lot simpler than Dutch auctions. This is not chauvinism, merely fact. In the English auction with two players, if player I will bid up to b_1 and player II up to b_2, then the payoffs are:

$$e_1(b_1,b_2) = \begin{cases} v_1 - b_2, & \text{if } b_1 > b_2, \\ \tfrac{1}{2}(v_1 - b_2), & \text{if } b_1 = b_2, \\ 0, & \text{if } b_1 < b_2, \end{cases} \tag{9.15}$$

$$e_2(b_1,b_2) = \begin{cases} 0, & \text{if } b_1 > b_2, \\ \tfrac{1}{2}(v_2 - b_2), & \text{if } b_1 = b_2, \\ v_2 - b_1, & \text{if } b_1 < b_2, \end{cases}$$

where v_1 and v_2 are the respective bidder's valuation of the object.

If we think of it as a cooperative game, then the maximin values are again 0, since if II plays v_1, all I can get is zero, whereas if I plays v_1 or less, he is sure of zero. Thus, $v_{\text{I}} = 0$ and similarly $v_{\text{II}} = 0$—the maximin values. Obviously the best thing to happen is when your opponent bids nothing, and so the Pareto optimal payoffs are $(v_1,0)$ and $(0,v_2)$ and the line joining them. The payoffs on the line occur when I and II have a jointly randomised strategy which decides with probability p that I must bid 0 and probability $(1 - p)$ that II bids 0. Problem 9.4 asks you to check that in this case the maximin bargaining solution is $(\tfrac{1}{2}v_1,\tfrac{1}{2}v_2)$. Notice that there is really no difference in the cooperative versions of the English and Dutch auctions.

Turning to the non-cooperative game, we can try to find the equilibrium pairs in the discrete version, where the bids are integer, just as in Section 9.2. Again put some upper bound B on the allowable bids. If $\mathbf{x} = (x_0,x_1,\ldots,x_B)$ and $\mathbf{y} = (y_0,y_1,\ldots,y_B)$ are the respective players' strategies, (9.15) gives:

$$e_1(\mathbf{x},\mathbf{y}) = \sum_{i=0}^{B} (v_1 - i)y_i\bar{X}_i, \tag{9.16}$$

$$e_2(\mathbf{x},\mathbf{y}) = \sum_{i=0}^{B} (v_2 - e)x_i\bar{Y}_i, \tag{9.17}$$

where $\bar{X}_i = (\tfrac{1}{2}x_i + x_{i+1} + \ldots + x_B)$ and $\bar{Y}_i = (\tfrac{1}{2}y_i + y_{i+1} + \ldots + y_B)$. Notice the $\bar{X}_i$ and $\bar{Y}_i$ are non-increasing in i. For $(\mathbf{x},\mathbf{y})$ to be an equilibrium pair we must have:

(a) if $(v_1 - i)y_i > 0$, then $\bar{X}_i = 1$;

(b) if $(v_1 - i)y_i = 0$, then $\bar{X}_i$ can be anything; (9.18)

(c) if $(v_1 - i)y_i < 0$, then $\bar{X}_i = 0$,

in order that **x** maximises (9.16). Similarly, if **y** is to maximise (9.17) we want:

(a) if $(v_2 - i)x_i > 0$, then $Y_i = 1$;

(b) if $(v_2 - i)x_i = 0$, then $\bar{Y}_i$ can be anything; (9.19)

(c) if $(v_2 - i)x_i < 0$, then $\bar{Y}_i = 0$.

We will again assume $v_1 \geqslant v_2$, and before going into the algebra, we solve two examples to see what is going on. They are the English auction equivalent of Examples 9.1 and 9.4. We will concentrate only on the equilibrium pairs where both players use a pure strategy—the pure equilibrium pairs.

Example 9.5—$v_1 = 4$, $v_2 = 1$. $v_1 > v_2$.

$$\begin{matrix} & II_0 & II_1 & II_2 & II_3 & II_4 \\ I_0 & (2,\tfrac{1}{2}) & (0,1) & (0,1) & (0,1) & (0,1)^* \\ I_1 & (4,0)^* & (1\tfrac{1}{2},0) & (0,0) & (0,0) & (0,0)^* \\ I_2 & (4,0)^* & (3,0)^* & (1,-\tfrac{1}{2}) & (0,-1) & (0,-1) \\ I_3 & (4,0)^* & (3,0)^* & (2,0)^* & (\tfrac{1}{2},-1) & (0,-2) \\ I_4 & (4,0)^* & (3,0)^* & (2,0)^* & (1,0)^* & (0,-1\tfrac{1}{2}) \end{matrix} \quad (9.20)$$

Just doing the easy check for pure equilibrium pairs—first component must be the maximum in the column, second component is the maximum in the row—gives the starred pairs in (9.20) as equilibrium pairs. These are of two kinds. If we are looking for pure equilibrium pairs of the form (I_{b_1}, II_{b_2}), then b_1 and b_2 must satisfy either

$$b_1 \geqslant \max\{v_2, b_2 + 1\}, \qquad b_2 \leqslant v_1, \quad (9.21)$$

or

$$b_2 \geqslant \max\{v_1, b_1 + 1\}, \qquad b_1 \leqslant v_2. \quad (9.22)$$

(9.21) gives the equilibrium pairs in the bottom left-hand corner of (9.20) and (9.22) those in the top right-hand corner. For any game with $v_1 > v_2$, the following analysis shows these conditions always give rise to pure equilibrium pairs.

If $b_1 > v_2$, then (9.19c) requires that $b_1 \geqslant b_2 + 1$, and if $b_2 < v_1$, then

$b_1 \geqslant b_2 + 1$ also ensures that (9.18a) holds. If $b_2 = v_1$, (9.18b) holds trivially. b_1 could equal v_2 in (9.21) and in this case (9.19b) holds and gives no requirement on the relation between b_1 and b_2. However, since now $b_1 = v_2 < v_1$, and $b_2 + 1 \leqslant v_2$, (9.18a) holds. So (9.21) gives rise to pure equilibrium pairs. (9.22) is similarly proved. It is messier but no harder to prove these are the only equilibrium pairs.

Now turn to the case $v_1 = v_2$ and the English auction equivalent of Example 9.4.

Example 9.6—$v_1 = v_2 = 2$.

$$\begin{array}{c} \\ \mathrm{I}_0 \\ \mathrm{I}_1 \\ \mathrm{I}_2 \\ \mathrm{I}_3 \\ \mathrm{I}_4 \end{array} \begin{array}{c} \begin{array}{ccccc} \mathrm{II}_0 & \mathrm{II}_1 & \mathrm{II}_2 & \mathrm{II}_3 & \mathrm{II}_4 \end{array} \\ \left(\begin{array}{ccccc} (1,1) & (0,2) & (0,2)^* & (0,2)^* & (0,2)^* \\ (2,0) & (\tfrac{1}{2},\tfrac{1}{2}) & (0,1)^* & (0,1)^* & (0,1)^* \\ (2,0)^* & (1,0)^* & (0,0)^* & (0,0)^* & (0,0)^* \\ (2,0)^* & (1,0)^* & (0,0)^* & (-\tfrac{1}{2},-\tfrac{1}{2}) & (0,-1) \\ (2,0)^* & (1,0)^* & (0,0)^* & (-1,0) & (-1,-1) \end{array}\right) \end{array}. \tag{9.23}$$

The starred pairs are again pure equilibrium ones, and again there are two sets of conditions (overlapping ones this time) each of which gives the pure equilibrium pairs. These are:

$$b_1 \leqslant v \leqslant b_2 \tag{9.24}$$

and

$$b_2 \leqslant v \leqslant b_1. \tag{9.25}$$

It is easy this time to check that these conditions satisfy (9.18) and (9.19) and there are no others (see Problem 9.5). Exactly the same results will occur in the continuous bid case, namely when $v_1 > v_2$, the pure equilibrium pairs satisfy (9.21) or (9.22), and when $v_1 = v_2$, they satisfy (9.24) or (9.25).

9.5 AUCTIONS WITH UNKNOWN VALUATIONS

Although we have spent a long time on auctions where bidders know the other bidder's valuation, such a model is not often realistic. What happens more often in practice is that you know your own valuation of the object, but only have an imprecise idea of the other bidders' valuations. We model this imprecision by saying a bidder knows the probability distribution of the other bidder's valuation.

Consider the case where there are two bidders, I and II, and each player believes the other's valuation is uniformly distributed between 0 and 1. So I knows his valuation v_1, $0 \leqslant v_1 \leqslant 1$, and II knows his valuation v_2, $0 \leqslant v_2 \leqslant 1$, but not each other's. A pure strategy B_1, for I, in this game is a function which maps the value v_1 to the bid he makes $B_1(v_1)$. For II, it is $B_2(v_2)$. These are the pure strategies, and we ask is there a pure strategy $B(v)$ which if both players play it gives an equilibrium pair? We will assume B is continuous and increasing as a function of v.

If $B(v)$ is the function that gives the bid you make when the item's value is v, we define the inverse function:

$$V(b) = \max_{v} \{v \mid B(v) = b\}. \tag{9.26}$$

and assume it is well defined. If I bids b, the chance the bid is successful if II is playing B, is the chance that II has valued the item at v_2, where $B(v_2) < b$. Using (9.26) this is the same as choosing a v_2 less than $V(b)$. Since all I knows is that v_2 is equally likely to be between 0 and 1, the chance it is less than $V(b)$ is $V(b)$.

So if I bids b when he values the item at v, his payoff is:

$$(v - b)V(b) + 0(1 - V(b)). \tag{9.27}$$

For I's strategy to be part of an equilibrium pair when II plays B, this choice of b must maximise his payoff. Hence the derivative of (9.27) with respect to b must be zero. That is

$$-V(b) + (v - b)V'(b) = 0. \tag{9.28}$$

For B to be the best response by I to II playing B, i.e. $e_1(B,B) \geqslant e_1(\tilde{B},B)$ for any $\tilde{B}$, then the b,v in (9.28) must be related by $b = B(v)$ or $v = V(b)$. Substituting this into (9.28), we have:

$$(V(b) - b)V'(b) = V(b). \tag{9.29}$$

Put $y(b) = V(b) - b$, then (9.29) becomes:

$$y(b)y'(b) = b. \tag{9.30}$$

Integrating both sides of (9.20) with respect to b leads to $\frac{1}{2}y(b)^2 = \frac{1}{2}b^2 + C$ or $y(b) = \sqrt{b^2 + 2C}$. Hence, $V(b) = b + \sqrt{b^2 + 2C}$. It is obvious that $B(0) = 0$, since you cannot bid less than zero, and you will not bid more for something worth zero. This means $V(0) = 0$ and substituting this into $V(b) = b + \sqrt{b^2 + 2C}$ gives $C = 0$ and $V(b) = 2b$. Thus, there is a symmetric equilibrium pair if both bidders use

$$B(v) = \tfrac{1}{2}v. \tag{9.31}$$

Using (9.31) in (9.27) shows if a bidder values the item at v, he will bid $\frac{1}{2}v$ and get it with probability v. So expected profit is $\frac{1}{2}v^2$. Since v varies uniformly between 0 and 1, the expected profit before I values the item is $\int_0^1 \frac{1}{2}v^2 dv = \frac{1}{6}$. This is also II's expected profit.

We can extend this analysis easily to the case where there are n bidders. Again we want to find an equilibrium n-tuple where all the bidders use the same strategy $B(v)$. We repeat the analysis leading to (9.27), but this time if I bids b when he values the item at v his payoff is:

$$(v - b)V(b)^{n-1}, \tag{9.32}$$

since all $n - 1$ other bidders must have chosen values of v less than $V(b)$.

Differentiating (9.32) gives:

$$-V(b)^{n-1} + (n-1)(v-b)V(b)^{n-2}V'(b) = 0, \tag{9.33}$$

and if we want $e_1(B,B,B,\ldots,B) \geqslant e_1(\tilde{B},B,B,\ldots,B)$ for any other strategy $\tilde{B}$, then the v and b in (9.33) are connected by $v = V(b)$. So

$$(n-1)(V(b) - b)V(b)' = V(b). \tag{9.34}$$

It is easy to show that the only solution to (9.34) that satisfies $V(0) = 0$ is $V(b) = nb/n - 1$ or $B(v) = (n-1)v/n$. Thus, bidder I makes a profit of $v - b = v - (n-1)v/n = v/n$ with probability v^{n-1} and 0 with probability $1 - v^{n-1}$. So expected profit is v^n/n, and as v varies from 0 to 1 the average profit I expects before he knows the value of the object is $\int_0^1 v^n dv/n = 1/n(n+1)$. v^n is the chance that the highest valuation of the n players is less than v, since this is equivalent to all n valuing it less than v. So the distribution function for the highest value is $F(v) = v^n$ and hence its density function is $f(v) = nv^{n-1}$. Thus, with n people, the expectation of their highest valuation is $\int_0^1 vnv^{n-1}dv = n/n + 1$. The average highest bid is $n - 1/n$ of this, i.e. $n - 1/n + 1$. Thus, obviously the more bidders, the higher the price the object will be sold at.

Let us apply the rules of the English auction to the same problem, and find if there are any symmetric pure strategy equilibrium pairs. With two bidders, suppose II chooses the bid strategy $B(v)$ which, as in (9.26), has inverse function $V(b)$. Again we can reinterpret $V(b)$ as the distribution function of II's bid—the probability he will bid less than b—and so $V'(b)$ is the density function. So, if I bids b_1 for the item which he values at v_1, under the English auction rules, his profit is:

$$\int_0^{b_1} (v_1 - b_2)V'(b_2)db_2. \tag{9.35}$$

Since $V'(b_2)$ is always non-negative, this integral will be maximised when $b_1 = v_1$, as this guarantees $v_1 - b_2$ is positive over the range 0 to b_1. So for an English auction the strategy $B(v) = v$ is the best response to any of your opponent's strategies and thus (B,B) is a pure symmetric equilibrium pair.

Exactly the same analysis will apply in the n-bidder case, the only difference being that we will need $V(b)^{n-1}$, the probability that all $n - 1$ other bidders bid less than b. This has density function $(n - 1)V(b)^{n-2}V'(b)$, and so the expected profit to player I is:

$$\int_0^{b_1} (v_1 - b_2)(n-1)V(b_2)^{n-2}V'(b_2)db_2. \tag{9.36}$$

Again $B(v) = v$ is optimal, and on substituting this into (9.36) we find the expected profit when $b_1 = v_1$ is:

$$\int_0^{v_1} (v_1 - b_2)(n-1)b_2^{n-2}db_2 = v_1^n/n. \tag{9.37}$$

This is exactly the same result as the Dutch auction and so the average profit before a player knows his valuation is $1/n(n + 1)$. To find the expected selling price, we need the distribution function of the second highest of n bids. We call this B_2 and note:

$$\text{Prob}(B_2 \leqslant b) = b^n + n(1-b)b^{n-1}, \tag{9.38}$$

since the chance it is less than b is the chance all n are less than b, plus the chance that any one of the n bids is greater than b and the rest are less than b. The corresponding density function on differentiating (9.38) is $f(b) = nb^{n-1} + n(n - 1)(1 - b)b^{n-2}$, and to find the average selling price we integrate $\int_0^1 bf(b)db$. The answer is $n - 1/n + 1$, which is also the corresponding result in the Dutch auction case. However, although the mean values of the characteristics of these auctions agree, there is considerable difference in their variation (Problem 9.7).

9.6 AUCTIONS OF TWO OR MORE OBJECTS

So far we have always assumed there is only one item for sale at the auction, but often in sale-room auctions a bidder's strategy is affected by the items yet to be auctioned. Similarly, in the sealed bid auctions for oil exploration rights, the price an oil company will pay for a particular area could well be affected by which other areas are, or will be, available at that round or subsequent rounds of exploration leasing. It is no good spending all the money set aside for exploration on one round of leasing, if potentially even more favourable sites are likely to be auctioned off subsequently.

To investigate this effect, consider the simple model where there are only two bidders, who only have c_1 and c_2 available to spend on buying items. Suppose there are two items available at the auction and both value the first at a_1, and the second at a_2. We will assume $c_1 \geqslant c_2$, $a_i \geqslant c_1$, $i = 1,2$, bids can go up continuously, and the English auction rules apply. Once the first object has been auctioned, then the second object will go to whoever has the more money left at a cost equal to the other player's remaining amount plus an infinitesimal amount ε. (If they have the same amount left, we toss a coin again.) Thus, the only decision for each player is how much to bid for the first object.

Suppose, during the auction for the first object, player II has just bid b, which is fractionally more than I's last bid. Should I continue bidding? We will ignore the trivial case when $c_1 \geqslant 2c_2$, so I can get both objects. Then if I decides to bid fractionally more than II and this is the winning bid, he wins the first object with profit $(a_1 - b)$. It seems that if $c_1 - b > c_2$, he could also win the second object, but II's equilibrium strategy must be to force him up sufficiently to more than $c_1 - c_2$ so that II can get at least one of the objects. If I does not bid, II will get this object at b, have $c_2 - b$ left, and so I will make a profit of $a_2 - (c_2 - b)$ from getting the second object. Thus, I will bid b provided $b \leqslant c_1$ and $(a_1 - b) > a_2 - (c_2 - b)$, i.e.

$$b \leqslant \min(c_1, \tfrac{1}{2}(a_1 - a_1 + c_2)). \tag{9.39}$$

A similar analysis for II leads to the criterion that II will continue bidding until:

$$b \leqslant \min(c_2, \tfrac{1}{2}(a_1 - a_2 + c_1)). \tag{9.40}$$

Thus, (9.39) and (9.40) give the limits of the bidding for the two bidders, and an equilibrium pair will occur when one of the players reaches this limit. The first object will go to the other player at essentially this price. Let us look at two numerical examples.

Example 9.7—$a_1 = 60$, $a_2 = 50$, $c_1 = 40$, $c_2 = 30$. I will bid b_1 provided $b_1 \leqslant \min(40, \tfrac{1}{2}(60 - 50 + 30)) = 20$. II will bid b_2 provided $b_2 \leqslant \min(30, \tfrac{1}{2}(60 - 50 + 40) = 25$. So II gets the first object for 20 and I will get the second one for $30 - 20 = 10$, and the payoff is 40 to both.

Reversing the order in which the objects are sold, i.e. $a_1 = 50$, $a_2 = 60$, gives a different result. I bids up to $\min(40, \tfrac{1}{2}(50 - 60 + 30) = 10$. II bids up to $\min(30, \tfrac{1}{2}(50 - 60 + 40)) = 25$. So II gets the first object for 10 and payoffs are 40 to I and 40 to II.

Example 9.8—$a_1 = 100$, $a_2 = 60$, $c_1 = 60$, $c_2 = 40$. Again I bids up to $\min(60, \tfrac{1}{2}(110 - 60 + 40)) = 45$ and II bids up to $\min(40, \tfrac{1}{2}(110 - 60 + 60)) = 40$. So I gets the first object for 40 and II gets the second for $60 - 40 = 20$. The payoffs are 70 to I and 40 to II.

This time if we reverse the ordering we find that I would only want to bid $\min(60, \tfrac{1}{2}(60 - 100 + 40)) = -5$. Obviously, he can't bid -5, but this means he does not bid at all and II who would bid $\min(40, \tfrac{1}{2}(60 - 110 + 60) = 5$ gets the first object for essentially nothing. I then gets the second for 40, and the profits are 60 and 60, respectively.

Thus the order in which the objects are presented is very important for both the buyers and the sellers. In the next section we look at another auctioning situation, where this is even more so.

9.7 HORSE MARKET AUCTIONS

In the nineteenth century, and even today in certain areas, horse fairs were annual events where horses were bought and sold. In fact, there were also fairs where one bought and sold all types of farm animals, and the hiring fairs where one acquired servants were run on the same lines. Consider a fair where there are M buyers each of whom wants one horse and there are N sellers each with one horse to sell. Let b_i be the value the ith buyer puts on owning a horse (it does not matter which horse) and c_j be what the jth seller thinks his horse is worth. The ith buyer's payoff is $b_i - p_i$ if p_i is the price he paid for the horse, and 0 if he did not buy one. The jth seller's payoff is $p_j - c_j$ is he sells his horse for p_j and 0 if he does not sell it. The problem is to find what trades take place. This was first looked at as a market game by Shapley and Shubik (1972) and subsequently in terms of an auction by Schotter (1974). We will follow both papers and first describe the market game results, and then concentrate on what happens under auctioning rules.

In order to 'solve' the horse auction as an n-person market game, we need to find its characteristic function v. The vital point here is that if you have a choice of deals to make, the set that gives the greatest profit is when the buyer with the highest value buys the horse from the seller with the lowest value; the second highest valued buyer buys the horse of the second lowest valued seller, and so on. To see this, look at the case with two buyers with values $b_1 > b_2$ and two sellers with values $c_1 < c_2$. If $c_2 \leqslant b_2$, then no matter which pair of trades occur, the overall profit is $b_1 + b_2 - c_1 - c_2$. However, if $b_1 > c_2 > b_2$, then the way suggested will lead to only one horse being sold at an overall profit of $(b_1 - c_1)$, whereas if the highest-priced buyer buys the higher-priced seller's horse and the lower-priced buyer buys the lower-priced seller's horse, both horses will be sold but the overall profit will be $(b_1 - c_1) + (b_2 - c_2)$, which is less than $b_1 - c_1$. This result still holds no matter how many buyers and sellers are involved.

For any coalition S of buyers and sellers, we let:

$$\alpha(S) = (b_1^S, b_2^S, \ldots, b_m^S; c_1^S, c_2^S, \ldots, c_n^S), \tag{9.41}$$

where

$$b_i^S \geqslant b_{i+1}^S, \qquad i = 1,\ldots,m;$$

$$c_i^S \leqslant c_{i+1}^S, \qquad i = 1,\ldots,n,$$

$$m + n = |S|,$$

be the values b_i^S the buyers in S put on a horse, written in decreasing order, and the values, c_i^S, the sellers in S put on their horse written in increasing order. From the argument in the previous paragraph, the highest overall profit S can get is when the b_1^S buyers buys from the c_1^S seller; the b_2^S buyers buys from the c_2^S seller, and so on, provided of course $b_i^S \geqslant c_i^S$. Eventually we run out of buyers or sellers or we get to an i so that $b_i^S < c_i^S$ and there will be no exchange. Thus, the characteristic function value of S is:

$$v(S) = \sum_{i:\{i \leqslant \min\{m,n\}, b_i^S \geqslant c_i^S\}} (b_i^S - c_i^S). \tag{9.42}$$

Having calculated $v(S)$ it is relatively easy to find the core for this $M + N$ game. From now on we assume the b_i's are ordered in decreasing order and the c_j's in increasing order.

Lemma 9.1. If the players are ordered so that $b_i \geqslant b_{i+1}$ and $c_j \leqslant c_{j+1}$, then the core is the set of imputations:

$$\{x|x = (b_1 - p, b_2 - p, \ldots, b_k - p, 0, 0, \ldots, 0, p - c_1, p - c_2, \ldots, p - c_k, 0, 0, \ldots, 0),$$

where $k = \max[k\colon b_k \geqslant c_k]$ and p satisfies $b_k \geqslant p \geqslant c_k$ and

$$c_{k+1} \geqslant p \geqslant b_{k+1}\}. \tag{9.43}$$

This looks a little complicated, but it is not. It says that trades take place at a fixed price p between the highest-valued buyers and the lowest-valued sellers until the next highest-valued buyer has a lower value than the next lowest-valued seller. The price will be somewhere between these two values.

Proof. If you recall Theorem 4.2 in Chapter 4, an imputation x is in the core if and only if $\sum_{i \in S} x_i \geqslant v(S)$, for all coalitions S. We first show all the imputations in (9.43) satisfy this condition and then that no others can do so. Take any coalition S of m buyers and n sellers ordered according to $\alpha(S)$ in (9.41). The buyers are actually $i_1, i_2, \ldots, i_m$ and the sellers $j_1, \ldots, j_n$, where for the k defined in (9.43) we have $i_r \leqslant k \leqslant i_{r+1}$ and $j_s \leqslant k \leqslant j_{s+1}$. Then (9.42) gives:

$$v(S) = \sum_{t=1}^{\min(r,s)} (b_{i_t} - c_{j_t}). \tag{9.44}$$

For any such S the imputations in (9.43) satisfy:

$$\begin{aligned} \sum x_i &= \sum_{t=1}^{r} (b_{i_t} - p) + \sum_{t=1}^{s} (p - c_{j_t}) \\ &= v(S) + \sum_{t=\min(r,s)}^{r} (b_{i_t} - p) + \sum_{t=\min(r,s)}^{s} (p - c_{j_t}) \\ &\geqslant v(S), \end{aligned} \tag{9.45}$$

since $b_{i_t} \geqslant p$, for $i_t \leqslant k$, and $p \geqslant c_{j_t}$, for $j_k \leqslant k$. Hence, all the imputations in (9.43) are in the core.

To prove that any imputation in the core must satisfy (9.43), look at a coalition of buyer 1 and seller i, $i \leqslant k$. For such a coalition S, $v(S) = b_1 - c_i$, and if $M + N$ is the coalition of everyone, $v(M + N - S) = v(M + N) - v(S)$. This is almost like saying the game is constant sum, but it only holds for these special coalitions S. If x is to be an imputation in the core, these conditions require:

$$\sum_{i \in S} x_i = x_1 + x_{M+i} \geqslant v(S) = b_1 - c_1 \tag{9.46}$$

and

$$\sum_{i \in M+N-S} x_i \geqslant v(M + N) - v(S). \tag{9.47}$$

Since $v(M + N) = \sum_{i \in M+N} x_i$ for any imputation, (9.47) says $v(S) \geqslant x_1 + x_{M+i}$, and so with (9.46) gives $x_1 + x_{M+i} = b_1 - c_i$. This holds for seller j as well, so we get:

$$x_{M+i} + c_i = x_{M+j} + c_j = b_1 - x_1 = p, \qquad i,j \leqslant k, \tag{9.48}$$

and so $x_1 = b_1 - p$, $x_{M+j} = p - c_j$, $1 \leqslant j \leqslant k$. A similar analysis for a coalition of buyer i and a seller j, $2 \leqslant i \leqslant k$, gives $x_i = b_i - p$. Hence,

$$\sum_{i=1}^{k} x_i + \sum_{j=1}^{k} x_{M+j} = \sum_{i=1}^{k} b_i - \sum_{i=1}^{k} c_i = v(M + N). \tag{9.49}$$

For x to be an imputation, all the remaining x_i must be 0. In order that

$$x_1 + x_{M+k+1} = b_1 - p \geqslant v(\{1,M + k + 1\}) = b_1 - c_{k+1},$$

we require $c_{k+1} \geqslant p$, and similarly for

$$x_{k+1} + x_{M+1} = p - c_1 \geqslant v(\{k + 1,M + 1\}) = b_{k+1} - c_1,$$

we want $p \geqslant b_{k+1}$. We have now proved that (9.43) gives all possible imputations in the core.

Let us relax after all this algebra to see what this means in a numerical example.

Example 9.9. Consider a horse market with four sellers and six buyers where $b_1 = 300$, $b_2 = 270$, $b_3 = 230$, $b_4 = 200$, $b_5 = 180$, $b_6 = 160$, $s_1 = 140$, $s_2 = 200$, $s_3 = 240$, and $s_4 = 260$. The characteristic function v for this game can be obtained by substituting in (9.42), and leads to the set of imputations:

$$E(v) = \{\mathbf{x} | \mathbf{x} = (x_1, x_2, x_3, \ldots, x_{10}), \text{ where } x_i \geqslant 0,\ 1 \leqslant i \leqslant 10,$$

and

$$\sum_{i=1}^{10} x_i = (300 - 140) + (270 - 200) = 230\}. \tag{9.50}$$

If buyer 1 buys seller 1's horse and buyer 2 buys seller 2's horse, buyer 3, who values a horse at 230, will not buy from seller 3 who values it at 240. Thus, the core is:

$$\{\mathbf{x} = (300 - p,270 - p,0,0,0,0,p - 140,p - 200,0,0),$$
$$230 \leqslant p \leqslant 240\}.$$

We have departed a long time from bidding, but now we return to ask: What would happen if the horses were auctioned using the English auction rule? Well, as Professor Joad said: 'It all depends.' In this case it depends on the order in which the horses are auctioned. We will use Example 9.9 to illustrate this.

If they are auctioned with lowest-valued horse first and highest-valued horse last, i.e. s_1, s_2, s_3, s_4, only two horses will be sold. We assume that the seller sets a reserve price equal to his valuation and the auctioneer will bid for the seller up to this price, if there is still a bidder in the market. Thus, the first horse is sold to buyer 1 at $270 + \varepsilon$, where 270 is the second highest buyer's valuation. (We will ignore the ε's from now on.) The second horse goes to bidder 2 for 230. However, none of the remaining bidders will pay 240 for the third horse, which is its seller's reserve price. Thus, the payoffs to the players are:

$$x = (30,40,0,0,0,0,130,30,0,0). \tag{9.51}$$

This is an imputation, i.e. Pareto optimal, but the sellers do far better than the imputations that are in the core.

However, if the horses are auctioned in decreasing order of seller's valuation, i.e. s_4, s_3, s_2, s_1, we have a completely different result. The first horse again goes to buyer 1 for 270. The second horse is bought by buyer 2 at 240, after the auctioneer has bid up to this—the seller's reserve price. The third horse is bought by buyer 3 for 200 and the fourth horse by buyer 4 at 180. This time the payoffs are given by:

$$x = (30,30,30,20,0,0,40,0,0,10). \tag{9.52}$$

This is not an imputation since the total payoff is 160 which is less than the Pareto optimal value of 230. Also the sellers do far worse under this system than any other.

One of the most interesting results about this game is that the standard market game analysis leads to a fixed price for all the transactions that occur, whereas in the auction game all the transactions occur at different prices. Finally, we are left with the auctioneer's dilemma. Which order should he sell the horses? Well, 'it all depends', whether he is paid a percentage of the seller's profit or a fixed amount for each horse sold.

9.8 FURTHER READING

Auctions and bidding, especially sealed bidding problems, is a very active area of research in Operations Research. In 1979, Stark and Rothkopf (1979) published a bibliography with over 500 works mentioned, the majority of which had appeared since 1970. Most of the work on bidding, however, thinks of it as a one decision-maker problem, and the other bidders only come in through some sort of probability distribution of the chance of a bid at a particular price being the highest. Even in the simplest case, this is a difficult thing to model, as the controversy between Gates and Friedman shows (see Rosenshine, 1972, for a summary of the controversy).

However, a substantial minority of the papers on the subject do take the game theoretical point of view, and the survey paper by Engelbrecht-Wiggans (1980) concentrates on these. Although the obvious results about bidding up to your value in English auctions and as little as possible above your opponent's valuation in sealed bid and Dutch auctions have been part of the folklore for a long time, the game theory approach to bidding started with a series of papers by Griesmer and Shubik (1963a, 1963b, 1963c), and a paper by Vickery (1961a). The Griesmer and Shubik papers look at a number of different models of auctions and try to solve them as games. These include situations which are akin to the discrete English and Dutch auctions with known valuations (Sections 9.2 and 9.4). Vickery, on the other hand, analysed two cases of the problem, where a bidder does not know his opponent's valuation—one where bidder I has a known value and bidder II has a uniform valuation on [0,1], and the second where I's valuation is uniform on [a,b] and II's is uniform on [0,1]. Several authors then extended these results by looking at other types of probability distributions. Griesmer, Shubik and Levitan (1967) did it for any two uniform distributions, Wilson (1967) where one bidder has a known value and the other's distribution is arbitrary, Rothkopf (1969) for essentially the Weibull distribution, and Beckman (1974) found equations that give the symmetric equilibrium pairs in the general case. Wilson (1969) also looked at the case where the bidders acquire different information about the object and update using Bayes's rule, while Smith and Case (1975) looked at the problem of finding equilibrium pairs when the bidders don't know their true valuations of the object. Most, if not all, of these papers have used the sealed bid version of the auction, but some other types of auction rules have also been studied. One of these is where all bidders have to pay the price they bid whether they get the object or not. Shubik (1971) discusses this in the context of the Dollar Auction (Problem 9.9), but it is essentially the same game that is involved in the 'war of attrition' model of evolutionary stable strategies (recall Chapter 8, Section 8.6).

There is hardly any work on the problem of a sequence of objects to be auctioned, apart from Griesmer and Shubik (1963c), even though this problem appeared in Sasieni, Yaspan and Friedman's standard text on Operations Research (1959). Vickery (1961b) looked at the problem where M identical objects are bid for at the same time, and an object is awarded to each of the M highest bidders.

The horse trading auction with several buyers and sellers, although first looked at by Böhm-Bawerk (1888), seems only to have been looked at in game theory terms in the two papers cited in the text—Shapley and Shubik (1972) and Schotter (1974).

PROBLEMS FOR CHAPTER 9

9.1. Consider the Dutch auction with two bidders, where one bidder values the item at v_1 and the other at v_2, as a two-person non-zero-sum game. If only integer bids are allowed and if both players bid the same amount, each is equally likely to get the item. The Pareto optimal boundary is given by Fig. 9.1. Find the maximin bargaining solution.

9.2. Find all the equilibrium pairs in the two-person Dutch auction non-zero-sum game with integer bids and an equal chance of winning the auction if bids are equal (a) when player I values the item at $v_1 = 4$ and player II values it at $v_2 = 2$, and (b) when $v_1 = 4$, $v_2 = 3$.

9.3. In the two-person Dutch auction with integer bids, where both players value the item at v, $v > 2$, and equal bids have the same chance of winning the auction, find all the symmetric equilibrium pairs. A symmetric pair is one where both players have the same strategy.

9.4. Consider the two-person English auction as a cooperative game. One bidder values the item at v_1, the other values it at v_2. Only integer bids are accepted and if the two bidders try to bid the same amount the auctioner is equally likely to take either bid. Find the maximin bargaining solution for the game.

9.5. Consider the two-person English auction with integer bids where both players value the object at v. Show that $(\mathrm{I}_{b_1}, \mathrm{II}_{b_2})$ is a pure equilibrium pair if and only if either $b_1 \geqslant v \geqslant b_2$ or $b_2 \geqslant v \geqslant b_1$.

9.6. Consider an auction involving three bidders as a cooperative three-person game. Since they are cooperating it does not matter whether it is an English or Dutch auction. Player I values the item at 10, player 2 at 8, and player 3 at 4. Find the set of imputations, the core and the nucleolus of this game if no one will bid higher than their valuation of the object.

9.7. In the auction with n bidder who do not know their opponents valuation of the object, only that it is uniformly distributed between 0 and 1, the expected profit to a bidder is $1/n(n + 1)$ before he knows the value of the object in both the English and Dutch auctions (see Sections 9.5). Also, the expected highest bid is $n - 1/n + 1$ in both types of auctions. Find the variance in the highest bid and the expected variance in the profit to a bidder before he knows the value of the object to him and show they are different for the two types of acutions.

9.8. Suppose three items are to be auctioned in sequence. The two bidders both value the first item at 70, the second at 100 and the third item at 80. Bidder 1 has 60 to spend and bidder 2 has 30. Find out how much each should bid up to for each item.

9.9. (Dollar Auction.) The auctioneer auctions off a dollar to the highest bidder with the understanding that both the highest bidder and the second highest bidder will pay what they bid. Bids must be made in multiples of five cents, say. Explain what is likely to happen in a non-cooperative game with two bidders. Are there any equilibrium pairs? (Think of the game in extensive as well as normal form.) Now consider it as a three-person game with the auctioneer involved, as he takes any profits or losses that occur. What is the characteristic function and the Shapley value?

CHAPTER 10

Gaming

'Gaming is the mother of lies and perjuries'—John of Salisbury

---

10.1 WHAT IS GAMING?

Gaming is one of those words which has several different but related meanings. In the quotation above, John of Salisbury was castigating gambling in the Middle Ages, which, however interesting, is not the content of this chapter. Even though the standard dictionary definition of gaming is also gambling, it has become used to describe any example of playing a game. Shubik (1975a) defines gaming as 'an exercise that employs human beings or robots acting as themselves or playing simulated roles in an environment, which is either actual or simulated and which contains elements of potential conflict or cooperation'. This definition also casts its net very wide and would seem to include many exercises that would normally be called simulation. For example, some of the simulations of queues, where customers have choices of which queue to join, could fall under this definition, but are usually considered as simulations.

We will content ourselves by saying that **gaming** is the playing of a situation involving two or more decision-makers which can be modelled as a game. Even among researchers in the area, there are wide differences in what they consider as gaming. A good example of this is given by the dialogue between Duke (1981) and Bowen (1981), resulting from the former's review of the latter's book. Duke describes games where there is considerable interaction and communication between the players, and the 'controller' of the game intervenes as little as possible, whereas Bowen looks at highly structured games where all communication between the players goes through the game controller and so one has a record of all moves explicit or implicit made by the players. The reason for these differences in gaming situations is that they have different objectives when setting up the gaming experiment in the first place. In the next section we look at some of the reasons one gets people to play these games.

10.2 WHY DO GAMING?

As has been suggested above, the reason there are so many different types of experiments classified as gaming is that there are often completely different objectives behind the setting up of these games. Let us look at some of these objectives, though of course they need not be exclusive, and a gaming situation is often set up in the hope of satisfying several objectives. We classify these objectives into four groups: **Teaching, Operational, Research,** and **Entertainment** (a similar classification is given by Shubik, 1972a). A mnemonic is useful for learning the names of these groups by **ROTE**. We will examine the four categories in more detail.

10.2.1 Teaching Games

This is the main objective behind the majority of exercises we think of as gaming. In universities it is used as a way of training students in the basic principles of business, as well as teaching them the 'skills' of management. It is an attractive alternative to case-study analysis with the advantage that mistakes and bad decisions are more readily remembered and corrected if they are your own, rather than those of some now-defunct firm. Some of these business games involve a lot of students and have been built up for many years, and it has become a hallmark of distinction to have played in them, e.g. the Carnegie Tech game and the Harvard Business School game. In fact, playing a business game is part of the curriculum in over half the American business schools.

If the students also participate in the construction of these games, it gives them an excellent opportunity to learn how to model complex decision problems, and those who do any analysis of the results learn some of the ideas underlying game theory far faster than by formal teaching. Two criticisms are sometimes made about such games. The first is that the participants become too involved in the game-playing and their objectives are not those of real players in the real situation that is being modelled. The players are more interested in winning, i.e. having the largest profit, whereas in reality firms are more interested in surviving. The second criticism is that one must be careful that the principles the game is using are true. It is all too easy to simplify building a game, by using a relationship between two variables which in reality are not connected. Then, when it is played, the players will go away with mistaken beliefs of how the real world works.

These teaching games are also being used by firms in much the same way to teach prospective managers the possible problems they are likely to face, and the way certain decisions cause matters to develop. For example, the National Coal Board in Britain has a game where future colliery managers have to decide where and when to develop new coal faces. The other players in the game are the other specialists used by that company, so in the N.C.B. game there will be geologists, engineers, accountants, as well as players representing the 'workers'. One of the objects of these games is for the players to realise the problems and responsibilities of other types of experts by getting inside their skins.

Gaming as an educational tool is starting earlier and earlier in the schools. It is used as a way of reinforcing recently learned facts. However, my definition of gaming would rule out many of the computer learning programs which led the student through one of a number of possible programs, depending on his answers. I would argue there is only one decision-maker here (the student) and so it is not a game.

10.2.2 Operational Gaming

The objectives in these games is to model complex real decision problems and find out what outcomes are likely. In teaching games, the participants are 'students' of some sort and the game-builders have analysed the model they built fairly well. In operational games, the players are often the actual potential decision-makers themselves and the game-builders try to build a game as near to reality as is feasible, with little knowledge of the analysis of the game. Such games are used mainly in military or diplomatic problems, and though there are opportunities for judging the decision-making style of a player, the main stated object in these games is to explore the consequences of a new diplomatic initiative, a new military weapon, or a new means of its deployment. Again one of the side advantages of doing such an exercise is the camaraderie it builds between the participants, but since they are often ordered to participate in such games, there might be a lack of enthusiasm.

Operational gaming is even more open to in-built biases than teaching games. It would be so easy to weight the rules of the game in favour of or against the concept being tested, that a great deal of care and honesty is needed by those building the game, and a great deal of scepticism is needed by those analysing it.

10.2.3 Research Gaming

These are games designed to test whether people's decisions in conflict or cooperative situations actually agree with the analysis of those situations made by game theorists or psychologists. As we mentioned previously, there has been a tremendous number of experiments done with the Prisoners' Dilemma—and many other 2×2 games—to see what people actually do in such situations. The idea is to make the game as near a repeatable scientific experiment as possible. Hence, everything is recorded, communication between players is kept to a minimum, and as few outside stimuli as possible are allowed. The game-builders want the players to concentrate as much as possible on the decision they have set before them. The difference between these highly restrained games and the operational games, where one is allowing if not encouraging new ideas and communication, exemplifies the wide range of gaming discussed by Duke and Brown.

One even more extreme development along this road is when the player does not actually play a game, but only thinks he does. Cooper (1979) describes a method called 'Superior Commander' where a player believes he is advising a game-player in playing a game, but in fact all the moves reported to him are predetermined to test his reactions to particular situations. This is again bordering on the edges of what is considered gaming.

One of the great disadvantages with the work on research games is that in the vast majority of cases the players are university students. Given that most game theory research is done at universities and students are an intelligent, cheap, and usually enthusiastic body of potential participants, it is only natural that this should happen. However, when their decisions are compared with other more experienced groups of decision-makers the results are often diverse (see Cooper, Klein, McDowell and Johnson, 1980, and references therein).

Most research games are set up to test a specific solution concept or hypothesis, but occasionally such gaming leads to new ideas of what should be analysed in such games.

10.2.4 Entertainment Gaming

This is the oldest form of gaming, as our discussion on the origins of the word 'gaming' indicated. It also includes some of the most popular pastimes of the human race. Most people play some sort of game for relaxation or enjoyment. Obviously in the recreational games like soccer, cricket, rugby, squash, basketball, and tennis, it is the athletic skill of the players that is more important than any strategic consideration. There are athletic games though, like American football, in which choice of strategy is very important.

It is more the board games and the gambling games which are what we think of as entertainment gaming. Poker (recall the simplified version of Example 1.2 in Chapter 1) is easily modelled in game theoretic terms and is a shining example of a zero-sum game. Other betting situations which we call games, such as Roulette and, to a certain extent, horse-betting, are not examples of gaming, because they involve only one decision-maker and the outcome depends purely on chance rather than skill. This is true for Roulette if the wheel is completely fair. However, most Roulette wheels have a small bias and it is an interesting problem to devise strategies that take advantage of this (see Downton, 1983).

Some of the board games have a timeless quality about them. The origins of Chess, Draughts, Go, and Mah-Jong are lost in ancient history. All these games are zero-sum perfect information games and so have an optimal pure strategy. The fascination is that their complications make them unsolvable. They each have a mixture of the strategic (long-run) and the tactical (immediate) interwoven in a fascinating way. Other board games seem much more of an age or a generation. It is interesting to notice that in the last forty years, the trend in these has been from Monopoly to Diplomacy, to Dungeons and Dragons. Monopoly is a non-cooperative n-person game, where the players do not form coalitions—and has a high degree of chance in the outcome. Diplomacy, as its name suggests, deals with coalition formations and reformations. There are no chance moves as the real skill in the game is the choice of which players to join with in a coalition, and when to double cross them. Dungeons and Dragons has moved even farther away from what we think of as a board game, and though there is a book full of rules, the real essence of the game is for the players to develop their 'characters' and invent new situations in the game. It is almost more an improvised play than a game. It is interesting to note this change in entertainment games from conflict to coalition to almost therapeutic, and to speculate whether the trend can continue in the future.

The great advance in entertainment games in the last two years has been the video game. Whether an arcade machine, or on a home computer there can be few children, or adults, in the Western world who have not played Space Invaders or Pac-Man. Although strictly these would not be considered examples of gaming, as again there seems to be only one player, the actual construction of such games is one of the areas where game theory can be profitably applied. What is the aim of the programmer who invents these games? Obviously to provide an entertainment, but should he choose a strategy that tries to minimise the player's score, maximise or minimise the length of the game. At present, the only way the game programs deal with good players is 'more and faster'. If a player essentially 'wins' the first game, then the second time around the program repeats its moves only faster or with more missiles. Should programmers try to introduce self-learning systems which will take advantage of any repetitions in the way the player plays the game? Such developments would bring these games close to the work on artificial intelligence.

Having described some of the objectives one tries to satisfy in a gaming experiment, let us examine how they are set up.

10.3 RUNNING A GAMING EXPERIMENT

Like an iceberg, 90% of the work involved in constructing a gaming experiment is never seen. All the hard work is done before the game starts. We can divide the performing of a gaming experiment into four sections: development of the game, preparation for playing the game, playing the game, and analysis of the game.

10.3.1 Development of the Game

This is the most difficult of the four stages, and it is probably impossible to do in one attempt. You should start off with a very simple version of what you are aiming at, play it several times and analyse the results. Then you can iron out any difficulties and add further detail to the game and repeat the process. The two main questions to answer immediately are: What are the objectives of the gaming exercise you want to construct, and what environment do you want it set in? Is it to be a business game about competing manufacturers, a strategic diplomacy game about the Middle East, or a tactical war game about the deployment of submarines given advances in their detection? Then you need to decide how many 'blocks' of players you want—a block is one firm in a business game, one chain of command in a war game—and what moves describe the strategic aspect of the game between the blocks. Having settled on that, you decide how many players are in each block, what their interactions are to be, and develop the tactical moves allowed in the game.

One consideration you should make is what type of players are going to play the game—students or professionals—and whether the players are to take on 'roles' or play themselves. Often, in operational war gaming, the players representing the host country can play as they like, whereas those who represent other countries must try to adopt the aims and responses of those countries.

Another aspect of the game that has to be decided in the earliest stages is how much and what type of communication is allowed between the players, and whether it is all to be recorded for post-game analysis. This slows down the game, but gives more insight into what went on. Finally, one has to settle how the game will end. Normally it is usual to put a time horizon on the game time—each move in the game is supposed to represent a certain passage of real time. It is necessary then to put limits on the time taken for each move in the game in order for its length of playing to be known beforehand.

10.3.2 Preparation For Playing

Having developed the game to a point where it is playable, the next step is to prepare for the gaming exercise itself. A critical feature, over which the game-builder often has little control, is the length of time the players can give to the game. Ths usual situation is that there are two or three full days which the participants can allot to or be allotted for the game. It becomes like a mini-course and is often held away from their normal environment. This has the advantage that it fosters a spirit of companionship and often generates considerable enthusiasm towards the game. It also has the less desirable effect of making 'winning the game' the objective of the participants, even if that is not really the objective assigned to their persona in the game. The alternative solution is when moves of the game are played at weekly intervals, say over an appropriate period. This generates less enthusiasm among the players, and is more susceptible to postponements through illness or changes in circumstances. It also means the players tend to 'forget' the rules and previous moves of the game and have to be given a quick résumé at the start of each playing period. It has the advantage, however, that the game takes its place with all the other problems the players are working on each week, and so their motives when playing it are more likely to be those they have in their workaday decision situations.

Whichever method is employed, it is necessary to prepare a written general scenario which tells each player the context of the game and describes whatever rules the game designer feels should be known. There should also be a pre-game discussion to ensure the players understand the scenario, and have some common understanding of what is going on. Each player should also be supplied with a brief which details his individual role and the options open to him, and maybe what characteristics should be pre-eminent in his decision-making.

Finally, the place where the game is to be played should be prepared. Whether all the players should be in the same room with easy access, or in separate areas with no access, depends on how much communication is to be allowed in the game. It is surprising how much information can be passed between players in a non-verbal fashion.

10.3.3 Playing The Game

The running of the game is supervised by a **controller** or group of controllers. They ensure the rules are being obeyed, and keep the game on its time schedule. They are also the arbitrators of chance events in the game, i.e. whether a certain missile exploded on impact, what the demand for a certain product in this quarter of the year will be. These decisions can be made by chance, and often are, or can be predetermined by the game-builders to create certain situations. The controllers also can introduce external events which impinge on the game. For example, in a war game that an influenza epidemic cuts the available manpower of one side by 20%, or in a business game that the government has imposed an extra tax on company profits for the year. In some business games, players operate as the government. In others, the controllers do.

In research games and most war games where communication between players is very limited, all communication is usually routed through the controllers, who thus have a record of it, for post-game analysis, and in some games are allowed to suppress or adjust the communication before passing it on.

A more recent development has been the use of a computer or a set of micro-computers as a controller. This speeds up the time taken to play the game, and is useful for storing the relevant information for post-game analysis. However, it lacks the catalysing reaction that can be gained from 'human controllers' in the way that they can make the game scenario come to life. One way they do this is to display 'newspaper headlines' giving lurid versions of the main events in the last round of moves, as well as fictional and extraneous events, which seem to set the scene.

One point we have not dealt with so far is what are the actual payoffs the players get in these games. In the two-person conflict research games, like Prisoners' Dilemma, it has become standard to give the players monetary rewards related to the payoffs they get in the game. This seems an effective way, especially among university students, of guaranteeing that they really do try to maximise their payoffs. In the larger games, monetary payoffs are less successful because the players are either in coalitions—the same firm, etc.—or have formed coalitions. Thus, a monetary payoff would add the bargaining problem of how to split the reward. There are many research games which deal purely with such bargaining problems but it is a complication one wants to avoid when dealing with a teaching or operational game. Thus, normally, the only incentives offered for 'doing one's best' is an appeal to inner motivation and a direct or indirect indication that doing well in the game will aid promotional prospects or count toward obtaining your degree.

10.3.4 Post-game Analysis

This should be the *raison d'être* for holding a gaming exercise, but is sometimes viewed as an anti-climax by the participants. It is very useful to have a meeting of all the players to describe the course of the game, and for them to explain the reasoning behind some of their actions. Thus, it is more important to have the meeting while the game is fresh in their memories than wait until all the statistics have been analysed. In teaching games, this is the time for emphasising the skills and principles which the game was designed to display. It might be possible to give a short course interweaving the results of the game and restatements of these principles.

In operational games, the players, controllers and builders all have much to contribute to the post-game analysis, as opposed to teaching games where the 'teacher–student' relationship reasserts itself at this point. Since these operational games are much more open ended and allow innovations, each player's view of the game and his role in it is very important as well as the overall course of the game. All connected with the game should have a general debriefing session on which views can be thrown back and fro, and a private debriefing to comment on their own role. Often the results of these sessions are written up, so that the players can make further comments on the report, after having had time for reflection.

The meeting of players and game-builders is less important in research games, where the clinical conditions of the gaming experiment usually mean the controllers have a record of all the players' actions. However, it is always useful to find out the motivations behind the players' actions. There is an apocryphal story told about one experiment on Prisoners' Dilemma. For the first half of the experiment the two players consistently played the individually rational 'confess–confess' pair of strategies, but halfway through turned abruptly to the 'don't confess–don't confess' group rational outcome. On quizzing the players afterwards the experimenter was told: 'Well you let us have coffee halfway through and I happened to meet the other player. I liked her so much, I had to be nice to her after.' This emphasises the need for post-game analysis, as well as ensuring your rules on communication—such as having no idea who are the other players—are adhered to both inside and outside the experiment. Mostly the post-game analysis of research games consists of the designer checking to see if the results confirm or conflict with the ideas proposed by game theory.

Post-game analysis of entertainment games is usually a literary venture, with books written analysing Chess games, and Bridge, Backgammon and Chess columns in most newspapers. If popular mythology is to be believed, post-game analysis of Bridge games between husband and wife is the quickest way to the divorce court.

10.4 FURTHER READING

The first reported gaming experiment was by Chamberlain (1948), but even prior to the Second World War the military in some countries were using war gaming. Since then more than two thousand references have appeared in the literature. The two books published by Shubik in 1975 (1975a, 1975b) are a good introduction to the subject. The first, *The Uses and Methods of Gaming* is well-described by its title, whereas the second, *Games for Society, Business and War*, looks at some specific games and details the basics of game theory needed to understand gaming. For readers who have not time to wade through the two books, the two articles by the same author, Shubik (1972a, 1972b), give a summary of the main points discussed.

Teaching games split into two classes: the large and complex games run by business schools or smaller games which are devoted to teaching one or a handful of principles. The large games have books written about them—see Cohen, Dill, Kuehn and Winters (1964) for the Carnegie Tech Management game; McKenney (1967) for the Harvard School Business game; and Shubik and Levitan (1980) for the Yale game. Details of the smaller games are scattered through the literature, but in Inbar and Stoll (1972) a dozen of the social science type games are described.

As we mentioned above, operational games are predominantly if not exclusively of the war gaming type. To get an idea of the work in war gaming, you should consult the bibliography by Riley and Young (1957), which notes the war gaming exercises in the immediate post-war period or the book by Brewer and Shubik (1979) which gives a much more recent perspective on the subject.

The amount of work done on research gaming with matrix games is phenomenal. Rapoport, Guyer and Gordon (1976) give a good summary of the results in 2×2 matrix games, including Prisoners' Dilemma. An interesting 'gaming' exercise on Prisoners' Dilemma was performed by Axelrod (1980) when he held a tournament of the strategies that people had suggested for playing the game. The analysis is somewhat similar to what occurs in evolutionary games (Chapter 8). There is a much smaller number of papers on gaming experiments with n-person games in characteristic function form. For a survey of this work see Shubik (1975c).

Entertainment gaming is too much fun to spoil by giving you a list of books to study, but if you want to look at some new simple mathematical games, read Conway (1976).

Solutions to problems

CHAPTER 1

1.1. (a) Monopoly, Ludo, Backgammon.
(b) Chess, Draughts, Noughts and Crosses, Go.
(c) Bridge, Poker, Dominoes.
(d) Scissors–Stone–Paper, Blitzkreig.

1.2. (a) Chess, Draughts.
(b) Prisoners' Dilemma, marriage (?).
(c) Poker, Blackjack.
(d) Business conflicts, international problems.

1.3. Because of symmetry we can always turn the grid around so I's first 0 is in the top left square. II's first X goes either diagonally opposite (2) or in the same row (1) or same column (3), as the 0. By twisting the grid on its diagonal, these two can be thought of as the same choice. So II has two strategies:

0	1
3	2

II_1, X in square 1, II_2, X in square 2. If II plays II_1, I's real choice is either I_1 square 3 or I_2 square 2. If II plays II_2, then I's reply of putting 0's in either 1 or 3 can be viewed as the same strategy under diagonal symmetry. So II's strategies are: II_1 put X in same row or column as 0, II_2 put X diagonally opposite 0. I's strategies are: I_1 if II plays II_1 put next 0 diagonally opposite first 0; I_2 if II plays II_1 put 0 in same row or column as first 0.

1.4. Each player has N strategies where the ith strategy is fire when $2i$ steps apart. In the noisy duel, this changes to fire when $2i$ steps apart, unless opponent has already fired, in which case fire when as near as possible.

1.5 $0.5u(100) + 0.5u(10) = u(40)$. So $u(40) = 50$. Next gamble says $u(0) < \frac{1}{4}u(50 - 10) + \frac{3}{4}u(0 - 10)$. Therefore, $0 < \frac{1}{4}\cdot 50 + \frac{3}{4}u(-10)$, or $u(-10) > -\frac{50}{3}$.

CHAPTER 2

2.1.

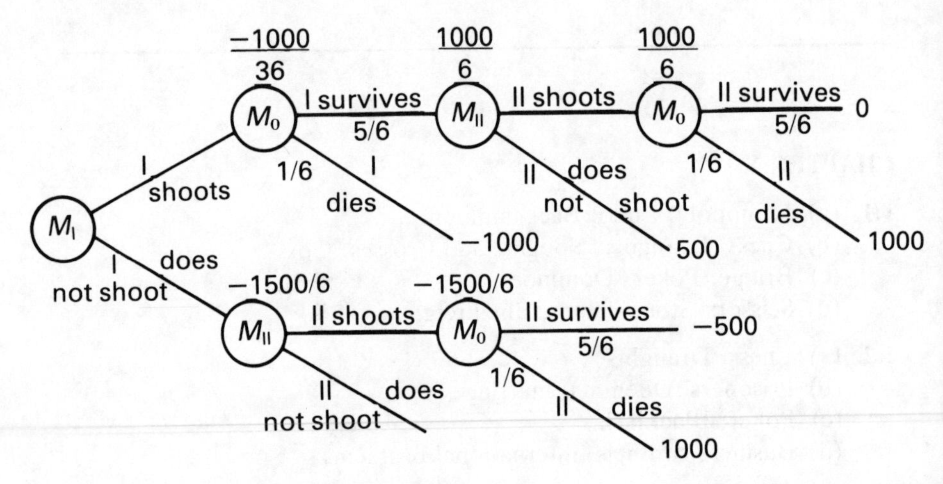

$v = \frac{1000}{6}$, I shoots and II shoots (!)

2.2.

	Send in Man	Dog	Mouse	Min for I
Send in Woman	0.5	0.6	0.1	0.1
Send in Lion	0.6	0.7	0.8	0.6
Send in Cat	0.2	0.5	0.9	0.2
Max for II	0.6	0.7	0.9	

$v_{\mathrm{I}} = 0.6$

$v_{\mathrm{II}} = 0.6$

So $v = v_{\mathrm{I}} = v_{\mathrm{II}} = 0.6$ and I should send in Lion, II should send in Man.

2.3. (a) I_1 dominates I_2, II_2 dominates II_1, so solution is $v = 1$; I_1 plays II_2.

(b) No domination, so by (2.41) and (2.42) $v = (4.6 - 0.2)/(4 + 6 - 0 - 2) = 3$. $\mathbf{x}^* = (\frac{1}{2},\frac{1}{2})$, $\mathbf{y}^* = (\frac{3}{4},\frac{1}{4})$.

(c) I_3 dominates I_2 leaving the matrix

$$\begin{pmatrix} 4 & 1 & 3 \\ 1 & 7 & 6 \end{pmatrix}$$

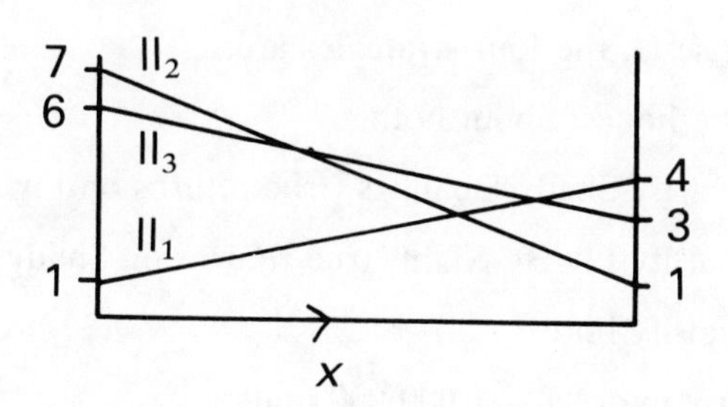

II_1 and II_2 worthwhile, so if $\mathbf{x}^* = (x,1-x)$ is optimal, $e(\mathbf{x}^*,II_1) = 4x + 1 - x = e(\mathbf{x}^*,II_2) = x + 7(1-x) = v$. So $v = 3$, $\mathbf{x}^* = (2/3,1/3)$. If $\mathbf{y}^* = (y,1-y,0)$ is optimal for II, $e(II_1,\mathbf{y}^*) = 4y + 1 - y = v = 3$. $\therefore$ $\mathbf{y}^* = (2/3,1/3,0)$.

2.4. No domination so by (2.41), (2.42), $v = (0.95.0 - 1.1)/(0.95 - 1 - 1) = 20/21$, $\mathbf{x}^* = (20/21,1/21)$, $\mathbf{y}^* = (20/21,1/21)$.

2.5. I_1 — split attacking forces 1 on each road.
I_2 — put all 2 attacking divisions on same road.
II_1 — put all 3 defending divisions on the same road.
II_2 — split forces 2 on one road, one on the other.

$$\begin{array}{c} \\ I_1 \\ I_2 \end{array} \begin{array}{c} \begin{array}{cc} II_1 & II_2 \end{array} \\ \begin{pmatrix} 1 & 1/2 \\ 1/2 & 3/4 \end{pmatrix} \end{array}$$

No domination as $v = 2/3$, $\mathbf{x}^* = (1/3,2/3)$, $\mathbf{y}^* = (1/3,2/3)$.

2.6. I_1 — make 200 colour sets. II_1 — make 400 colour sets.

I_2 — make 200 B/W sets. II_2 — make 200 colour, 200 B/W sets.

II_3 — make 400 B/W sets.

$$\begin{array}{c} \\ I_1 \\ I_2 \end{array} \begin{array}{c} \begin{array}{ccc} II_1 & II_2 & II_3 \end{array} \\ \begin{pmatrix} 1333\frac{1}{3} & 2000 & 4000 \\ 2000 & 2000 & 1333\frac{1}{3} \end{pmatrix} \end{array}.$$

II_1 dominates II_2 and applying (2.41), (2.42) to remaining game gives $v = 1866\frac{2}{3}$, $\mathbf{x}^* = (1/5,4/5)$, $\mathbf{y}^* = (4/5,0,1/5)$.

2.7. For I, internal revenue service the strategies are:

I_1 — investigate Fly-By-Night.

I_2 — investigate Shady Dealings.

For enterprise II, the four strategies are:

II_1 — be honest about both.

II_2 — falsify Shady Dealings, true returns on Fly-By-Night.

II_3 — falsify Fly-By-Night, true returns on Shady Dealings.

II_4 — falsify both.

Payoff matrix in dollars, 1,000,000 units:

$$\begin{array}{c} \\ I_1 \\ I_2 \end{array} \begin{array}{c} \begin{array}{cccc} II_1 & II_2 & II_3 & II_4 \end{array} \\ \begin{pmatrix} 16 & 4 & 18 & 6 \\ 16 & 22 & 12 & 18 \end{pmatrix} \end{array}$$

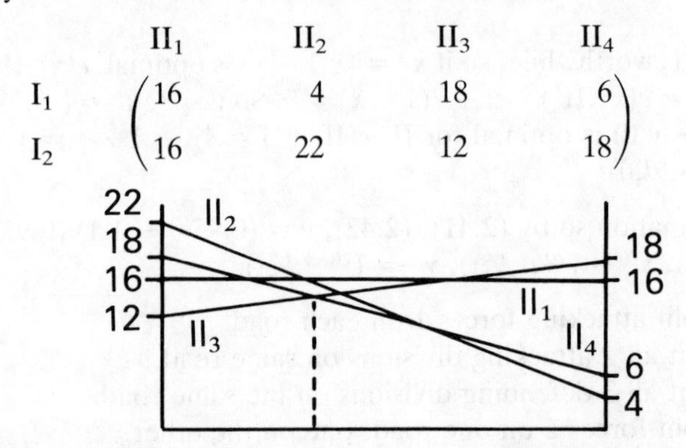

II_3, II_4 are worthwhile, If $\mathbf{x}^* = (x, 1 - x)$ is optimal, $e(\mathbf{x}^*, II_3) = 18x + 12(1 - x) = v = e(\mathbf{x}^*, II_4) = 6x + 18(1 - x)$. $v = 14$, $\mathbf{x}^* = (1/3, 2/3)$. If $\mathbf{y}^* = (0,0,y,1 - y)$ is optimal for II, $e(I_1, \mathbf{y}^*) = 18y + 6(1 - y) = 14$. $\mathbf{y}^* = (0,0,2/3,1/3)$.

Changing the penalty gives a new payoff matrix:

$$\begin{array}{c} \\ I_1 \\ I_2 \end{array} \begin{array}{c} \begin{array}{cccc} II_1 & II_2 & II_3 & II_4 \end{array} \\ \begin{pmatrix} 16 & 4 & 24 & 12 \\ 16 & 40 & 12 & 36 \end{pmatrix} \end{array}.$$

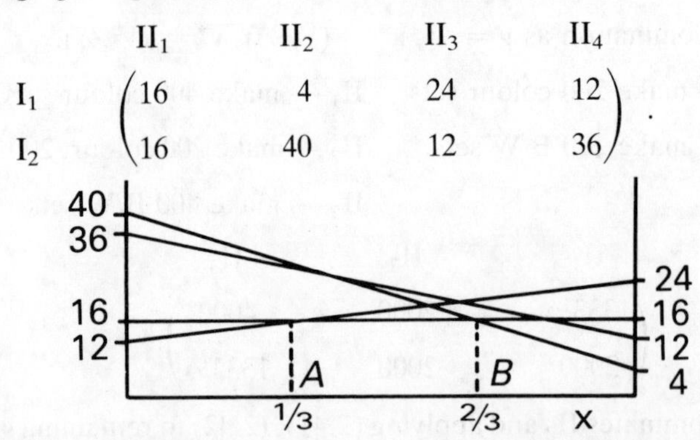

Maximin value is 16 from A to B. This corresponds to a non-unique solution. Solution has value $v = 16$, and optimal strategies can be any of:

(a) $\mathbf{x}^* = (x, 1 - x)$, $1/3 < x < 2/3$, $\mathbf{y}^* = (1,0,0,0)$, i.e. II_1
This corresponds to the points between A and B.

2.8. I_1 — believe II when he says 'Ace'.
I_2 — don't believe II when he says 'Ace'.
II_1 — say 'Queen' when dealt a Queen; say 'King' when dealt a King.
II_2 — say 'Ace' when dealt a Queen; say 'King' when dealt a King.
II_3 — say 'Queen' when dealt a Queen; say 'Ace' when dealt a King.
II_4 — say 'Ace' when dealt a Queen; say 'Ace when dealt a King.

$$\begin{array}{c} \\ I_1 \\ I_2 \end{array} \begin{array}{c} \begin{array}{cccc} II_1 & II_2 & II_3 & II_4 \end{array} \\ \begin{pmatrix} 0 & -\frac{2}{3} & -\frac{1}{3} & -1 \\ -\frac{1}{3} & 0 & \frac{1}{3} & \frac{2}{3} \end{pmatrix} \end{array}$$

II_3 dominated by II_2.

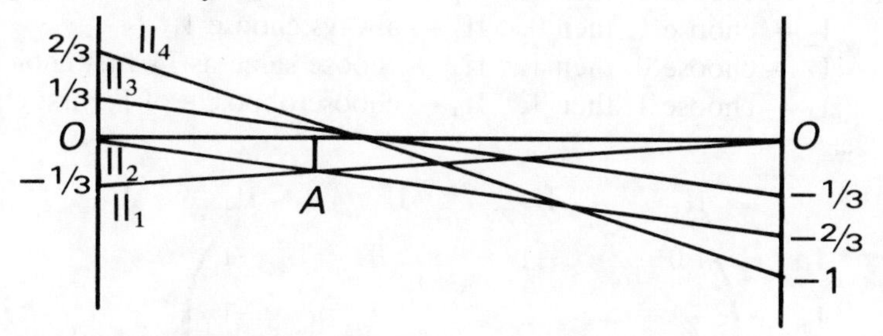

II_1, II_2 are worthwhile, which leads to $v = -\frac{2}{9}$, $\mathbf{x}^* = (\frac{1}{3},\frac{2}{3})$, $\mathbf{y}^* = (\frac{2}{3},\frac{1}{3},0,0)$, i.e. never lie when dealt a King.
If saying 'King' costs a, this only affects payoffs against II_1 and II_2 which becomes:

$$\begin{array}{c} \\ I_1 \\ I_2 \end{array} \begin{array}{c} \begin{array}{cccc} II_1 & II_2 & II_3 & II_4 \end{array} \\ \begin{pmatrix} 0+\frac{1}{3}a & -\frac{2}{3}+\frac{1}{3}a & -\frac{1}{3} & -1 \\ -\frac{1}{3}+\frac{1}{3}a & \frac{1}{3}a & \frac{1}{3} & \frac{2}{3} \end{pmatrix} \end{array}.$$

From the diagram above, it follows easily that for a close to 0, the optimal solutions stay the same: $\mathbf{x}^* = (\frac{1}{2},\frac{2}{3})$, $\mathbf{y}^* = (\frac{2}{3},\frac{1}{3},0,0)$ and $v = -\frac{2}{9} + \frac{1}{3}a$, since the II_1 and II_2 lines just get raised by $\frac{1}{3}a$. So it is worth lying only when A cuts II_3 or II_4, i.e. $a = 1$.

2.9. I_2 dominates I_3; II_4 dominates II_3, leaving

$$\begin{pmatrix} 1 & 4 & 2 \\ 5 & 2 & 4 \end{pmatrix}.$$

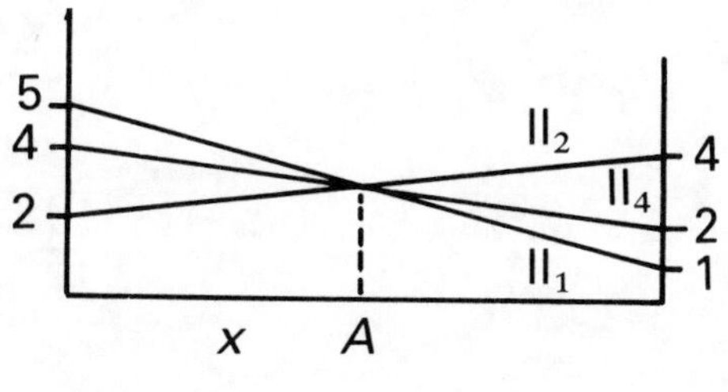

II_1, II_2, II_4 are all worthwhile. $e(\mathbf{x}^*,\text{II}_1) = e(\mathbf{x}^*,\text{II}_2) = e(\mathbf{x}^*,\text{II}_4) = 4 \Rightarrow v = 3$, $\mathbf{x}^* = (\frac{1}{2},\frac{1}{2})$. If $\mathbf{y}^*$ is optimal for II, $e(\text{I}_1,\mathbf{y}^*) = v$ gives $\mathbf{y}^* = (\frac{1}{2} + \frac{1}{2}y, y, y, \frac{1}{2} - \frac{3}{4}y)$, $0 \leqslant y \leqslant \frac{1}{3}$. So II's optimal strategy is not unique.

2.10. $(\mathbf{x}_1,\mathbf{y}_1)$ equilibrium pair $\Rightarrow e(\mathbf{x}_2, \mathbf{y}_1) \leqslant e(\mathbf{x}_1,\mathbf{y}_1) \leqslant e(\mathbf{x}_1,\mathbf{y}_2)$.
$[\mathbf{x}_2,\mathbf{y}_2)$ equilibrium pair $\Rightarrow e(\mathbf{x}_1,\mathbf{y}_2) \leqslant e(\mathbf{x}_2,\mathbf{y}_2) \leqslant e(\mathbf{x}_2,\mathbf{y}_1)$.
So $e(\mathbf{x}_2,\mathbf{y}_1) = e(\mathbf{x}_1,\mathbf{y}_1) = e(\mathbf{x}_1,\mathbf{y}_2) = e(\mathbf{x}_2,\mathbf{y}_2)$, and for any $\mathbf{x},\mathbf{y}$ $e(\mathbf{x},\mathbf{y}_2) \leqslant e(\mathbf{x}_2,\mathbf{y}_2) = e(\mathbf{x}_1,\mathbf{y}_2) = e(\mathbf{x}_1,\mathbf{y}_1) \leqslant e(\mathbf{x}_1,\mathbf{y})$.
So $(\mathbf{x}_1,\mathbf{y}_2)$ is an equilibrium pair.

2.11. I_1 — choose 0, then 0. II_1 — always choose 0.
I_2 — choose 1, then 0. II_2 — always choose 1.
I_3 — choose 0, then 1. II_3 — choose same as I's first choice.
I_4 — choose 1, then 1. II_4 — choose opposite of I's first choice.

$$\begin{array}{c} \\ \text{I}_1 \\ \text{I}_2 \\ \text{I}_3 \\ \text{I}_4 \end{array} \begin{array}{c} \begin{array}{cccc} \text{II}_1 & \text{II}_2 & \text{II}_3 & \text{II}_4 \end{array} \\ \begin{pmatrix} 0 & -1 & 0 & -1 \\ -1 & 2 & 2 & -1 \\ -1 & 2 & -1 & 2 \\ 2 & -3 & -3 & 2 \end{pmatrix} \end{array}$$

$$\max\ y_1 + y_2 + y_3 + y_4$$

$$\begin{aligned} \text{subject to } -y_2 - y_4 &\leqslant 1 \\ -y_1 + 2y_2 + 2y_3 - y_4 &\leqslant 1, \\ -y_1 + 2y_2 - y_3 + 2y_4 &\leqslant 1, \\ 2y_1 - 3y_2 - 3y_3 + 2y_4 &\leqslant 1, \\ y_i \geqslant 0, \quad i &= 1,2,3,4. \end{aligned}$$

n	I plays	II plays	Pay off	Total pay offs for I against II's strategies	Total pay off for II against I's strategies	Total actual pay off	I_1	I_2	I_3	I_4	II_1	II_2	II_3	II_4
1	I_1	II_1	0	(0,−1,−1,2)	−(0,−1,0,−1)	0	1	0	0	0	1	0	0	0
2	I_4	II_2^*	−3	(−1,1,1,−1)	−(2,−4,−3,1)	−3	1	0	0	1	1	1	0	0
3	I_2^*	II_2	2	(−2,3,3,−4)	−(1,−2,−1,0)	−1	1	1	0	1	1	2	0	0
4	I_2^*	II_2	2	(−3,5,5,−7)	−(0,0,1,−1)	1	1	2	0	1	1	3	0	0
5	I_2^*	II_4	−1	(−4,4,7,−5)	−(−1,2,3,−2)	0	1	3	0	1	1	3	0	1
6	I_3	II_4	2	(−5,3,9,−3)	−(−2,4,2,0)	2	1	3	1	1	1	3	0	2
7	I_3	II_1	−1	(−5,2,8,−1)	−(−3,6,1,2)	1	1	3	2	1	2	3	0	2
8	I_3	II_1	−1	(−5,1,7,1)	−(4,8,0,4)	0	1	3	3	1	3	3	0	2
9	I_3	II_1	−1	(−5,0,6,3)	−(−5,10,−1,6)	−1	1	3	4	1	4	3	0	2
10	I_3	II_1	−1	(−5,−1,5,5)	−(−6,12,−2,8)	−2	1	3	5	1	5	3	0	2
11	I_4^*	II_1	2	(−5,−2,4,7)	−(−4,9,−5,10)	0	1	3	5	2	6	3	0	2
12	I_4	II_3	−3	(−5,0,3,4)	−(−2,6,−8,12)	−3	1	3	5	3	6	3	1	2

So estimate is $v = -\,3/12 = -\,1/4$, $\mathbf{x}^* = (1/12, 3/12, 5/12, 3/12)$, $\mathbf{y}^* = (6/12, 3/12, 1/12, 2/12)$.

[True value is $v = 1/8$, $\mathbf{x}^* = (0, 5/8, 0, 3/8)$, $\mathbf{y}^* = (5/8, 3/8y, 3/8(1-y), 0)$ $0 \leqslant y \leqslant 1$.]

CHAPTER 3

3.1. (a) $e_1(\mathbf{x},\mathbf{y}) = x(2y - 1) - 3y + 4$. $e_2(\mathbf{x},\mathbf{y}) = y(2x - 3) - x + 4$.

To maximise if $y < 1/2$, $x = 0$. To maximise if x = anything, $y = 0$

$y = 1/2$, x = anything

$y = 1/2$, $x = 1$.

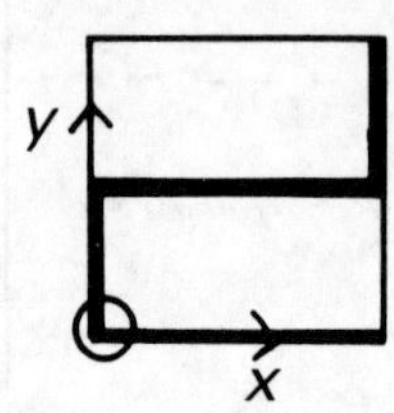

Solution is $x = 0$, $y = 0$, i.e. $(\mathrm{I}_2,\mathrm{II}_2)$.

(b) $e_1(\mathbf{x},\mathbf{y}) = x(5y - 4) - 4y + 4$. $e_2(\mathbf{x},\mathbf{y}) = y(5x - 1) - x + 1$.

$y < 4/5 \Rightarrow x = 0$.	$x < 1/5 \Rightarrow y = 0$.
$y = 4/5 \Rightarrow x$ = anything.	$x = 1/5 \Rightarrow y$ = anything.
$y > 4/5 \Rightarrow x = 1$.	$x > 1/5 \Rightarrow y = 1$.

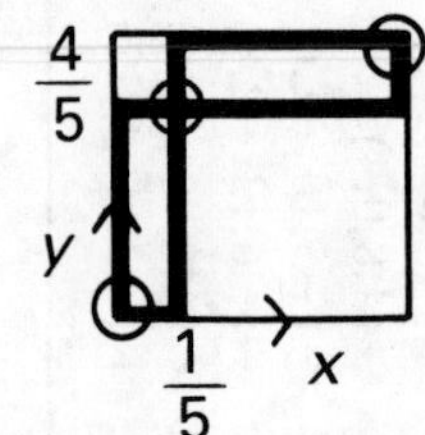

Solutions are
$x = 0, y = 0$ $(\mathrm{I}_2,\mathrm{II}_2)$,

$x = 1, y = 1$ $(\mathrm{I}_1,\mathrm{II}_1)$.

$x = 1/5$, $y = 4/5$, $(1/5, 4/5)$ v. $(4/5, 1/5)$.

(c) $e_1(\mathbf{x},\mathbf{y}) = x - 9y - 1$. $e_2(\mathbf{x},\mathbf{y}) = y - 9x - 1$.

To maximise this $x = 1$, any y. To maximise this $y = 1$, any x.

So equilibrium point is $x = 1$, $y = 1$ $(\mathrm{I}_2,\mathrm{II}_2)$.

3.2. (a) I's payoff matrix is:

$$\begin{pmatrix} 2 & 4 \\ 6 & 3 \end{pmatrix}$$

II's payoff matrix is:

$$\begin{array}{cc} & \begin{array}{cc} \mathrm{I}_1 & \mathrm{I}_2 \end{array} \\ \begin{array}{c} \mathrm{II}_1 \\ \mathrm{II}_2 \end{array} & \begin{pmatrix} 1 & 2 \\ 3 & 1 \end{pmatrix} \end{array}$$

Solution is $\mathbf{x} = (3/5, 2/5)$, $v = 3.6$.

Solution is $\mathbf{y} = (2/3, 1/3)$, $v = 5/3$.

Maximin pair is $(3/5, 2/5)$ v. $(2/3, 1/3)$.

$e_1(\mathbf{x},\mathbf{y}) = x(1 - 5y) + 3y + 3.$ $e_2(\mathbf{x},\mathbf{y}) = y(1 - 3x) + 2x + 1.$

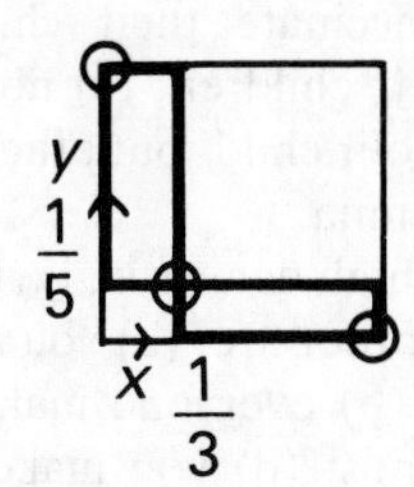

Solution is
$x = 1, y = 0\ (\mathrm{I}_1,\mathrm{II}_2)$,
$x = 0, y = 1\ (\mathrm{I}_2,\mathrm{II}_1)$,
$x = 1/3, y = 1/5, (1/3, 2/3)\ v.\ (1/5, 4/5)$.

(b) I's payoff matrix is:

$$\begin{array}{c} \\ \mathrm{I}_1 \\ \mathrm{I}_2 \end{array} \begin{array}{c} \mathrm{II}_1 \quad \mathrm{II}_2 \\ \begin{pmatrix} 4 & 10 \\ 12 & 5 \end{pmatrix} \end{array}$$

$\mathbf{x}^* = (7/13, 6/13),\ v = 7\tfrac{9}{13}$.

II's matrix is:

$$\begin{array}{c} \\ \mathrm{II}_1 \\ \mathrm{II}_2 \end{array} \begin{array}{c} \mathrm{I}_1 \quad \mathrm{I}_2 \\ \begin{pmatrix} -30 & 8 \\ 6 & 4 \end{pmatrix} \end{array}$$

$\mathbf{y}^* = (1/20, 19/20),\ v = 4.2$

Maximin pair is $(7/13, 6/13)\ v.\ (1/20, 19/20)$.

$e_1(\mathbf{x},\mathbf{y}) = x(5 - 13y) + 7y + 5.$ $e_2(\mathbf{x},\mathbf{y}) = y(4 - 40x) + 2x + 4.$

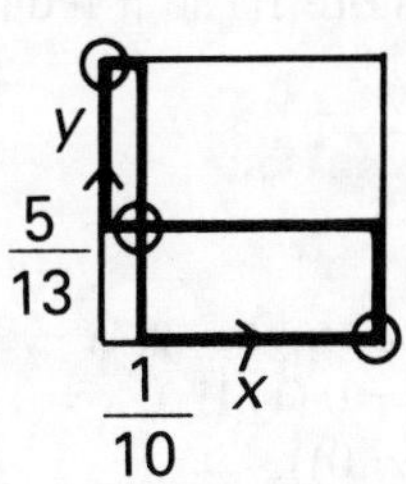

Solution is
$x = 0, y = 1\ (\mathrm{I}_2,\mathrm{II}_1)$,
$x = 1, y = 0\ (\mathrm{I}_1,\mathrm{II}_2)$,
$x = 1/10, y = 5/13\ (1/10, 9/10)\ v.\ (5/13, 8/13)$.

3.3. Each country has two strategies: I_1 (II_1) is to keep weapons; while I_2 (II_2) is to disarm. By symmetry the payoff matrix is:

$$\begin{array}{c} \\ \mathrm{I}_1 \\ \mathrm{I}_2 \end{array} \begin{array}{c} \mathrm{II}_1 \qquad \mathrm{II}_2 \\ \begin{pmatrix} (c,c) & (a,d) \\ (d,a) & (b,b) \end{pmatrix} \end{array}$$

Suppose a country prefers (a) having nuclear weapons when the other country does not, to (b) neither having weapons; (b) neither having weapons to (c) both having weapons; and (c) both having weapons to (d) not having them, when the other country does have them. Then this is exactly the preference order of the Prisoners' Dilemma.

For the whooping cough vaccine problem, let I_1 (II_1) be not to vaccinate your child, while I_2 (II_2) is to vaccinate. If the preferences

are, in decreasing order of preference: (a) do not vaccinate your own child, but other people vaccinate their children; (b) you and everyone else vaccinates their children; (c) no one vaccinates their children; (d) you vaccinate your child, but other people do not, then we have the Prisoners' Dilemma.
For wage demands I_1 (II_1) is high demands, I_2 (II_2) are low demands. If preferences in decreasing order are: (a) you make a high demand, others make low demands; (b) everyone makes low demands; (c) everyone makes high demands; (d) you make a low demand, but others make high demands, then again we have a game with the same ordering of strategies as the Prisoners' Dilemma.

3.4 If $x_1^* \neq 0$, let $\bar{\mathbf{x}} = (0, x_1^* + x_2^*, x_3^*, \ldots, x_n^*)$. Then:

$$\begin{aligned} e_1(\mathbf{x}^*,\mathbf{y}^*) &= x_1^* e_1(\mathrm{I}_1,\mathbf{y}^*) + \sum_{j=2}^{n} x_j^* e_1(\mathrm{I}_j,\mathbf{y}^*) \\ &< x_1^* e_1(\mathrm{I}_2,\mathbf{y}^*) + \sum_{j=2}^{n} x_j^* e_1(\mathrm{I}_j,\mathbf{y}^*) \\ &= e_1(\bar{\mathbf{x}},\mathbf{y}^*), \end{aligned}$$

which contradicts $(\mathbf{x}^*,\mathbf{y}^*)$ being an equilibrium pair. Since II_2 strictly dominates II_3 we can now ignore II_3 so it reduces to the game with payoff matrix:

$$\begin{pmatrix} (2,4) & (0,2) \\ (1,1) & (3,3) \end{pmatrix}.$$

$e_1(\mathbf{x},\mathbf{y}) = x(4y - 3) - 2y + 3$, $e_2(\mathbf{x},\mathbf{y}) = y(4x - 2) - x + 3$. This leads to equilibrium pairs $x = 0$, $y = 0$ $(\mathrm{I}_2,\mathrm{II}_2)$, $x = 1$, $y = 1$ $(\mathrm{I}_1,\mathrm{II}_1)$, and $x = \frac{1}{2}$, $y = \frac{3}{4}$ $((\frac{1}{2},\frac{1}{2})\ v.\ (\frac{3}{4},\frac{1}{4},0))$.
Maximin pair is $(\frac{1}{2},\frac{1}{2})\ v.\ (\frac{1}{4},\frac{3}{4},0)$.

3.5. If typical strategies are $\mathbf{x} = (x_1,x_2,1 - x_1 - x_2)$, $\mathbf{y} = (y_1,y_2,1 - y_1 - y_2)$, then:

$$e_1(\mathbf{x},\mathbf{y}) = x_1(4y_1 + 4y_2 - 1) + x_2(4y_1 + 6y_2 - 3) + 3 - 3y_1 - 3y_2,$$

$$e_2(\mathbf{x},\mathbf{y}) = y_1(-3x_2) + y_2(2x_2 - 2) + 2 + x_2.$$

Notice that $4y_1 + 4y_2 - 1 \geqslant 4y_1 + 6y_2 - 3$, and $0 \geqslant -3x_2$, $0 \geqslant 2x_2 - 2$. To maximise $e_1(\mathbf{x},\mathbf{y})$ over x we have:

if $4y_1 + 4y_2 - 1 > 0$, then $x_1 = 1$, $x_2 = 0$;

if $4y_1 + 4y_2 - 1 = 0$, then x_1 is anything $x_2 = 0$;

if $4y_1 + 4y_2 - 1 < 0$, then $x_1 = x_2 = 0$;

if $y_2 = 1$, then x_1, x_2 can be anything $x_1 + x_2 = 1$.

To maximise $e_2(\mathbf{x},\mathbf{y})$ over y we have:

if $x_2 = 0$, then $y_2 = 0$, y_1 can be anything;

if $0 < x_2 < 1$, then $y_1 = y_2 = 0$;

if $x_2 = 1$, then $y_1 = 0$, y_2 can be anything.

The pairs that satisfy both sets of conditions are:

$x_1 = 1, x_2 = 0, y_1 > 1/4, y_2 = 0$, i.e. I_1 *v.* $(y,0,1 - y), y > 1/4$.

$x_1 =$ anything, $x_2 = 0; y_1 = 1/4, y_2 = 0$, i.e. $(x,0,1 - x)$ *v.* $(1/4,0,3/4)$.

$x_1 = 0, x_2 = 0, y_1 < 1/4, y_2 = 0$, i.e. I_3 *v.* $(y,0,1 - y), y < 1/4$.

$x_1 = 0, x_2 = 1, y_1 = 0, y_2 = 1$, i.e. I_2 *v.* II_2.

Notice that (I_2,II_2) is an equilibrium pair but I_1 dominates I_2, II_3 dominates II_2, so the result in Problem 3.4 does not hold with dominated replacing strictly dominated.

3.6. For (a) of 3.2 we found the equilibrium pairs to be (I_1,II_2), (I_2,II_1), and $(1/3,2/3)$ *v.* $(1/5,4/5)$. These are not interchangeable, so there is no Nash solution. The Pareto optimal equilibrium pairs are (I_1,II_2) and (I_2,II_1), which do not have the same payoffs, so there is no solution in the strict sense. Finally, no strategy dominates another strategy, so $X^r = X$, $Y^r = Y$ and since there was no solution in the strict sense, there is none in the completely weak sense.
For problem (b) of 3.2 the equilibrium pairs are (I_1,II_2), (I_2,II_1) and $(1/10,9/10)$ *v* $(5/13,8/13)$. These are not interchangeable, so no solution in the Nash sense exists. The only Pareto optimal equilibrium pair is (I_2,II_1) so trivially this is the solution in the strict sense. Since there is no domination of strategies, it is also the solution in the completely weak sense.
For Problem 3.5, the equilibrium pairs are I_1 *v.* $(y,0,1 - y), 1 \geqslant y > 1/4$; $(x,0,1 - x)$ *v.* $(1/4,0,3/4), 0 \leqslant x \leqslant 1$; I_3 *v.* $(y,0,1 - y), 0 < y < 1/4$, and I_2 *v.* II_2. These are not interchangeable, so no solution in the Nash sense exists. The Pareto optimal equilibrium pair is (I_2,II_2), so the game has a solution in the strict sense. However I_2 is dominated by I_1, II_2 by II_3 and so $X^r = (I_1,I_3)$, $Y^r = (II_1,II_3)$. The Pareto optimal equilibrium pairs in this reduced game are (I_1,II_1) and (I_3,II_3) which are not interchangeable. So this game has no solution in the completely weak sense!

3.7. For game (a) of Problem 3.2, $v_{\mathrm{I}} = 3\frac{3}{5}$, $v_{\mathrm{II}} = 1\frac{2}{3}$. From the diagram:

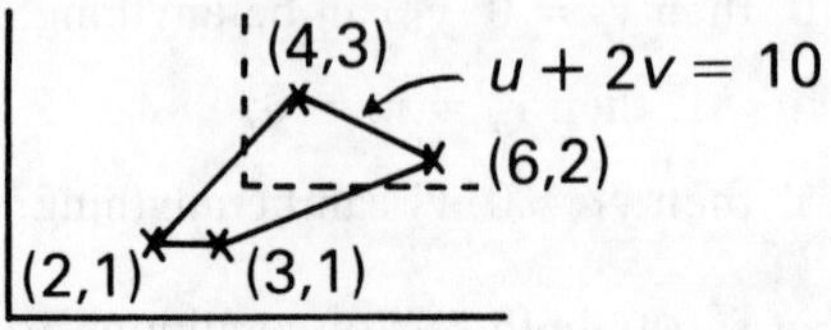

The negotiation set is $\{(u,v)\ \frac{1}{2}u + v = 5,\ 4 \leqslant u \leqslant 6\}$.
To find the maximin bargaining solution, maximise $(u - 3\frac{3}{5})(5 - \frac{1}{2}u - 1\frac{2}{3})$, i.e. $u = 5\frac{2}{15}$, $v = 2\frac{13}{30}$. For the threat solution, look at game $\frac{1}{2}e_1(\ ,\ ) - e_2(\ ,\ )$, i.e.

$$\begin{pmatrix} 0 & -1 \\ 1 & \frac{1}{2} \end{pmatrix}$$

This has the solution $w = \frac{1}{2}$, $(\mathrm{I}_2,\mathrm{II}_2)$, which leads to the threat solution, $u^* = 5\frac{1}{2}$, $v^* = 2\frac{1}{4}$ by substituting in (3.52) or otherwise.
For game (b) of Problem 3.2, $v_{\mathrm{I}} = 7\frac{9}{13}$, $v_{\mathrm{II}} = 4\frac{1}{5}$, but the negotiation set is just the point (12,8). So both threat and maximin bargaining solutions are (12,8).

3.8 The only equilibrium pair is $(\mathrm{I}_1,\mathrm{II}_1)$ with payoff (3,8). For maximin bargaining solution $v_{\mathrm{I}} = 3$, $v_{\mathrm{II}} = 4\frac{4}{5}$, so the negotiation set is $\{(u,v)|4u + v = 20,\ 3 \leqslant u \leqslant 3\frac{4}{5}\}$.
The maximin bargaining solution is $u = 3\frac{2}{5}$, $v = 6\frac{2}{5}$.
For the threat bargaining solution, look at $4e_1(\ ,\ ) - e_2(\ ,\ )$, which has the value $w^* = \frac{60}{11}$. Substituting in (3.52) gives $u^* = 3\frac{2}{11}$, $v^* = 7\frac{3}{11}$. Both bargaining solutions give II less than he gets in the only equilibrium pair, so he would prefer the non-cooperative game.

3.9 The maximin values for this game are:

$$v_{\mathrm{I}} = \begin{cases} 2, & \text{if } a \leqslant 2, \\ a, & \text{if } 2 < a < 3, \\ 3, & \text{if } a \geqslant 2, \end{cases} \qquad v_{\mathrm{II}} = 1,$$

and the negotiation set is:

$$((u,v)|2u + v = 6,\ 2 \leqslant u \leqslant 3), \quad \text{if } a \leqslant 2,$$
$$((u,v)|2u + (3 - a)v = 6,\ a \leqslant u \leqslant 3), \quad \text{if } 2 < a < 3,$$
$$(a,2), \quad \text{if } a \geqslant 3.$$

The maximin bargaining solution is then:

$(2\frac{1}{4}, 1\frac{1}{2}), \quad a \leqslant 2,$

$(\frac{3}{4} + \frac{3}{4}a, 1\frac{1}{2}), \quad 2 < a < 3,$

$(a, 2), \quad a \geqslant 3.$

It is not continuous at $a = 3$.

3.10. If I_1 is make 200 colour TVs, I_2 make 100 colour, 100 black and white TVs, II_1 is make 50 colour TVs and II_2 make 100 black and white TVs, the payoff matrix is:

$$\begin{array}{c} \\ I_1 \\ I_2 \end{array} \begin{array}{c} \begin{array}{cc} II_1 & \qquad II_2 \end{array} \\ \begin{pmatrix} (2{,}000,0) & (10{,}000,1{,}500) \\ (9{,}000,2{,}000) & (4{,}000, -3{,}500) \end{pmatrix}. \end{array}$$

Maximin values are $v_I = £6308$, $v_{II} = £429$.
The negotiation set is $\{(u,v) | \frac{1}{2}u + v = 6{,}500, 9{,}000 \leqslant u \leqslant 10{,}000\}$.
The maximin bargaining solution to the nearest integer is (9,225,1,888).
The threat solution looks at $\frac{1}{2}e_1(\ , \ ) - e_2(\ , \ )$ and leads to $w^* =$ 2500 and solution (9,000,2,000).
II's threat of going into the colour market changes the status quo point to his advantage.

CHAPTER 4

4.1. The maximin values of the game are $v_I = 1\frac{1}{2}$, $v_{II} = 1$ (domination). So $v(1) = 1\frac{1}{2}$, $v(2) = 1$. The payoffs with maximum sum are (4,1) and (3,2), so $v(1,2) = 4 + 1 = 3 + 2 = 5$. Hence, $E(v) = \{(x, 5 - x),$ $1\frac{1}{2} \leqslant x \leqslant 4\}$. The set of imputations allows for side-payments, while in the negotiation set the players only get their payoffs in the game. So $(1\frac{1}{2}, 3\frac{1}{2})$ is not a payoff in the game, but can be got from (2,3) if 1 gives a side payment of $\frac{1}{2}$:

$$\phi_1 = \frac{0!1!}{2!} v(1) + \frac{1!0!}{2!} (v(1,2) - v(2)) = \tfrac{1}{2} \times 1\tfrac{1}{2} + \tfrac{1}{2}(5 - 1)$$
$$= 1\tfrac{1}{4}.$$

$$\phi_2 = \frac{0!1!}{2!} v(2) + \frac{1!0!}{2!} (v(1,2) - v(1)) = \tfrac{1}{2} \times 1 + \tfrac{1}{2}(5 - 1\tfrac{1}{2})$$
$$= \tfrac{9}{4}.$$

By Theorem 4.2, $x = (x_1, x_2) \in C(v)$ if $x_1 \geqslant 1\frac{1}{2}$, $x_1 + x_2 = 5$, so $C(v) = E(v)$. For the nucleolus, if $x = (x, 5 - x)$, the excesses $v(S) - x(S)$

are $1\frac{1}{2} - x$, $1 - (5 - x)$, $5 - 5$, $0 - 0$; to minimise the maximum of $1\frac{1}{2} - x, x - 4$, choose x where they are equal, $1\frac{1}{2} - x = x - 4$. So $x = 1\frac{1}{4}$.

4.2. If $v(1) = a$, $v(2) = b$, $v(1,2) = c$, $E(v) = \{(x, c - x), a \leqslant x \leqslant c - b\}$. By Theorem 4.2, $x = (x_1, x_2) \in C(v)$ if $x_1 \geqslant a$, $x_2 \geqslant b$, $x_1 + x_2 = c$, which are just the conditions for it to be an imputation. So $C(v) = E(v)$. Since for any stable set $S(v)$, $C(v) \subseteq S(v) \subseteq E(v)$, this means $E(v)$ is the only stable set.

4.3. A typical imputation is $x = (x, c - x)$, $a \leqslant x \leqslant c - b$, and excesses $v(S) - x(S)$ are $a - x$, $b - (c - x)$, $c - c$, $0 - 0$. The maximum of $a - x, x + b - c$ is minimised when $a - x = x + b - c$, i.e. $x = \frac{1}{2}(a + c - b)$:

$$\phi_1 = \frac{0!1!}{2!}(v(1)) + \frac{1!0!}{2!}(v(1,2) - v(2)) = \tfrac{1}{2}a + \tfrac{1}{2}(c - b).$$

$$\phi_2 = \frac{0!1!}{2!}(v(1,2)) + \frac{1!0!}{2!}(v(1,2) - v(1)) = \tfrac{1}{2}b + \tfrac{1}{2}(c - a).$$

For the maximin bargaining solution, we know by definition that $v_{\text{I}} = a$, $v_{\text{II}} = b$, and we are given that the negotiation set is $u + v = c$. So we need to maximise $(u - a)(b - v) = (u - a)(c - u - b)$. Differentiating with respect to u and setting the derivative equal to 0 gives $u = \frac{1}{2}(a + c - b)$. This only works if the slope of the negotiation set is -1. For the three-person game a typical imputation is $(x_1, x_2, 1 - x_1 - x_2)$, $x_2 \geqslant 0$, $x_1 + x_2 \leqslant 1$, and so excesses for the coalitions S are $0 - x_1$, $0 - x_2$, $0 - x_3$, $a - (x_1 + x_2)$, $b - (1 - x_2)$, $c - (1 - x_1)$, $1 - 1$, 0. The maximum of these is either $a - x_1 - x_2$, $b - 1 + x_2$ or $c - 1 + x_1$, and to minimise the maximum of these three, we set them all equal. This gives $x_1 = \frac{1}{3}(a + b - 2c + 1)$, $x_2 = \frac{1}{3}(a + c - 2b + 1)$. The Shapley value has $\phi_1 = \frac{1}{6}(a + b + 2 - 2c)$, $\phi_2 = \frac{1}{6}(a + c + 2 - 2b)$, $\phi_3 = \frac{1}{6}(b + c + 2 - 2a)$. $x_1 = \phi_1$, $x_2 = \phi_2$, $x_3 = \phi_3$ only if $a = b = c$.

Let:

$$c'(S) = \Big(v(S) - \sum_{i \in S} a_i\Big) \Big/ (a_{123} - a_1 - a_2 - a_3),$$

so:

$$v'(1,2) = (a_{12} - a_1 - a_2)/(a_{123} - a_1 - a_2 - a_3),$$

$$v'(1,3) = (a_{13} - a_1 - a_3)/(a_{123} - a_1 - a_2 - a_3),$$

$$v'(2,3) = (a_{23} - a_2 - a_3)/(a_{123} - a_1 - a_2 - a_3).$$

From above, we only get the nucleolus and Shapley value equal if $a_{12} - a_1 - a_2 = a_{13} - a_1 - a_3 = a_{23} - a_2 - a_3$.

4.4. $\phi_1 = 4\frac{2}{3}$, $\phi_2 = 2\frac{1}{6}$, $\phi_3 = 3\frac{1}{6}$.
If $x = (x_1,x_2,x_3) \in C(v)$, $x_1 \geqslant 4$, $x_2 \geqslant 0$, $x_3 \geqslant 0$, $x_1 + x_2 \geqslant 5$, $x_1 + x_3 \geqslant 7$, $x_1 + x_3 \geqslant 6$, $x_1 + x_2 + x_3 = 10 \rightarrow x_1 = 4$, $x_2 + x_3 = 6$, i.e. $\{(4,3 - x, 3 + x), 0 \leqslant x \leqslant 2\}$.
Since the nucleolus is in the core, look only at $(4,3 - x, 3 + x)$, $0 \leqslant x \leqslant 2$. Excesses $v(S) - x(S)$ are $4 - 4$, $0 - (3 - x)$, $0 - (3 + x)$, $5 - (7 - x)$, $7 - (7 + x)$, $6 - (3 - x + 3 + x)$, $10 - 10$, 0, i.e. $x - 2$, $x - 3$, $-x$, $-x - 3$, $0, 0, 0, 0$. The maximum is either $-x$ or $x - 2$, and to minimise the maximum of these choose $-x = x - 2$, i.e. $x = 1$ and so $(4,2,4)$ is the nucleolus.
To show $\{(4,6 - x,x), 0 \leqslant x \leqslant 10\}$ is a stable set we must prove internal and external stability. Since domination can only occur on (1,2), (1,3) or (2,3), internal stability is trivial because $(4,6 - x,x)$ cannot dominate $(4,6 - y,3y)$ on (1,2) or (1,3) since both give player 1 the same and on (2,3) if $6 - x > 6 - y$, then $x < y$, so no domination. For external stability notice any imputation not in this set must be $(4 + x,6 - y,y - x)$, $x > 0; 0 \leqslant y < 6)$ and if we choose $\bar{x}$ so that $\bar{x} = y - (x/2)$, $\bar{x} > y - x$, $6 - \bar{x} > 6 - y$ and $(4,6 - \bar{x},\bar{x})$ dominates it on (2,3).

4.5. $E(v) = \{(x_1,x_2,x_3), x_1 \geqslant 1, x_2 \geqslant 2, x_3 \geqslant 3, x_1 + x_2 + x_3 = 12\}$. If $x = (x_1,x_2,x_3) \in C(v)$, $x_1 \geqslant 3$, $x_2 \geqslant 2$, $x_3 \geqslant 3$, $x_1 + x_2 \geqslant 3$, $x_1 + x_3 \geqslant 10$, $x_2 + x_3 \geqslant 6$, $x_1 + x_2 + x_3 = 12 \Rightarrow x_2 = 2$, $x_1 + x_3 = 10$, $x_1 \geqslant 1$, $x_3 \geqslant 4$, i.e. $\{(x,2,10 - x, 1 \leqslant x \leqslant 6\}$. Take $x \in C(v)$ and look at $v(S) - x(S)$ for each coalition S. These are $1 - x$, $2 - 2$, $3 - (10 - x)$, $3 - (x + 2)$, $10 - (x + 10 - x)$, $6 - (2 + 10 - x)$, $12 - 12$, i.e. $1 - x$, $x - 7$, $x - 6$, 0. So the maximum is either $1 - x$ or $x - 6$. At the nucleolus these are the same, i.e. $1 - x = x - 6$. So $x = 3\frac{1}{2}$, $x = (3\frac{1}{2},2,6\frac{1}{2})$. The bargaining set is:

{1}, {2}, {3};	(1,2,3),
{12}, {3};	(1,2,3),
{23}, {1};	(1,2,4),
{13}, {2};	$(x,2,10 - x)$, $1 \leqslant x \leqslant 6$,
{1,2,3};	$(x,2,10 - x)$, $1 \leqslant x \leqslant 6$.

The kernel is:

{1}, {2}, {3};	(1,2,3),
{12}, {3};	(1,2,3),
{2}, {1};	(1,2,4),
{13}, {2};	$(3\frac{1}{2},2,6\frac{1}{2})$,
{123};	$(3\frac{1}{2},2,6\frac{1}{2})$.

4.6. $v(1) = 2$, Bungo keeps his mattress; $v(2) = 0$, Wellington has no mattress; $v(3) = 4$, Orinoco has one; $v(1,2)$, Bungo sells Wellington the mattress; $v(1,3) = 7$, Orinoco has both mattresses; $v(2,3) = 6$, Wellington has Orinoco's mattress; $v(1,2,3) = 10$, Wellington has one, Orinoco has one.

$\phi_1 = 3\frac{1}{2}$, $\phi_2 = 2$, $\phi_3 = 4\frac{1}{2}$. $E(v) = \{(x_1,x_2,x_3), x_1 \geqslant 2, x_2 \geqslant 0, x_3 \geqslant 4, x_1 + x_2 + x_3 = 10\}$.
$C(v) = \{(3 + x, 3 - x, 4), 0 \leqslant x \leqslant 1\}$
Look at $\{(2 + x, 4 - x, 4), 0 \leqslant x \leqslant 4\}$ for internal stability. By comments in the text we can only have domination on $\{1,2\}$, $\{1,3\}$ or $\{2,3\}$, and since player 3's value is constant, we cannot have domination on $\{1,3\}$ or $\{2,3\}$. On $\{1,2\}$ if $2 + x > 2 + y$, then $4 - x < 4 - y$ and so there is no domination.
For external stability, notice that any imputation outside this set is of the form $\{(2 + x, 4 - x - y, 4 + y), y > 0, 0 \leqslant x \leqslant 4 - y\}$. For such an imputation choose $\bar{x} = x + \frac{1}{2}y$, so $2 + \bar{x} > 2 + x$, $4 - \bar{x} > 4 - x - y$ and so $(2 + \bar{x}, 4 - \bar{x}, 4)$ dominates it on $\{1,2\}$.

4.7. If 1 is Arthur, 2 is Buck, 3 is Chuck, and 4 is Donald, the characteristic function is $v(1) = \frac{1}{4}$; $v(2) = v(3) = v(4) = 0$; $v(1,2) = v(1,3) = v(2,3) = \frac{3}{4}$; $v(1,4) = v(2,4) = v(3,4) = \frac{1}{4}$; $v(1,2,3) = v(1,2,4) = v(1,3,4) = 1$, $v(2,3,4) = \frac{3}{4}$, $v(1,2,3,4) = 1$. $E(v) = \{(x_1,x_2,x_3,x_4), x_1 \geqslant \frac{1}{4}, x_2, x_3, x_4 \geqslant 0, x_1 + x_2 + x_3 + x_4 = 1\}$. If $x = (x_1,x_2,x_3,x_4) \in C(v)$, $x_1 \geqslant \frac{1}{4}$, $x_2 \geqslant 0$, $x_3 \geqslant 0$, $x_4 \geqslant 0$, $x_1 + x_2 \geqslant \frac{3}{4}$, $x_1 + x_3 \geqslant \frac{3}{4}$, $x_2 + x_3 + x_4 \geqslant 1$, $x_2 + x_3 + x_4 \geqslant \frac{3}{4}$, $x_1 + x_2 + x_3 + x_4 = 1$ $\Rightarrow x_2 = x_3 = x_4 = 0$, which contradicts $x_2 + x_3 + x_4 \geqslant \frac{3}{4}$. So the core is empty.
S_1 is a stable set. Internal stability follows since it is only on $\{2,3\}$ that imputations give different rewards and if $x > y$, $\frac{3}{4} - x < \frac{3}{4} - y$, so there is no domination. For $x \notin S_1$, $x_1 + x_4 > \frac{1}{4}$, and so $x_2 + x_3 < \frac{3}{4}$. There is always an element of S_1 that dominates it on $\{2,3\}$.
S_2 is not a stable set because it does not satisfy external stability. $x = (\frac{1}{4},\frac{1}{2},\frac{1}{8},\frac{1}{8}) \notin S_2$ and the players who get more under S_2 are 1 and 4. Since $v(1,4) = \frac{1}{4} < x_1 + x_4 = \frac{1}{4} + \frac{1}{8}$, no element of S_2 can dominate x on $\{1,4\}$, and so it is not dominated by S_2. (There are many other counter-examples.)
S_3 is not a stable set because it does not satisfy internal stability. $(\frac{1}{2} - x, x, x, \frac{1}{2} - x) > (\frac{1}{2} - y, y, y, \frac{1}{2} - y)$ on $\{2,3\}$ provided, $y < x < \frac{1}{2}v\{2,3\} = \frac{3}{8}$.
For the three-player game, $v(1) = v(2) = \frac{1}{4}$, $v(3) = 0$, $v(1,2) = 1$, $v(1,3) = \frac{3}{4}$, $v(2,3) = \frac{3}{4}$, $v(1,2,3) = 1$. As v is essentially a constant sum game, there is no core. (A similar proof holds in a four-player game.) Obviously $S_1 = \{(x, 1 - x, 0), \frac{1}{4} \leqslant x \leqslant \frac{3}{4}\}$ is a good example.

4.8. $v(S) = 1$ if $\#S \geqslant 10$, otherwise $v(S) = 0$.
If $x = (x_1,x_2,\ldots,x_{12}) \in C(v)$, Theorem 4.2 gives $x_1 + x_2 + \ldots + x_{11} \geqslant 1$, $x_{12} \leqslant 0$, $x_1 + x_2 + \ldots + x_{12} = 0$, i.e. $x_{12} = 0$. Similarly, we can prove $x_i = 0$, $1 \leqslant i \leqslant 12$, which contradicts $x_1 + x_2 + \ldots + x_{12} = 1$. So, no core.
For player 1, only time $v(S)$ and $v(S - \{1\})$ are different is when $\#S = 10$.
There are 11!/2!9! different S with 1 in it. So:

$$\phi_1 = \frac{9!2!}{12!} \cdot \frac{11!}{2!9!} \cdot (1 - 0) = \tfrac{1}{12}.$$

By symmetry, $\phi_i = \tfrac{1}{12}$, $1 \leqslant i \leqslant 12$. The symmetry of the game gives the nucleolus as also symmetric, i.e. $x = (\tfrac{1}{12},\tfrac{1}{12},\ldots,\tfrac{1}{12})$. With judge, $v(S) = 1$, if $1 \in S$ and $\#S \geqslant 11$, otherwise $v(S) = 0$. If $x = (x_1,\ldots,x_{13}) \in C(v)$, $x_1 + \ldots + x_{12} \geqslant 1$, $x_{13} \geqslant 0$, $x_1 + \ldots + x_{13} = 1 \Rightarrow x_{13} = 0$. Similarly, we show $x_i = 0$, $2 \leqslant i \leqslant 13$, which leaves $x = (1,0,0,\ldots,0)$:

$$\phi_1 = {}_{12}C_{10}\,\frac{10!2!}{13!} + {}_{12}C_{11}\,\frac{11!1!}{13!} + \frac{12!0!}{13!} = \frac{18}{78}\,.$$

$$\phi_i = {}_{11}C_9\,\frac{10!2!}{13!} = \frac{5}{78}\,, \quad 2 \leqslant i \leqslant 13.$$

Since the nucleolus must be in the core if it exists, $N(v) = (1,0,\ldots,0)$. This game is simple, so stable sets are of the form V_s, where $V_s = \{x \in E(v),\ x_i = 0, \text{ if } i \notin S\}$, for all minimum winning coalitions S. The minimum winning coalition S, are those with $1 \in S$, $\#S = 11$.

4.9. If n is odd, the chairman's vote is only decisive under either rule if the other members split evenly for and against the motion. This is also true of any other committee member. So $\phi_i = 1/n$ for them all. If n is odd, then under rule (a) since there will be no ties, the chairman never votes and so $\phi_1 = 0$. Under (b) if the chairman joins a coalition of $\tfrac{1}{2}n$ or $\tfrac{1}{2}n - 1$ players he changes it from a losing to a winning coalition. There are $(n - 1)!/(\tfrac{1}{2}n)!\,(\tfrac{1}{2}n - 1)!$ coalitions of $\tfrac{1}{2}n$ players not including the chairman and a similar number of $\tfrac{1}{2}n - 1$ player coalitions. Thus:

$$\phi_1 = \frac{(n-1)!}{(\tfrac{1}{2}n)!(\tfrac{1}{2}n - 1)!} \cdot \frac{(\tfrac{1}{2}n)!(\tfrac{1}{2}n - 1)!}{n!} + \frac{(n-1)!}{(\tfrac{1}{2}n)!(\tfrac{1}{2}n - 1)!} \cdot \frac{(\tfrac{1}{2}n - 1)!(\tfrac{1}{2}n)!}{n!} = \frac{2}{n}\,.$$

The winning coalitions are (1,2), (2,3), (1,2,3), (1,2,4), (1,2,5), (1,3,4), (1,3,5), (2,3,4), (2,3,5), (2,4,5), (1,2,3,4), (1,2,3,5), (1,2,4,5), (1,3,4,5), (2,3,4,5) and (1,2,3,4,5). To find the Shapley value we must list the coalitions that win with him, but not without him.

For 1: (1,2), (1,2,4), (1,2,5), (1,3,4), (1,3,5), (1,3,4,5):

$$\phi_1 = 1 \cdot \frac{1!3!}{5!} + 4 \cdot \frac{2!2!}{5!} + 1 \cdot \frac{3!1!}{5!} = 1 \cdot \frac{1}{20} + 4 \cdot \frac{1}{30} + 1 \cdot \frac{1}{20} = \frac{7}{30}$$

For 2: (1,2), (2,3), (1,2,3), (1,2,4), (1,2,5), (2,3,4), (2,3,5), (2,4,5), (1,2,4,5), (2,3,4,5):

$$\phi_2 = 2 \cdot 1/20 + 6 \cdot 1/30 + 2 \cdot 1/20 = 2/5 = 12/30.$$

For 3: (2,3), (1,3,4), (1,3,5), (2,3,4), (2,3,5), (1,3,4,5). So: $\phi_3 = 1 \cdot 1/20 + 4 \cdot 1/30 + 1 \cdot 1/20 = 7/20$.

For 4: (1,3,4), (2,4,5) $\phi_4 = 2 \cdot 1/30 = 1/15 = 2/30$.

For 5: (1,3,5), (2,4,5) $\phi_5 = 2 \cdot 1/30 = 1/15 = 2/30$.

4.10. For any imputation x, let the excesses $v(S) - x(S)$ be written $e_i(x,S_i)$, where $e_1(x,S_1) \geqslant e_2(x,S_2) \geqslant \dots$. If $e_i(y,S_i')$ is written similarly with the extra property that if $e_i(y,S_i') = e_{i+1}(y,S_{i+1}')$, we choose S_i', S_{i+1}' to have the same order as they did in S_1, S_2, ... Then $\theta(x) = \theta(y)$ means $e_i(x,S_i) = e_i(y,S_i')$, for all i. However, $S_i \neq S_i'$, for all i, as this would mean $x = y$. Let i^* be the smallest i so $S_i \neq S_i'$. For any S, $e(1/2(x + y),S) = 1/2 e(x,S) + 1/2 e(y,S)$, and so for all $i < i^*$, $e_i(1/2(x + y),S_i) = e_i(x,S_i)$. But if $\bar{S_i}$ is the next coalition for $1/2(x + y)$:

$$e_{i^*}(1/2(x + y),\bar{S}_i) = 1/2 e_i(x,\bar{S}_i) + 1/2 e_i(y,\bar{S}_i) < 1/2 e_i(x,S_{i^*}) + 1/2 e_i(y,S_{i^*}'),$$

and so $\theta(1/2(x + y) < \theta(x)$.

CHAPTER 5

5.1. By concavity of the utility function, $v(1) = u(a,0)$, $v(2) = u(0,b)$ and $v(1,2) = 2u(a/2,b/2)$. By Problem 4.3, the result holds. The maximin values are $v_{\mathrm{I}} = u(1,0) = 2$, $v_{\mathrm{II}} = u(0,1) = 2$. At any Pareto optimal point the tangents to the payoffs $u(x,y)$, $u(1 - x,1 - y)$ must be parallel. For $u(x,y) = (1 + x)(1 + y)$, $\partial u/\partial x = 1 + y$, $\partial u/\partial y = 1 + x$, so $-(\partial u/\partial y)(\partial u/\partial x) = -(1 + x)/(1 + y)$ is the gradient of the tangent. For player 2, $u(1 - x, 1 - y) = (2 - x)(2 - y)$, so $-(\partial u/\partial y)(\partial u/\partial x) = -(x - 2)/(y - 2)$:

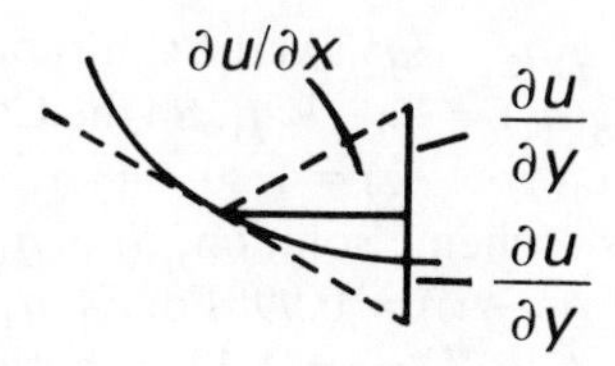

Tangents are parallel if $x = y$. On this curve payoffs are $(1 + x)^2$, $(2 - x)^2$, respectively, so to be greater than the maximin values we require $(1 + x)^2 \geq 2$, i.e. $x \geq \sqrt{2} - 1$ and $(2 - x)^2 \geq 2$, i.e. $x \leq 2 - \sqrt{2}$.

For the maximin bargaining solution, since the payoff region and the status quo point are symmetric, axiom N6 says the solution is symmetric, i.e. $(x,x) = (1 - x, 1 - x)$ or $x = ½$ with $u(½,½) = 9/4$.

5.2. $v(1) = u(0,0)$; $v(2) = v(3) = u(0,b)$; $v(1,2) = v(1,3) = 2u(a/2,b/2)$; $v(2,3) = 2u(0,b)$; $v(1,2,3) = 3u(a/3,2b/3)$. By Theorem 4.2 $(x_1,x_2,x_3) \in C(v)$ if $x_1 \geq u(a,0)$, $x_2 \geq u(0,b)$, $x_3 \geq u(0,b)$, $x_1 + x_2 \geq 2u(a/2,b/2)$, $x_1 + x_3 \geq 2u(a/2,b/2)$, $x_2 + x_3 \geq 2u(0,b)$, $x_1 + x_2 + x_3 = 3u(a/3,2b/3)$. Since $2u(a/2,b/2) \geq u(a,0) + u(0,b)$, the result follows.
Shapley values are $\phi_1 = u(a/3,2b/3) + ⅔u(a/2,b/2) + ⅓u(a,0) - u(0,b)$, $\phi_2 = \phi_3 = u(a/3,2b/3) - ⅓u(a/2,b/2) + ½u(0,b) - ⅙u(a,0)$. By symmetry of 2 and 3 the nucleolus is $(3u(a/3,2b/3) - 2x,x,x)$, where $u(0,b) \leq x \leq 3u(a/3,2b/3) - 2u(a/2,b/2)$. The excesses $v(S) - x(S)$ are $u(0,b) - x$, $2u(a/2,b/2) + x - 3u(a/3,2b/3)$, $2(u(0,b)-x)$, $2x + u(0,a) - 3u(a/3,2b/3)$, 0 and the maximum variable excess is one of the first two terms. To minimise the maximum of these, make them equal, i.e. $x = 3/2u(a/3,2b/3) + ½u(0,b) - u(a/2,b/2)$.

5.3. If $(x_1,x_2,\ldots,x_n) \in C(v)$ by Theorem 4.2, $x_1 + x_2 \ldots + x_n = v(N)$ and $x_2 + x_3 + \ldots + x_n \geq v(2,3,\ldots,n) > v(N)(N - 1)/N$. Hence, $x_1 < v(N)/N$. We can repeat this for all x_i and so $x_1 + x_2 + \ldots + x_n < N(V(N)/N) = v(N)$, which contradicts the definition of a imputation. Hence, no imputation in the core.

5.4. $v(1) = v(2) = 4$; $v(3) = v(4) = 3$; $v(1,2) = 8$, $v(3,4) = 6$; $v(1,3) = v(1,4) = v(2,3) = v(2,4) = 7½$; $v(1,2,3) = v(1,2,4) = 11⅔$; $v(1,3,4) = v(2,3,4) = 10⅔$; $v(1,2,3,4) = 15$.
$E(v) = \{(x_1,x_2,x_3,x_4), x_1 \geq 4, x_2 \geq 4, x_3 \geq 3, x_4 \geq 3, x_1 + x_2 + x_3 + x_4 = 15\}$.
$C(v) = \{(x,x, 7½ - x, 7½ - x), 4⅙ \leq x \leq 4⅓\}$.
$\phi_1 = \phi_2 = 4¼$; $\phi_3 = \phi_4 = 3¼$.
$N(v) = (4¼, 4¼, 3¼, 3¼)$.

5.5. If $q_1 = 1 + ⅓p_2 - ½p_1$, and $q_2 = 1 + ¼p_1 - ½p_2$, then $p_1 = 5 - 3q_1 - 2q_2$, $p_2 = 4½ - 1½q_1 - 3q_2$ and so $e_1(q_1,q_2) = 5q_1 - 3q_1^2 - 2q_1q_2$,

$e_2(q_1,q_2) = 4\frac{1}{2}\, q_1 - 1\frac{1}{2}\, q_1 q_2 - 3q_2^2$. At the Cournot equilibrium, $q_1 = g(q_2) = 5/6 - 1/3 q_2$, $q_2 = g(q_1) = 3/4 - 1/4 q_1$. So the Cournot equilibrium C is $q_1 = 7/11$, $q_2 = 13/22$, $e_1(7/11, 13/22) = 1.20$, $e_2(7/11, 13/22) = 1.047$. If 1 is leader in the Stackleberg scheme solution, S_1 is $q_1 = 7/10$, $q_2 = 23/40$, $e_1(7/10, 23/40) = 1.225$, $e_2(7/10, 23/40) = 0.99$. For S_2, $q_1 = 37/60$, $q_2 = 13/20$, $e_1(37/60, 13/20) = 1.056$, $e_2\ (37/60, 13/20) = 1.14$. If both try to be leaders, we get the solution S where $q_1 = 7/10$, $q_2 = 13/20$, $e_1(7/10, 13/20) = 1.12$, $e_2(7/10, 13/20) = 0.975$. Joint maximisation of profit, J, leads to $q_1 = 3/5$, $q_2 = 2/5$, $e_1(3/5, 2/5) = 1.44$, $e_2(3/5, 2/5) = 0.96$ (i.e. same as the price model). But maximin results are when an opponent makes an infinite amount, so it is best for the player to make zero, then $e_1(0,\infty) = 0$, $e_2(\infty,0) = 0$.

5.6. Cournot reaction functions are $p_1 = g_1(p_2) = p_2 + 5$, and $p_2 = g_2(p_1) = (20 + p_1)/6$. Hence, the Cournot equilibrium is $p_1 = 10$, $p_2 = 5$, $e_1(10,5) = 100$, $e_2(10,5) = 75$. S_1 is $p_1 = 12\frac{1}{2}$, $p_2 = 5\frac{5}{12}$, with $e_1(12\frac{1}{2}, 5\frac{5}{12}) = 104.17$, $e_2(12\frac{1}{2}, 5\frac{5}{12}) = 88.02$. S_2 is $p_1 = 11\frac{1}{4}$, $p_2 = 6\frac{1}{4}$, $e_1(11\frac{1}{4}, 6\frac{1}{4}) = 126.56$, $e_2(11\frac{1}{4}, 6\frac{1}{4}) = 79.68$.
For joint maximisation of profit, $e_1(p_1,p_2) + e_2(p_1,p_2) = 10p_1 + 20p_2 + 3p_1p_2 - p_1^2 - 3p_2^2$, and maximum of this expression is $p_1 = 40$, $p_2 = 70/3$. This contradicts $q_2 = 20 + p_1 - 3p_2 \geqslant 0$ and so the maximum is on the boundary $q_2 = 20 + p_1 - 3p_2 = 0$. Substituting this gives $p_1 = 35$, $p_2 = 55/3$, $e_1(35, 55/3) = 443\frac{1}{3}$, $e_2(35, 55/3) = 0$. Maximin points are M_1, $p_1 = 5$, $p_2 = 0$ guarantees 1, 25; M_2, $p_2 = 10/3$, $p_1 = 0$ guarantees 2, $33\frac{1}{3}$.

5.7. $e_1(p_1,p_2,p_3) = 10p_1 + p_1p_3 - p_1^2$, $e_2(p_1,p_2,p_3) = 10p_2 + p_2p_3 - p_2^2$; $e_3(p_1,p_2,p_3) = 20p_3 + p_1p_3 + p_2p_3 - 4p_3^2$. The Cournot reaction functions are $p_1 = g_1(p_2,p_3) = 5 + \frac{1}{2}p_3$, $p_2 = g_2(p_2,p_3) = 5 + \frac{1}{2}p_3$, $p_3 = 2\frac{1}{2} + \frac{1}{8}p_1 + \frac{1}{8}p_2$. The Cournot equilibrium is $p_1 = p_2 = 50/7$, $p_3 = 30/7$, with profits 50.02, 50.02, 73.4, respectively. Joint maximisation of profit is $p_1 = p_2 = 15$, $p_3 = 10$, with profits 75, 75, 100, respectively. Maximin payoffs are M_1, $25(p_1 = 5, p_2 = p_3 = 0)$; M_2 is $25(p_2 = 5, p_1 = p_3 = 0)$; M_3 is 25 ($p_3 = 2.5$, $p_1 = p_2 = 0$). For Stackleberg strategy, note that 3 plays $p_3 = g_3(p_1,p_2) = 2\frac{1}{2} + \frac{1}{8}p_1 + \frac{1}{8}p_2$. 2 tries to maximise $e_2(p_1,p_2,g_3(p_1,p_2))$. The maxima lie on $p_2 = h_2(p_1) = 50/7 + 1/14\, p_1$. 1 maximises $e_1(p, h_2(p_1), g_3(p_1, h_2(p_1)))$, which leads to $p_1 = 7\frac{71}{97}$, $p_2 = 7\frac{472}{679}$, $p_3 = 4\frac{2327}{5432}$.
As a three person game, the characteristic function is $v(1) = v(2) = 25$, $v(3) = 25$, $v(1,2,3) = 250$ from maximin payoffs and joint maximisation of profit. Maximising profits for two firms with the price of the other set at zero gives the characteristic function of the two player coalition: $v(1,2) = 50$, $v(1,3) = v(2,3) = 100$.

$E(v) = \{(x_1,x_2,x_3),\ x_1 \geqslant 25,\ x_2 \geqslant 25,\ x_3 \geqslant 25,\ x_1 + x_2 + x_3 = 250\}.$
$C(v) = \{(x_1,x_2,250 - x_1 - x_2),\ 25 \leqslant x_i \leqslant 150,\ i = 1,2,\ x_1 + x_2 \leqslant 225\}.$
$\phi_1(v) = \phi_2(v) = 75,\ \phi_3(v) = 100.$
$N(v) = (83\tfrac{1}{3},83\tfrac{1}{3},83\tfrac{1}{3}).$

CHAPTER 6

6.1. In $12G$, 2's strategies are: f_1, always go to theatre; f_2, always go to football; f_3, do same as 1; f_4, do opposite to 1. Whereas 1's sixteen strategies are labelled (t,t,f,t) which describes its response to each of 2's strategies. Payoff matrix is as follows where the first payoff is to 1, second to 2.

1

2	(t,t,t,t)	(t,t,t,f)	(t,t,f,t)	(t,f,t,t),	(f,t,t,t)
f_1	$(1,4)^{12}$	$(1,4)^{12}$	$(1,4)^{12}$	$(1,4)^{12}$	$(0,0)$
f_2	$(0,0)$	$(0,0)$	$(0,0)$	$(4,1)^{1}$	$(0,0)$
f_3	$(1,4)^{2}$	$(1,4)^{2}$	$(4,1)^{1}$	$(1,4)^{2}$	$(1,4)^{2}$
f_4	$(0,0)^{1}$	$(0,0)^{1}$	$(0,0)^{1}$	$(0,0)^{1}$	$(0,0)^{1}$

	(t,t,f,f)	(t,f,t,f)	(t,f,f,t)	(f,t,t,f)	(f,t,f,t)
f_1	$(1,4)^{12}$	$(1,4)^{12}$	$(1,4)^{12}$	$(0,0)$	$(0,0)$
f_2	$(0,0)$	$(4.1)^{1}$	$(4,1)^{1}$	$(0,0)$	$(0,0)$
f_3	$(4,1)^{1}$	$(1,4)^{2}$	$(4,1)^{1}$	$(1,4)^{2}$	$(4,1)^{12}$
f_4	$(0,0)^{1}$	$(0,0)^{1}$	$(0,0)^{1}$	$(0,0)^{1}$	$(0,0)^{1}$

	(f,f,t,t)	(t,f,f,f)	(f,t,f,f)	(f,f,t,f)	(f,f,f,t)	(f,f,f,f)
f_1	$(0,0)$	$(1,4)^{12}$	$(0,0)$	$(0,0)$	$(0,0)$	$(0,0)$
f_2	$(4,1)^{1}$	$(4,1)^{1}$	$(0,0)$	$(4,1)^{1}$	$(4,1)^{12}$	$(4,1)^{12}$
f_3	$(1,4)^{2}$	$(4,1)^{1}$	$(4,1)^{12}$	$(1,4)^{2}$	$(4,1)^{12}$	$(4,1)^{12}$
f_4	$(0,0)^{1}$	$(0,0)^{1}$	$(0,0)^{1}$	$(0,0)^{1}$	$(0,0)^{1}$	$(0,0)^{1}$

$\hat{E}(12G) = \{(t,t),(f,f)\}$ with payoffs (4,1) and (1,4).

6.2. $\hat{R}_1(1G) = \{(b,c);\ (a,d)\}$, $\hat{R}_2(1G) = \{(a,c);\ (a,d);\ (b,d)\}$.
$\hat{E}(1G) = (a,d)$.
$\hat{R}_1(2G) = \{(a,c),(a,d),(b,c)\}$, $\hat{R}_2(2G) = \{(a,c),(b,d)\}$.
$\hat{E}(2G) = (a,c)$.
$\hat{R}_1(21G) = \{(b,c),(a,d)\}$, $\hat{R}_2(21G) = \{(a,c),(a,d),(b,d)\}$.
$\hat{E}(21G) = (a,d)$.
$\hat{R}_1(12G) = \{(a,c),(a,d),(b,c)\}$, $\hat{R}_2(12G) = \{(a,c),(b,d)\}$.
$\hat{E}(12G) = (a,c)$.
No symmetric metaequilibria.

6.3. $\hat{R}_1(1G) = \{(a,b),(b,a)\}$, $\hat{R}_2(1G) = \{(a,b),(b,a),(b,b)\}$, $\hat{E}(1G) = \{(a,b),(b,a)\}$, $\hat{R}_1(2G) = \{(a,b),(b,a),(b,b)\}$, $\hat{R}_2(2G) = \{(a,b),(b,a)\}$, $\hat{E}(2G) = \{(a,b),(b,a)\}$, $\hat{R}_1(21G) = \hat{R}_2(21G) = \hat{E}(21G) = \hat{R}_1(12G) = \hat{R}_2(12G) = \hat{E}(12G) = \{(a,b),(b,a),(b,b)\}$.

6.4. If player 1 is North Vietnam, player 2 is the U.S.
$\hat{R}_1(1G) = \hat{R}_1(2G) = \hat{R}_1(12G) = \hat{R}_1(21G) = \{(c,w),(c,c)\}$, $\hat{R}_2(2G) = \{(w,w),(c,w)\}$, $\hat{R}_2(1G) = \hat{R}_2(21G) = \hat{R}_2(12G) = \{(w,w),(w,c),(c,w)\}$. So $\hat{E}(G) = (c,w)$. Player 1 says we will continue fighting no matter what. Player 2 would rather withdraw than continue.

6.5. If player 1 is North Vietnam, player 2 is the U.S. and player 3 is South Vietnam, for all games $k_1k_2k_3G$.
$\hat{R}_1(k_1k_2k_3G) = \{(c,w,c),(c,c,c),(s,w,s),(c,w,s),(c,c,s)\}$.
$\hat{R}_2(k_1k_2k_3G) = \{(s,w,c),(s,c,c),(c,w,c),(c,c,c),(s,w,s),(s,c,s),(c,w,s)\}$.
$\hat{R}_3(k_1k_2k_3G) = \{(s,w,c),(s,c,c),(c,w,c),(c,c,c),(s,w,s),(s,c,s)\}$.
Therefore, $\hat{E}(G) = \{(c,w,c),(c,c,c),(s,w,s)\}$.

6.6.

	C	L	F
C	metarational 1 metarational 2	metarational 1 rational 2	— metarational 2
L	metarational 1 rational 2	metarational 1 rational 2	— metarational 2
F	metarational 1 —	rational 1 —	rational 1 rational 2

So (F,F) is equilibrium; {(C,C),(C,L),(L,C),(L,L)} are metaequilibrium and analysis is very pessimistic.

6.7. Analysis depends on preferences chosen.

6.8. Analysis depends on preferences chosen.

CHAPTER 7

7.1. Γ_i is the subgame corresponding to the ith move from the end, so Γ_i has payoff matrix:

$$\text{Seeker}\quad \begin{matrix} \\ I_1 \\ I_2 \end{matrix} \begin{matrix} \text{Hider} \\ \begin{matrix} II_1 & II_2 \end{matrix} \\ \begin{pmatrix} \tfrac14 \,\&\, \tfrac34\Gamma_{i-1} & \Gamma_{i-1} \\ \Gamma_{i-1} & \tfrac12 \,\&\, \tfrac12\Gamma_{i-1} \end{pmatrix} \end{matrix}, \qquad \Gamma_0 = (0).$$

If $v_i = \text{val}\,\Gamma_i$, then $v_0 = 0$, $v_1 = 1/6$, $v_2 = 11/36$, $v_3 = 91/216$.

7.2. From the payoff matrices, if $v_{ij} = \text{val}\,\Gamma_{ij}$, we have $v_{11} = v_{12} = v_{13} = 0$; $v_{21} = v_{23} = \frac{1}{2}$; $v_{22} = \frac{1}{3}$; $v_{32} = \frac{7}{11}$; $v_{31} = v_{33} = \frac{5}{7}$.
For the reappearing monster game, the payoff matrices are:

$$\Gamma_{ij}, i = 2,3, j = 1,3 \qquad \begin{pmatrix} 1 \,\&\, \Gamma_{32}, & \Gamma_{i-12} \\ \Gamma_{i-1j}, & 1 \,\&\, \Gamma_{32} \\ \Gamma_{i-1j}, & \Gamma_{i-12} \end{pmatrix}.$$

$$\Gamma_{i2}, i = 2,3 \qquad \begin{pmatrix} 1 \,\&\, \Gamma_{32}, & \Gamma_{i-12}, & \Gamma_{i-13} \\ \Gamma_{i-11}, & 1 \,\&\, \Gamma_{32}, & \Gamma_{i-13} \\ \Gamma_{i-11}, & \Gamma_{i-12}, & 1 \,\&\, \Gamma_{32} \end{pmatrix}.$$

If $v_{ij} = \text{val}\,\Gamma_{ij}$, then we get $v_{21} = v_{23} = \frac{1}{2}(1 + v_{32})$, $v_{22} = \frac{1}{3}(1 + v_{32})$ and $v_{32} = ((1 + v_{32})^2 - 2v_{22}\, v_{23} + v_{23}(1 + v_{32}))/(3 + 3v_{32} - 2v_{22} - v_{23})$. This is satisfied by $v_{21} = v_{23} = \frac{11}{8}$, $v_{22} = \frac{11}{12}$, $v_{32} = \frac{7}{4}$.

7.3. Following calculations similar to (7.26)–(7.29) gives $\mathbf{w}^0 = (0,0)$, $\mathbf{w}^1 = (16,-8)$ and $\mathbf{w}^2 = (14.4, -4.8)$. The bound (7.33) then has $\beta = 0.8$, $\|w^2 - w^1\| = \max\{1.6.3.2\} = 3.2$, so $14.4 - 12.8 \leqslant v_1 \leqslant 14.4 + 12.8$, $-4.8 - 12.8 \leqslant v_2 \leqslant -4.8 + 12.8$. If you do more iterations the bounds become tighter.

7.4. Using value iteration starting with $\mathbf{w}^0 = (0,0)$ gives $\mathbf{w}^1 = (0,2)$, $\mathbf{w}^2 = (0.2,2.4)$ and $\mathbf{w}^3 = (0.32,2.54)$. Use (7.33) with $\beta = 0.5$ and $\|w^3 - w^2\| = \max\{0.12,0.14\} = 0.14$ to get $0.18 \leqslant v_1 \leqslant 0.46$, $2.4 \leqslant v_2 \leqslant 2.68$. Notice that in the third iteration the optimal strategies were $(\mathrm{I}_1,\mathrm{II}_1)$ in Γ_1 and $(\mathrm{I}_2,\mathrm{II}_1)$ in Γ_2. Suppose these were optimal in the infinite horizon game, then v_1 and v_2 would satisfy $v_1 = 0.5\,(0.8v_1 + 0.2v_2)$, $v_2 = 2 + 0.5(0.6v_1 + 0.4v_2)$. This gives $v_1 = \frac{4}{9}$, $v_2 = 2\frac{2}{3}$, and on substituting back into the payoff matrices we have:

$$v_1 = \text{val}\begin{pmatrix} \frac{4}{9} & 4\frac{22}{45} \\ -1\frac{7}{15} & 2\frac{2}{45} \end{pmatrix} = \frac{4}{9}, \qquad v_2 = \text{val}\begin{pmatrix} -1 & 2\frac{14}{5} \\ 2\frac{2}{3} & 4\frac{38}{45} \end{pmatrix} = 2\frac{2}{3}.$$

So this is the solution to the game.

7.5. The payoff matrices are:

$$\Gamma_1 \begin{pmatrix} 2 \,\&\, (\frac{1}{4}\Gamma_1 + \frac{1}{4}\Gamma_2), & 5 \,\&\, (\frac{1}{4}\Gamma_1 + \frac{1}{2}\Gamma_2) \\ -1 \,\&\, (\frac{3}{4}\Gamma_1), & 0 \,\&\, (\frac{1}{2}\Gamma_1 + \frac{1}{2}\Gamma_2) \end{pmatrix},$$

$$\Gamma_2 \begin{pmatrix} -2, & \frac{1}{2}\Gamma_2 \\ -2 \,\&\, \frac{1}{2}\Gamma_1, & -2 \,\&\, (\frac{1}{4}\Gamma_1 + \frac{1}{2}\Gamma_2) \end{pmatrix},$$

with $\beta = \tfrac{2}{3}$. So value iteration gives $\mathbf{w}^0 = (0,0)$, $\mathbf{w}^1 = (2,-2)$, $\mathbf{w}^2 = (2,-1.62)$, $\mathbf{w}^3 = (2.06, -1.58)$. In the bound (7.33) $\beta = \tfrac{2}{3}$, $\|w^3 - w^2\| = 0.06$, so $1.94 \leqslant v_1 \leqslant 2.18$, $-1.70 \leqslant v_2 \leqslant -1.46$. Using I_1, II_1 in Γ_1 always gives $v_1 = 2 + \tfrac{2}{3}(\tfrac{1}{4}v_1 + \tfrac{1}{4}v_2)$, whereas using mixed strategies in Γ_2 gives by (2.42):

$$v_2 = ((-2)(-2 + \tfrac{2}{3}(\tfrac{1}{4}v_1 + \tfrac{1}{2}v_2)) - (\tfrac{1}{3}v_2)(-2 + \tfrac{1}{3}v_1))/$$
$$-4 + \tfrac{1}{6}v_1 + \tfrac{1}{3}v_2 - \tfrac{1}{3}v_2 + 2 - \tfrac{1}{6}v_1.$$

This leads to $v_1 = (-81 + 3\sqrt{929})/5 = 2.09$, $v_2 = -93 + 3\sqrt{929} = -1.56$, which, on checking back in the payoff matrices, are the payoffs.

7.6. Starting with $\mathbf{w}^0 = (0,0)$ payoff matrices are:

$$\Gamma_1 \quad \begin{pmatrix} 2 & 5 \\ -1 & 0 \end{pmatrix} \quad \text{and} \quad \Gamma_2 \quad \begin{pmatrix} -2 & 0 \\ -2 & 2 \end{pmatrix}$$

So the optimal strategy for II is $y = (1,0)$ in both cases. If II always plays this, then the games become:

$$\Gamma_1 \quad \begin{pmatrix} 2\ \&\ (\tfrac{1}{4}\Gamma_1 + \tfrac{1}{4}\Gamma_2) \\ -1\ \&\ \tfrac{3}{4}\Gamma_1) \end{pmatrix}, \qquad \Gamma_2 \quad \begin{pmatrix} -2 \\ -2\ \&\ \tfrac{1}{2}\Gamma_1 \end{pmatrix}$$

So $\mathbf{w}^1 = (w_1^1, w_2^1)$ is the value of such a game, with discount factor $\beta = \tfrac{2}{3}$, satisfies $w_1^1 = \max\{2 - \tfrac{2}{3}(\tfrac{1}{4}w_1^1 + \tfrac{1}{4}w_2^1), -1 + \tfrac{2}{3}\cdot\tfrac{3}{4}w_1^1\}$ and $w_2^1 = \max\{-2, -2, + \tfrac{2}{3}\cdot\tfrac{1}{2}w_1^1\}$. Solving this gives $\mathbf{w}^1 = (2,-2)$ and substituting in payoff matrices leads to:

$$\Gamma_1 \quad \begin{pmatrix} 2 & 4\tfrac{2}{3} \\ 0 & 0 \end{pmatrix}, \quad \Gamma_2 \quad \begin{pmatrix} -2 & -\tfrac{2}{3} \\ -1\tfrac{1}{3} & -2\tfrac{1}{3} \end{pmatrix}$$

Optimal strategies for II are $\mathbf{y} = (1,0)$ in Γ_1, $\mathbf{y} = (\tfrac{5}{7}, \tfrac{2}{7})$ in Γ_2. If II always plays these strategies, then the value of the subsequent game is $\mathbf{w}^2$, where

$$w_1^2 = \max\{2 + \tfrac{1}{6}w_1^2 + \tfrac{1}{6}w_2^2, -1 + \tfrac{1}{2}w_1^2\},$$
$$w_2^2 = \max\{-\tfrac{10}{7} + \tfrac{2}{21}w_2^2, -2 + \tfrac{2}{7}w_1^2 + \tfrac{2}{21}w_2^2\}.$$

This gives $\mathbf{w}^2 = (2.09, -1.55)$, and repeating the procedure gives $\mathbf{w}^3 = (2.09, -1.56)$.

7.7. If I_i (II_i) is defend with i divisions, the payoff matrix of Γ is:

$$\begin{array}{cccc} & II_0 & II_1 & II_2 \\ I_0 & -1 & -1 & 1 \\ I_1 & 1 & 1 & \Gamma \\ I_2 & 1 & \Gamma & \Gamma \\ I_3 & \Gamma & \Gamma & \Gamma \end{array}.$$

An ε'-optimal strategy for I is $x = (\varepsilon^2, \varepsilon, 1 - \varepsilon - \varepsilon^2, 0)$ which guarantees a payoff of $1 - 2\varepsilon - 2\varepsilon^2$ against II_0, $(1 - \varepsilon)/(1 + \varepsilon)$ against II_1, and 1 against II_2.

7.8. Following Example 2.1, let I_1 believe II when he says 'Ace'; I_2 is do not believe II when he says 'Ace', while II's strategies are: II_1, say 'Two' when you have a two, II_2 say 'Ace' when you have a two. If Γ_i is the game where I has £i, the payoff matrix of Γ_i is:

$$\begin{array}{ccc} & II_1 & II_2 \\ I_1 & \tfrac{1}{2}\Gamma_{i-1} + \tfrac{1}{2}\Gamma_{i+1} & \Gamma_{i-1} \\ I_2 & \tfrac{1}{2}\Gamma_{i-2} + \tfrac{1}{2}\Gamma_{i+1} & \tfrac{1}{2}\Gamma_{i-2} + \tfrac{1}{2}\Gamma_{i+2} \end{array},$$

and on substituting v_i for Γ_i we get $v_0 = 0$, $v_4 = 1$:

$$v_1 = \text{val}\begin{pmatrix} \tfrac{1}{2}v_2 & 0 \\ \tfrac{1}{2}v_2 & \tfrac{1}{2}v_3 \end{pmatrix}, \quad v_2 = \text{val}\begin{pmatrix} \tfrac{1}{2}v_1 + \tfrac{1}{2}v_3 & v_1 \\ \tfrac{1}{2}v_3 & \tfrac{1}{2} \end{pmatrix},$$

$$v_3 = \text{val}\begin{pmatrix} \tfrac{1}{2} + \tfrac{1}{2}v_2, & v_2 \\ \tfrac{1}{2} + \tfrac{1}{2}v_1, & \tfrac{1}{2} + \tfrac{1}{2}v_1 \end{pmatrix},$$

which with the inequalities given leads to (i), (ii) and (iii). [v_2 has mixed optimal strategies.] This leads to $v_1 = 1/6$, $v_2 = 1/3$, $v_3 = 7/12$.

7.9. (I_c, II_c) is supergame stable, because it is an equilibrium pair and so by Theorem 7.3 the strategies

$$X_1(n, H_{n-1}) = I_c, \quad X_2(n, H_{n-1}) = II_c, \quad \text{for all } n = 0,1,2,\ldots$$
and all H_{n-1},

are equilibrium pairs which result in (I_c, II_c) always being played. (I_n, II_n) is also supergame stable, by using the strategies:

$$X_1(n, H_{n-1}) = \begin{cases} I_n, & \text{if } h_k = (I_n, II_n), \quad \text{for all } k = 1,2,\ldots,n-1, \\ I_c & \text{otherwise,} \end{cases}$$

and a similar strategy for X_2. The analysis surrounding (7.54) and (7.55) proves these are an equilibrium pair which guarantees $(\mathrm{I}_n,\mathrm{II}_n)$. $(\mathrm{I}_c,\mathrm{II}_n)$ and $(\mathrm{I}_n,\mathrm{II}_c)$ are not supergame stable.

7.10. We can prove any set of prices $(\bar{p}_1,\bar{p}_2,\dots,\bar{p}_n)$ is supergame stable for some $\beta < 1$ if (7.61) holds by looking at the strategies:

$$X_i(n,H_{n-1}) = \begin{cases} \bar{p}_i, & \text{if } h_k = (\bar{p}_1,\bar{p}_2,\dots,\bar{p}_n), \qquad k = 1,2,\dots,n-1, \\ 0, & \text{otherwise,} \end{cases}$$

$i = 1,\dots,n$. It is easy to show, as in (7.53), that $(X_1,X_2,\dots,X_n)$ is an equilibrium n-tuple, which results in $(\bar{p}_1,\bar{p}_2,\dots,\bar{p}_n)$ always being played. As

$$\min_{p_1p_2\dots p_{i-1}p_{i+1}\dots p_n} \max_{p_i} e_i(p_1,p_2,\dots,p_n)$$

$$\geqslant \min_{p_{i_1}p_{i_2}p_{i_k}} \max_{p_i} \min_{p_{i_{k+1}}p_{i_{n-1}}} e_i(p_1,p_2,\dots,p_n),$$

it follows that any strategies $(\bar{p}_1,\bar{p}_2,\dots,\bar{p}_n)$ with

$$e_i(\bar{p}_1,\bar{p}_2,\dots,\bar{p}_n) \geqslant \min_{p_1,p_2\dots p_{i-1}p_{i+1}\dots,p_n} \max_{p_i} e_i(p_1,p_2,\dots,p_n)$$

is metarational for i in any complete metagame based on G. So if the above condition holds for all $i = 1,2,\dots,n$, we have all symmetric metaequilibria.

7.11. If $v_i = \text{val}\,\Gamma_i$ we have, using the discount factor β, that $v_2 = 0$, $v_3 = 1/1 - \beta$ and:

$$v_1 = \text{val} \begin{pmatrix} 1 + \beta v_1 & 0 \\ \beta v_1 & 1/1 - \beta \end{pmatrix},$$

which has mixed optimal strategies:

$$\mathbf{x} = \left(\frac{1}{2-\beta}, \frac{1-\beta}{2-\beta}\right),$$

$\mathbf{y} = (\tfrac{1}{2},\tfrac{1}{2})$, $v_1 = \tfrac{1}{2}(1 - \beta)$. So for $\beta = 0.9$, $\mathbf{x} = (\tfrac{10}{11},\tfrac{1}{11})$, $\mathbf{y} = (\tfrac{1}{2},\tfrac{1}{2})$, $v = 5$; for $\beta = 0.999$, $\mathbf{x} = (\tfrac{1000}{10001},\tfrac{1}{1001})$, $\mathbf{y} = (\tfrac{1}{2},\tfrac{1}{2})$, $v = 500$.

In the average reward case, once in Γ_2 the average reward is 0, and once in Γ_3 the average reward is 1. So if I plays $(x,1-x)$ in Γ_1 against $(y,1-y)$, he can guarantee himself $(1-x)$ if $y < 1$ and x if $y = 1$. So the guaranteed payoff is $\text{maximin}_x\,\{(1-x,x)\} = \tfrac{1}{2}$. II can stop I getting more than $(1-x)$ if $y < 1$ and x if $y = 1$, and as $\min_y \max_x \{(1-x,x)\} = 1$, he cannot guarantee to stop I getting more than 1. So no solution using stationary history-forgetting strategies.

CHAPTER 8

8.1. (a) By diagonal domination, both $\mathbf{x}_1 = (1,0)$ and $\mathbf{x}_2 = (0,1)$ are ESS and by Lemma 8.2, there are no others.

(b) $\mathbf{x} = (4/5, 1/5)$ is the only ESS.

(c) $\mathbf{x} = (1/3, 1/3, 1/3)$ is only the ESS.

8.2. Let fitness matrix be

$$\begin{pmatrix} a & b \\ c & d \end{pmatrix}$$

Then if I_1 dominates I_2, either $a > c$ and $b \geqslant d$, so by the diagonal domination result of Section 8.5.1, I_1 is ESS or $a \geqslant c$, $b > d$. In that case, $e(I_1,I_1) = e(x,I_1)$ for all x, but $e(I_1,x) > e(x,x)$, so I_1 is ESS.

8.3. Since

$$p_k - q_k = \sum_{i=1}^{k-1} (p_i - q_i):$$

$$\sum_{i=1}^{k} \sum_{j=1}^{k} (p_i - q_i)b_{ij}(p_j - q_j) = \sum_{i=1}^{k-1} \sum_{j=1}^{k-1} (p_i - q_i)b_{ij}(p_j - q_j)$$

$$- \sum_{i=1}^{k-1} \sum_{j=1}^{k-1} (p_i - q_i)b_{kj}(p_j - q_j) - \sum_{i=1}^{k-1} \sum_{j=1}^{k-1} (p_i - q_i)b_{ik}(p_j - q_j) +$$

$$\sum_{i=1}^{k-1} \sum_{j=1}^{k-1} (p_i - q_i)\, b_{kk}(p_j - q_j)$$

$$= \sum_{i=1}^{k-1} \sum_{j=1}^{k-1} (p_i - q_i)c_{ij}(p_j - q_j) = 1/2(p - q)^{\mathrm{T}} (C + C^{\mathrm{T}})(p - q).$$

Hence, if (8.67) holds, $(p - q)^{\mathrm{T}}B(p - q) < 0$.

8.4. $e(I_1,I_1) = e(\mathbf{q},I_1) = 0$, for all $\mathbf{q}$. If $\mathbf{q} = (1 - q_2 - q_3, q_2, q_3)$, $e(I_1,\mathbf{q}) = 2q_2 + 2q_3 > (1 - q_2 - q_3)(2q_2 - q_3)(2q_2 + 2q_3) + q_2^2 + q_3^2 = e(\mathbf{q},\mathbf{q})$, so I_1 is ESS.
Since $e(I_1,I_1) = e(I_2,I_1) = e(I_3,I_1) = 0$, then $k = 3$ and so

$$C = \begin{pmatrix} -1 & 1 \\ 1 & 2 \end{pmatrix}$$

and if $\mathbf{z} = (0,1)$, $\mathbf{z}^{\mathrm{T}}\mathbf{C}\mathbf{z} = 2$, which is not negative definite.

8.5. (a) $\mathbf{x} = (3/4, 1/4)$ is ESS.

(b) The payoff matrix of $1G$ is:

	II Hawk	Dove
f_1 always Hawk	(−1,−1)	(6,0)
f_2 always Dove	(0,6)	(3,3)
f_3 same as II	(−1,−1)	(3,3)
f_4 opposite to II	(0,6)	(6,0)

$\hat{E}(1G) = \{(\mathrm{h},\mathrm{d}),(\mathrm{d},\mathrm{h})\}$.

(c) The fitness matrix is:

	Hawk	Dove	Bully	Retaliator
Hawk	−1	6	6	−1
Dove	0	3	0	3
Bully	0	6	3	−1
Retaliator	−1	3	−1	3

If $\mathbf{p} = (\frac{3}{4},\frac{1}{4})$ $e(\mathbf{p},\mathbf{p}) = \frac{3}{4} = e(\mathrm{I}_1,\mathbf{p}) = e(\mathrm{I}_3,\mathbf{p})$, whercas $e(\mathrm{I}_2,\mathbf{p})$ and $e(\mathrm{I}_4,\mathbf{p})$ is less than $\frac{3}{4}$. So we need only apply (8.4) for $\mathbf{y} = (y,0,1-y,0)$. The same analysis as in part (a) now gives the result. Retaliator, I_4, satisfies $e(\mathrm{I}_4,\mathrm{I}_4) = e(\mathrm{I}_2,\mathrm{I}_4) = 3$. Since $e(\mathrm{I}_4,\mathrm{I}_2) = e(\mathrm{I}_2,\mathrm{I}_2)$ retaliator is not ESS as we need $e(\mathrm{I}_4,\mathrm{I}_2) > e(\mathrm{I}_2,\mathrm{I}_2)$.

8.6. (a) $\mathbf{p}^0 = \mathbf{p}^1 = \mathbf{p}^2 = \mathbf{p}^3 = \mathbf{p}^4 = \mathbf{p}^5 = \mathbf{p}^6 = (0.5,0.5)$.

(b) $\mathbf{p}^0 = (0.5,0.5)$, $\mathbf{p}^1 = (0.64,0.36)$, $\mathbf{p}^2 = (0.70,0.30)$, $\mathbf{p}^3 = (0.74,0.26)$, $\mathbf{p}^4 = (0.76,0.24)$, $\mathbf{p}^5 = (0.77,0.23)$, $\mathbf{p}^6 = (0.78,0.22)$.

(c) $\mathbf{p}^0 = \mathbf{p}^1 = \mathbf{p}^k = (\frac{1}{3},\frac{1}{3},\frac{1}{3})$, for all k.
$\mathbf{p}_0 = \mathbf{p}_2 = \mathbf{p}_4 = \mathbf{p}_6 = (0.4,0.4,0.2)$, $\mathbf{p}_1 = \mathbf{p}_3 = \mathbf{p}_5 = (0.25,0.5,0.25)$.

8.7. If $p_1 > p_2$, $e(\mathbf{p}_1,\mathbf{p}_2) = V(1-p_2) + \frac{1}{2}(V-D)p_2$, and if $p_1 = p_2$, $e(p_1,p_1) = \frac{1}{2}V(1-p_1) + \frac{1}{2}(V-D)p_1$. Hence, $e(p+h,p) - e(p,p) = \frac{1}{2}V(1-p) > 0$, and so choosing a fixed p is not ESS:

$$e(p,f) = \int_0^p (V(1-x) + \tfrac{1}{2}(V-D)x)\alpha(1-x)^{\alpha-1}\mathrm{d}x + \int_p^1 \tfrac{1}{2}(V-D)p\alpha(1-x)^{\alpha-1}\mathrm{d}x = 0.$$

So f is part of an equilibrium pair. It is the only such mixed strategy which you can check by solving $de(\mathrm{p},g)/d\mathrm{p} = 0$. Any mixed equilibrium pair g must satisfy that. It leads to $g(x) = \alpha(1-x)^{\alpha-1}$.

To find whether f is ESS we must prove $e(f,g) > e(g,g)$ for all other strategies g. The proof follows that of 'war of attrition'. Since $e(f,f) = e(g,f)$ it is enough to show that

$$T(f,g) = e(f,f) + e(g,g) - e(f,g) - e(g,f) < 0.$$

Since:

$$e(f,g) + e(g,f) = V - D\int_0^1 x(f(x)G(x) + g(x)\bar{F}(x))\mathrm{d}x,$$

we can show that:

$$T(f,g) = -D\int_0^1 (\bar{F}(x) - \bar{G}(x))^2\mathrm{d}x < 0.$$

8.8. Let $\bar{\mathbf{x}} = (1-h)\mathbf{x} + h\mathbf{r}$, $\bar{\mathbf{y}} = (1-h)\mathbf{y} + h\mathbf{s}$. (8.55) leads to:

$$(1-h)[\mathbf{x}^{\mathrm{T}}A\mathbf{x} + \mathbf{x}^{\mathrm{T}}B\mathbf{y} + \mathbf{y}^{\mathrm{T}}C\mathbf{x} + \mathbf{y}^{\mathrm{T}}D\mathbf{y}] + h[\mathbf{x}^{\mathrm{T}}A\mathbf{r} + \mathbf{x}^{\mathrm{T}}B\mathbf{s} + \mathbf{y}^{\mathrm{T}}C\mathbf{r} + \mathbf{y}^{\mathrm{T}}D\mathbf{s}]$$

$$> (1-h)[\mathbf{r}^{\mathrm{T}}A\mathbf{x} + \mathbf{r}^{\mathrm{T}}A\mathbf{y} + \mathbf{s}^{\mathrm{T}}C\mathbf{x} + \mathbf{s}^{\mathrm{T}}D\mathbf{y}] + h[\mathbf{r}^{\mathrm{T}}A\mathbf{r} + \mathbf{r}^{\mathrm{T}}B\mathbf{s} + \mathbf{s}^{\mathrm{T}}C\mathbf{r} + \mathbf{s}^{\mathrm{T}}D\mathbf{s}].$$

For this to hold for all small h, the $(1-h)$ term on the L.H.S. of the inequality is as great as the $(1-h)$ term on the R.H.S. of the inequality, while the h term on the L.H.S. must be greater than the h term on the R.H.S. This gives the result.

8.9. The first inequality follows from putting $\mathbf{s} = \mathbf{y}$ in (8.57) and the second from putting $\mathbf{r} = \mathbf{x}$ in (8.57).
Looking at $e_1(\mathbf{r},(\mathbf{x},\mathbf{y}))$, where $\mathbf{x} = (x,1-x)$, $\mathbf{y} = (y,1-y)$, $\mathbf{r} = (r,1-r)$, we have this as maximised at $r = x = 1$ if $7/12 > y$; $r = \mathrm{x} =$ anything if $6x = 13 - 12y$. Similarly, looking at $e_2(\mathbf{s},(\mathbf{x},\mathbf{y}))$, where $\mathbf{s} = (s,1-s)$, we have $s = y = 1$ if $x > 5/6$; $s = y =$ anything if $12x = 3y + 7$ and $s = y = 0$ if $x < 7/12$. The only possible values of x and y that satisfy both sets of conditions is $\tilde{x} = 41/54$, $\tilde{y} = 19/27$. $e_1(\mathbf{r},((41/54,13/54),(19/27,8/27)) + e_2(\mathbf{s}((41/54,13/54),(19/27,3/27))$ is constant for all r and s. So to check if this pair $(\tilde{\mathbf{x}},\tilde{\mathbf{y}})$ is ESS we must see if it satisfies inequality (8.58). Since $e_1(I_1,(\mathrm{I}_1,\mathrm{II}_1)) + e_2(\mathrm{II}_1,(\mathrm{I}_1,\mathrm{II}_1)) > e_1(\tilde{\mathbf{x}},(\mathrm{I}_1,\mathrm{II}_1)) + e_2(\tilde{\mathbf{y}},(\mathrm{I}_1,\mathrm{II}_1))$, $(\tilde{\mathbf{x}},\tilde{\mathbf{y}})$ is not ESS.

CHAPTER 9

9.1. Maximin strategies are if an opponent bids your value of the item, giving you zero reward. So $v_{\mathrm{I}} = v_{\mathrm{II}} = 0$. The Pareto optimal boundary is $A_1 \cup A_2$, where $A_1 = \{(u,v)|v_2u + (v_1-2)v = v_2(v_1-1),\ \tfrac{1}{2}v_1 \leqslant u \leqslant v_1 - 1\}$ and $A_2 = \{(u,v)|(v_2-2)u + v_1v = v_1(v_2-1),\ \tfrac{1}{2}v_2 \leqslant v \leqslant v_2 - 1\}$. The maximum of $(u-0)(v-0)$ on A_1 is at $u = v = \tfrac{1}{2}v_1$, since the derivative is zero at $u = \frac{1}{2}(v_1 - 1)$, which is outside the range. Similarly, uv is maximised at $u = v = \frac{1}{2}v_1$ on A_2, so this is the maximin bargaining solution.

9.2. (a) From the payoff matrix (9.8) we have $A_0 = 2y_0$, $A_1 = 3y_0 + 1\frac{1}{2}y_1$, $A_2 = 2y_0 + 2y_1 + y_2$, $A_3 = y_0 + y_1 + y_2 + \frac{1}{2}y_3$, $A_4 = 0$, $B_0 = x_0$, $B_1 = x_0 + \frac{1}{2}x_1$, $B_2 = 0$, $B_3 = -x_0 - x_1 - x_2 - \frac{1}{2}x_3$, $B_4 = -2x_0 - 2x_1 - 2x_2 - 2x_3 - x_4$. $B_3 > B_4 \Rightarrow y_4 = 0 \Rightarrow A_3 > A_4 \Rightarrow x_4 = 0 \Rightarrow B_2 > B_3 \Rightarrow y_3 = 0$. $\text{Max}(A_1,A_2) > A_0 \Rightarrow x_0 = 0$. $A_2 > A_3$ unless $y_2 = 1$. If $y_2 = 1 \Rightarrow x_1 = 0 \Rightarrow A_2 = A_3$ and says $(0,0,x,1-x,0)$ v. II_2, $0 \leqslant x \leqslant 1$ is an equilibrium point. If $A_2 > A_3 \Rightarrow y_2 < 1 \Rightarrow x_3 = 0$. Either $x_1 = 0 \Rightarrow \frac{1}{2}y_1 + y_2 \geqslant y_0$, which gives I_2 v. $(y_0,y_1,y_2,0,0)$, $\frac{1}{2}y_1 + y_2 \geqslant y_0$, or $y_1 = 1 \Rightarrow A_2 > \max(A_1,A_3) \Rightarrow x_2 = 1$. So I_2 v. II_1 is also an equilibrium point but is contained in a previous set of equilibrium points.

(b) $A_0 = 2y_0$, $A_1 = 3y_0 + 1\frac{1}{2}y_1$, $A_2 = 2y_0 + 2y_1 + y_2$, $A_3 = y_0 + y_1 + y_2 + \frac{1}{2}y_3$, $A_4 = 0$, $B_0 = 1\frac{1}{2}x_0$, $B_1 = 2x_0 + x_1$, $B_2 = x_0 + x_1 + \frac{1}{2}x_2$, $B_3 = 0$, $B_4 = -x_0 - x_1 - x_2 - x_3 - \frac{1}{2}x_4$. $B_3 > B_4 \Rightarrow y_4 = 0 \Rightarrow A_3 > A_4 \Rightarrow x_4 = 0$. $\text{Max}(A_1,A_2) > A_0 \Rightarrow x_0 = 0$. Either $x_3 = 1$ or $B_2 > B_3$. $x_3 = 1$ gives equilibrium pairs I_3 v. $(y_0,y_1,y_2,y_3,0)$, $\frac{1}{2}y_3 \geqslant y_0 + y_1$ and $\frac{1}{2}y_3 + y_2 \geqslant 2y_0 + \frac{1}{2}y_1$. If $B_2 > B_3 \Rightarrow y_3 = 0$, either $y_2 = 1$ or $x_3 = 0$. If $y_2 = 1 \Rightarrow x_1 = 0$ and shows, $(0,0,x_2,x_3,0)$ v. II_2 are equilibrium pairs. If $x_3 = 0 \Rightarrow y_0 = 0 \Rightarrow A_2 > A_1 \Rightarrow x_1 = 0 \Rightarrow B_2 > B_1 \Rightarrow y_2 = 1$. This is contained in the above set of equilibrium points.

9.3. The standard argument in Example 9.1 shows that an equilibrium strategy can only be a mixture of bids betweeen 0 and v. Using the notation of that example, $(\mathbf{x},\mathbf{x})$ is a symmetric equilibrium pair only if when $x_i \neq 0$, $A_i = \max_{1 \leqslant j \leqslant v} A_j$, where $A_j = (v - j)(\frac{1}{2}x_j + x_{j+1} + \ldots + x_B)$. Suppose $x_0 > 0$, then as $v > 2$, $A_1 = (v - 1)(x_0 + \frac{1}{2}x_1)) > A_0 = \frac{1}{2}vx_0$, which makes $x_0 = 0$. Similarly, if $x_0 = x_1 = x_{i-1} = 0$, then if $x_i > 0$, provided $(v - i - 1) > \frac{1}{2}(v - i)$, than $A_{i+1} > A_i$, which in turn makes $x_i = 0$. This holds for all $i < v - 2$. So $x = (0,0,\ldots,0,x_{v-2},x_{v-1},x_v)$ and $A_v = 0$, $A_{v-1} = \frac{1}{2}x_{v-1} + x_{v-2}$, $A_{v-2} = x_{v-2}$. If $x_v \neq 0 \Rightarrow A_v \geqslant A_{v-1} \Rightarrow x_{v-1} = x_{v-2} = 0$, so $x_v = 1$ is an equilibrium pair. If $x_v = 0$, either $x_{v-1} = 0$ so $x_{v-2} = 1$ which is an equilibrium pair, or $x_{v-1} \neq 0$. In that case we have $A_{v-1} > A_{v-2}$ and so $x_{v-1} = 1$. Thus equilibrium pairs are $x_v = 1$ (I_v,II_v), $x_{v-1} = 1$ $(\text{I}_{v-1},\text{II}_{v-1})$ and $x_{v-2} = 1$ $(\text{I}_{v-2},\text{II}_{v-2})$.

9.4. Analysis is identical with Dutch auction of Problem 9.1.

9.5. From (9.19a), $b_1 > v \Rightarrow b_1 \geqslant b_2 + 1$; from (9.19c), $b_1 < v \Rightarrow b_1 + 1 \leqslant b_2$; from (9.18a), $b_2 > v \Rightarrow b_2 \geqslant b_1 + 1$, while (9.18c) gives $b_2 < v \Rightarrow b_2$

$+ 1 \leqslant b_1$. Hence if $b_1 > v$ we must have $b_1 \geqslant b_2 + 1$ which then requires $b_2 \leqslant v$; if $b_1 < v$, then (9.19) requires $b_1 + 1 \leqslant b_2$, which can only be satisfied by (9.18) if $b_2 \geqslant v$. Lastly, if $b_1 = v$, any b_2 completes an equilibrium pair.

9.6. $v(1) = 2$, $v(2) = v(3) = 0$, $v(1,2) = 6$, $v(1,3) = 2$, $v(2,3) = 0$, $v(1,2,3) = 10$. The set of imputations $= \{(x_1,x_2,x_3): x_1 \geqslant 2, x_2,x_3 \geqslant 0\ x_1 + x_2 + x_3 = 10\}$. The core $= \{(2 + x,y,8 - x - y) | x,y \geqslant 0\ 4 \leqslant x + y \leqslant 8\}$. The nucleolus is (5,3,2).

9.7. If $P(v)$ is profit to a bidder if his value is v and $H(v)$ is highest bid, then for a Dutch auction:

$$\text{Var } P(v) = v^{n+1}/n - (v^n/n)^2, \text{ so Exp Var } P(v) = (2n^2-1)/(n+1)n^2(2n+1).$$

$$\text{Var } H(v) = (n-1)^2/n(n+1)^2(n+2).$$

For English auction:

$$\text{Var } P(v) = (2v^{n+1}/n) - (v^n/n)^2, \text{ so Exp Var } P(v) = (2n-1)/n^2(n+2).$$

$$\text{Var } H(v) = (2n^2 + 6n - 1)/(n+1)^2(n+2).$$

9.8. If with one item gone I has x left to spend and II has y to spend, then (9.39) and (9.40) say I should bid up to $\min\{x,\frac{1}{2}y + 10\}$, while II should bid up to $\min\{y,\frac{1}{2}x + 10\}$ for the second item provided $x > \frac{1}{2}y$ and $y > \frac{1}{2}x$. We can use this to find out what the outcome is to each player if, with the opponent having bid b they decide to bid more or not bid. Such an analysis shows II will not bid more than 20 for the first item, because then I can get both remaining items. So I gets the first item for 20, the second item for 35, giving a profit of 115, while II gets the third item for 5 and a profit of 75.

9.9. If your last bid was b cents and the other player has just bid $b + 5$ cents, you should always bid $b + 10$, as this involves a possible loss of $b - 100$, whereas no bid involves losing b cents! Such an analysis leads to escalation of bidding to well above a dollar. The only possible equilibrium pair is in the extensive form of a game where the first bidder bids a dollar and the second does not then bid. We really have a multi-stage game where Γ_i is the game when the last bid is i by the opponents. The payoffs are then $(-i + 5)$ if you do not bid and your last bid was $i - 5$, or Γ_{i+5} if you bid $i + 5$. Such a game does not have a value.

$v(1) = -95$, $v(2) = v(3) = v(1,2) = v(1,3) = v(1,2,3) = 0$, $v(2,3) = 95$,

where bidders 2 and 3 can get 95 cents if one bids 5 cents and the other zero. The Shapley value is $\phi_1 = -66.6^r$, $\phi_2 = \phi_3 = 33.3^r$.

References

Abakuks, A., 1980, 'Conditions for evolutionarily stable strategies', *J. Appl. Prob.*, **17**, 559–562.

Aubin, J-P., 1979, *Mathematical Methods of Game and Economic Theory*, North-Holland, Amsterdam.

Aumann, R.J., 1959, 'Acceptable points in general cooperative *n*-person games', in: *Contributions to the Theory of Games, Vol. IV* (Ann. Math. Studies No. 40), ed. Tucker, A.W. and Luce, R.D., Princeton University Press, Princeton, pp. 287–324.

Aumann, R.J., 1960, 'Acceptable points in games of perfect information', *Pacific J. of Maths.*, **10**, 381–417.

Aumann, R.J., 1967, 'A survey of cooperative games without side payments', in: *Essays in Mathematical Economics*, ed. M. Shubik, Princeton University Press, Princeton, pp. 3–27.

Aumann, R.J. and M. Maschler, 1964, 'The bargaining set for cooperative games', in: *Advances in Game Theory* (Ann. Math. Studies No. 52), eds. M. Dresher, L.S. Shapley and A.W. Tucker, Princeton University Press, Princeton, pp. 443–447.

Aumann, R.J. and M. Maschler, 1967a, 'Repeated games with incomplete information. The zero sum extensive case', in: *Report to the U.S. Arms Control and Disarmament Agency*, Mathematics Policy Research, Inc., Princeton, pp. 25–108.

Aumann, R.J. and M. Maschler, 1967b, 'Repeated games with incomplete information. A survey of recent results', in: *Report to the U.S. Arms Control and Disarmament Agency*, Mathematics Policy Research Inc., Princeton, pp. 287–403.

Axelrod, R., 1980, 'Effective choice in the prisoner's dilemma', *J. Conflict Resolution,* **24**, 3–26.

Bacharach, M., 1977, *Economics and the Theory of Games*, Macmillan Press, London.

Beckman, M.J., 1974. 'A note on cost estimation and the optimal bidding strategy', *Oper. Res.*, **22**, 510–513.

Bennet, P.G., 1977, 'Towards a theory of hypergames', *Omega*, **5**, 749–751.

Bennet, P.G. and M.R. Dando, 1979, 'Complex strategic analysis: A hypergame study of the "fall of France" ', *J. Oper. Res. Soc.,* **30**,. 23–32.

Bennet, P.G. and C.S. Huxham, 1982, 'Hypergames and what they do', *J. Oper. Res. Soc.,* **33**, 41–50.

Bennet, P.G., M.R. Dando, and R.G. Sharp, 1980, 'Using hypergames to model difficult social issues: An approach to the crisis of soccer hooliganisms', *J. Oper. Res. Soc.,* **31**, 621–634.

Bertrand, J., 1883, 'Review of "Théorie mathematique de la richesse sociale" ', *Journal des Savants*, 499–509, Paris.

Bewley, T. and E. Kohlberg, 1976, 'Asymptotic theory of stochastic games', *Mathematics of Oper. Res.,* **1**, 197–208.

Billera, L.J., 1981, 'Economic market games', in: '*Proceedings of Symposia in Applied Mathematics*, Vol. 24, *Game Theory and its Applications*', ed. W.F. Lucas, A.M.S. Rhode Island, pp. 55–68.

Bishop, D.T. and C. Canning, 1978, 'A generalised war of attrition', *J. Theor. Biol.,* **70**, 85–124.

Blackett, D.W., 1954, 'Some Blotto games', *Nav. Rev. Log. Quart.,* **1**, 55–60.

Blackwell, D. and T.S. Ferguson, 1968, 'The big match', *Am. Math. Stat.,* **39**, 159–163.

Böhm-Bawerk, E. von, 1888, *Interest and Capital*, South Holland, Amsterdam.

Borel, E., 1953, 'The theory of play and integral equations with skew symmetric kernels'; 'On games that involve chance and the skill of the players', and 'On systems of linear forms of skew symmetric determinants and the general theory of play', Translated by L.J. Savage, *Econometrica*, **21**, 97–117.

Bowen, K.C., 1981, Letter to the Editor, *Europ. J. Op. Res.* **8**, 88.

Brams, S.J., 1980, *Biblical Games*, M.I.T. Press, Cambridge, Massachusetts.

Brams, S.J. and P.D. Straffin, 1979, 'Prisoner's dilemma and professional sports draft', *Amer. Math. Monthly,* **86**, 80–86

Brewer, G. and M. Shubik, 1979, *The War Game,* Harvard Univ. Press, Cambridge, Mass., U.S.A.

Brown, G.W., 1951, 'Iterative solutions of games by fictitious play', in: *Activity Analysis of Production and Allocation*, ed. T.C. Koopmans, Wiley, New York, pp. 374–376.

Cassady, R., Jr, 1967, *Auctions and Auctioneering,* Univ. of California

Press, Berkeley, California.
Chamberlin, E.H., 1948, 'An experimental imperfect market', *J. Political Econ.*, **56**, 95–108.
Charnes, A. and R.G. Schroeder, 1967, 'On some stochastic tactical anti-submarine games', *Nav. Rev. Log. Quart.*, **14**, 291–311.
Cohen, K.J., W. Dill, A. Kuehn and P. Winters, 1964, *The Carnegie Tech Management Game: An experiment in business education,* Irwin, Homewood, Illinois.
Conway, J.H., 1976, *On Numbers and Games*, Academic Press, London.
Cooper, D.F., 1979, 'The superior commander: A methodology for the control of crisis games', *J. Oper. Res. Soc.*, **30**, 529–537.
Cooper, D.F., J. Klein, R.C. McDowell and P.V. Johnson, 1980, 'The development of a research game', *J. Oper. Res. Soc.*, **31**, 191–193.
Cournot, A.A., 1838, *Recherches sur les Principles Mathematiques de la theorie des Richesses* (translated 1897), Macmillan, New York.
Dantzig, G.B., 1951, 'A proof of the equivalence of the programming problem and the game problem', in: *Activity Analysis of Production and Allocation*, ed. T.C. Koopmans, Wiley, New York, pp. 330–338.
Davis, M. and M. Maschler, 1965, 'Kernel of a cooperative game', *Nav. Rev. Log. Quart.*, **12**, 223–259.
Deshmukh, S.D. and W. Winston, 1978, 'A zero-sum stochastic game model of duopoly', *Int. J. Game Theory*, **7**, 19–30.
Downton, F., 1983, 'Rational Roulette', to appear in *Bull. Australian Math. Soc.*
Drescher, M., A.W. Tucker and P. Wolfe (eds.), 1957, *Contributions to the Theory of Games, Vol. III* (Ann. Math. Studies No. 39), Princeton University Press, Princeton.
Duke, R.D., 1981, 'Letter to the Editor', *Europ. J. Oper. Res.*, **8**, 88–89.
Edgeworth, F.Y., 1881, *Mathematical Psychics*, Keegan, Paul, London.
Engelbrecht-Wiggans, R., 1980, 'Auctions and bidding models. A survey', *Mgmt. Sci.*, **26**, 119–142.
Everett, H., 1957. 'Recursive games', in: *Contributions to the Theory of Games, Vol. III* (Ann. Math. Studies No. 39), by M. Drescher, A.W. Tucker and P. Wolfe, Princeton University Press, Princeton, pp. 47–78.
Frazer, N.M. and K.W. Hipel, 1980, 'Metagame analysis of the Poplar River Conflict', *J. Oper. Res. Soc.*, **31**, 377–385.
Frazer, N.M. and K.W. Hipel, 1981, 'Computer assistance in Labor-management Negotiations', *Interfaces,* **11**, 22–29.
Friedman, J.W., 1977, *Oligopoly and the Theory of Games*, North-Holland, Amsterdam.
Gale, D., H.W. Kuhn and A.W. Tucker, 1951, 'Linear programming and the theory of games', in: *Activity Analysis of Production and*

Allocation, ed. T.C. Koopmans, Wiley, New York, pp. 317–329.
Gillette, D., 1957, 'Stochastic games with zero stop probabilities', in: *Contributions to the Theory of Games, III* (Ann. Math. Studies, No. 39), ed. M. Drescher, A.W. Tucker and P. Wolfe, Princeton University Press, Princeton, pp. 179–187.
Gillies, D.B., 1959, 'Solutions to general non-zero-sum games, in: *Contributions to the Theory of Games, Vol. IV* (Ann. Math. Studies, No. 40), ed. A.W. Tucker and D.R. Luce, Princeton University Press, Princeton, pp. 47–85.
Grafen, A., 1979, 'The Hawk–Dove game played between relatives', *Anim. Behav.,* **27**, 905–907.
Griesmer, J.H. and M. Shubik, 1963a, 'Towards a study of bidding processes. Some constant sum games', *Nav. Rev. Log. Quart.,* **10**, 11–21.
Griesmer, J.H. and M. Shubik, 1963b, 'Towards a study of bidding processes, II. Games with capacity limitations', *Nav. Rev. Log. Quart.,* **10**, 151–173.
Griesmer, J.H. and M. Shubik, 1963c, 'Towards a study of bidding processes, III. Some special models', *Nav. Rev. Log. Quart.,* **10**, 193–217.
Griesmer, J.H., M. Shubik and R.E. Levitan, 1967, 'Towards a study of bidding processes, IV. Games with unknown costs', *Nav. Rev. Log. Quart.,* **14**, 415–433.
Guyer, M. and B. Perkel, 1972. *Experimental Games. A Bibliography 1945–1971*, Communication No. 293, Mental Health Research Institute, Univ. of Michigan.
Haigh, J., 1975, 'Game theory and evolution', *Adv. Appl. Prob.,* **7**, 8–11.
Haigh, J. and M.C. Rose, 1980, 'Evolutionary game auctions', *J. Theor. Biol.,* **85**, 381–397.
Hamburger, H., 1979, *Games as Models of Social Phenomena*, W.H. Freeman, New York.
Hammerstein, P., 1981, 'The role of asymmetries in animal contests', *Anim. Behav.,* **29**, 193–205.
Harsanyi, J.C., 1959, 'A bargaining model for the cooperative *n*-person game', in: *Contributions to the Theory of Games, Vol. IV* (Ann. Math. Studies, No. 40). ed. A.W. Tucker and D.R. Luce, Princeton University Press, Princeton, pp. 325–356.
Harsanyi, J.C., 1963, 'A simplified bargaining model for the *n*-person cooperative game', *Int. Econ. Review,* **4**, 194–220.
Harsanyi, J.C., 1967, 'Games with incomplete information played by Bayesian players, I', *Mgmt. Sci.,* **14**, 159–182.
Harsanyi, J.C., 1968a, 'Games with incomplete information played by Bayesian players, II', *Mgmt. Sci.,* **14**, 320–334.

Harsanyi, J.C., 1968b, 'Games with incomplete information played by Bayesian players, III', *Mgmt. Sci.,* **14**, 486–502.

Harsanyi, J.C., 1977, *Rational Behavior and Bargaining Equilibrium in Games and Social Situations*, Cambridge Univ. Press, Cambridge, U.K.

Hines, W.G.S., 1977, 'Competition with an evolutionary stable strategy', *J. Theor. Biol.,* **67**, 141–153.

Hirsch, M.W. and S. Smale, 1974, *Differential Equations, Dynamic Systems and Linear Algebra*, Academic Press, New York.

Hofbauer, J., P. Schuster and K. Sigmund, 1979, 'A note on ESS and game dynamics', *J. Theor. Biol.,* **81**, 609–612.

Hotelling, H., 1929, 'Stability in competition', *Economic Journal,* **39**, 41–57.

Howard, N., 1966a, 'Theory of metagames', *General Systems,* **11**, 167–186.

Howard, N., 1966b, 'The mathematics of metagames', *General Systems,* **11**, 187–200.

Howard, N., 1968. 'Theory of metagames', Ph.D. Thesis, Univ. of London.

Howard, N., 1970, 'Some developments in the theory and applications of metagames', *General Systems,* **15**, 205–231.

Howard, N., 1971, *Paradoxes of Rationality*, M.I.T. Press, Cambridge, Massachusetts.

Inbar, M. and C.S. Stoll, 1972, *Simulation and Gaming in Social Science*, Free Press, New York.

Isaacs, R., 1965, *Differential Games*, Wiley, New York.

Jones, A.J., 1980, *Game Theory. Mathematical Models of Conflict*, Ellis Horwood, Chichester, England.

Karlin, S., 1959, *Mathematical Methods and Theory in Games, Programming and Economics, Vol. II. Theory of Infinite Games*, Addison Wesley, Reading, Massachusetts.

Kirman, A. and M. Sobel, 1974, 'Dynamic oligopoly with inventories', *Econometrica,* **42**, 279–287.

Kohlberg, E., 1975, 'Optimal strategies in repeated games with incomplete information', *Int. J. Game Theory,* **4**, 7–24.

Kuhn, H.W. and A.W. Tucker, 1950 (eds.), *Contributions to the Theory of Games, Vol. I* (Ann. Math. Studies, No. 24), Princeton University Press, Princeton.

Kuhn, H.W. and A.W. Tucker, 1953, *Contributions to the Theory of Games, Vol. II* (Ann. Math. Studies, No. 28), Princeton University Press, Princeton.

Lucas, W.F., 1967, 'A counterexample in game theory', *Mgmt. Sci.,* **13**, 766–767.

Lucas, W.F., 1968, 'A game with no solution', *Bull. Amer. Math. Soc.,* **74**,

237–239.

Lucas, W.F., 1969, 'Games with unique solutions that are non-convex', *Pac. J. Maths.,* **28**, 599–602.

Lucas, W.F., 1971, 'Some recent developments in *n*-person game theory', *S.I.A.M. Review,* **13**, 491–523.

Luce, R.D., 1954, 'A definition of stability for *n*-person games', *Ann. Maths.,* **59**, 357–366.

Luce, R.D. and E.W. Adams, 1956, 'Determination of subjective characteristic functions in games with misperceived payoff functions', *Econometrica,* **24**, 158–171.

Luce, R.D. and H. Raiffa, 1957, *Games and Decisions*, Wiley, New York.

McKenney, J.L., 1967, *Simulation Gaming for Management Development*, Harvard Business School, Cambridge, Massachusetts.

Maynard-Smith, J., 1974, 'The theory of games and the evolution of animal conflicts', *J. Theor. Biol.,* **47**, 209–221.

Maynard-Smith, J., 1978, 'Optimisation theory in evolution', *Ann. Rev. Ecol. Syst.,* **9**, 31–56.

Maynard-Smith, 1982, *Evolution and the Theory of Games*, Cambridge University Press, Cambridge, U.K.

Maynard-Smith, J. and G.A. Parker, 1976, 'Logic of asymmetric contests', *Anim. Behav.,* **24**, 159–175.

Maynard-Smith, J. and G.R. Price, 1973, 'The logic of animal conflict', *Nature,* **246**, 15–18.

Mertens, J.F. and A. Neyman, 1981, 'Stochastic games', *Int. J. Game Theory,* **10**, 53–66.

Mertens, J.F. and S. Zamir, 1971, 'Value of two person repeated games with lack of information on both sides', *Int. J. Game Theory,* **1**, 39–64.

Myerson, R.B., 1978, 'Refinements of the Nash equilibrium concept', *Int. J. Game Theory,* **7**, 73–80.

Nash, J.F., 1950a, 'The bargaining problem', *Econometrica,* **18**, 155–162.

Nash, J.F., 1950b, 'Equilibrium points in *n*-person games', *Proc. Nat. Acad. Sc.,* **36**, 48–49.

Nash, J.F., 1951, 'Non-cooperative games', *Annals of Maths.,* **54**, 286–295.

Nash, J.F., 1953, 'Two-person cooperative games', *Econometrica,* **21**, 128–140.

Orkin, M., 1972, 'Recursive matrix games', *J. Appl. Prob.,* **9**, 813–820.

Owen, G., 1968, *Game Theory*, Saunders, Philadelphia.

Parker, G.A., 1970, 'The reproductive behaviour and the nature of selection in Scatophaga stercororia, II', *J. Anim. Ecol.,* **39**, 205–228.

Parthasarathy, T. and T.E.S. Ragharan, 1971, *Some Topics in Two Person Games*, Elsevier, New York.

Parthasarathy, T. and T. Stern, 1977, 'Markov games—A Survey', in: *Differential Games and Control Theory, Vol. II*, ed. E.O. Roxin, P-T

Lui and R.L. Sternberg, Marcel Dekker Co., New York, pp. 1–46.

Pollateschek, M.A. and B. Avi-itzhak, 1969, 'Algorithms for stochastic games', *Mgmt. Sci.,* **15**, 399–415.

Radford, K.J., 1975, *Managerial Decision Making*, Resten, Resten, Virginia.

Radford, K.J., 1977, *Complex Decision Problems: An Integrated Strategy for Resolution*, Resten, Resten, Virginia.

Radford, 1980, 'Analysis of a complex business decision situation—The Simpsons/Simpsons–Sears merger proposal', *Omega,* **8**, 421–431.

Raiffa, H., 1953, 'Arbitration schemes for generalised two person games', in: *Contributions to the Theory of Games, Vol. II* (Ann. Math. Studies, No. 28), ed. H. Kuhn and A.W. Tucker, Princeton University Press, Princeton, pp. 361–387.

Rao, S.S., R. Chandrasekaran and K.P. Nair, 1973, 'Algorithms for discounted games', *J. Opt. Theory and Appl.,* **11**, 627–637.,

Rapoport, A., 1970, *N-person Game Theory*, Univ. of Michigan Press, Ann Arbor.

Rapoport, A., 1974, 'Prisoner's dilemma—Recollections and observations', in: *Game Theory as a Theory of Conflict Resolution*, ed. A. Rapoport, Reidel, Dordecht, Holland, pp. 17–34.

Rapoport, A. and A.M. Chammak, 1965, *Prisoner's Dilemma*, Univ. of Michigan Press, Ann Arbor.

Rapoport, A., and M. Guyer, 1966, 'A taxonomy of 2 × 2 games', *General Systems,* **11**, 203–214.

Rapoport, A., M. Guyer and D.G. Gordon, 1976, *The 2 × 2 Game*, Univ. of Michigan Press, Ann Arbor.

Riley, J.G., 1979, 'Evolutionary equilibrium strategies', *J. Theor. Biol.,* **76**, 109–123.

Riley, V. and J.P. Young, 1957, *Bibliography on War Gaming*, John Hopkins, Operations Research Office, Chevy Chase, U.S.A.

Rogers, P.D. 1969, *Non-Zero-Sum Stochastic Games*, Publication ORC-69–8, Operations Research Centre, University of California, Berkeley.

Rosenshine, M., 1972, 'Bidding models. Resolution of a controversy', *J. Const. Division ASCE,* **98**, 143–148.

Roth, A.E., 1976, Subsolutions and the supercore of cooperative games, *Maths. of Oper. Res.,* **1**, 43–49.

Rothkopf, M.A., 1969, 'A model of rational competitive bidding', *Mgmt. Sci.,* **15**, 362–373.

Sasieni, M., A. Yaspan and L. Friedman, 1959, *Operations Research—Methods and Problems*, Wiley, New York.

Schmeidler, D., 1969, 'The nucleolus of a characteristic function game, *S.I.A.M. J. Appl. Math.,* **17**, 1163–1170.

Schotter, A., 1974, 'Auctioning Böhm-Bawerk's horses', *Int. J. Game Theory,* **3**, 195–215.

Schuster, P. and K. Sigmund, 1981, 'Coyness, philandering and stable strategies', *Anim. Behav.,* **29**, 186–192.

Selten, R., 1975, 'Reexamination of the perfectness concept for equilibrium points in extensive games', *Int. J. Game Theory,* **4**, 25–55.

Selten, R., 1980, 'A note on evolutionary stable strategies in asymmetrical animal conflicts', *J. Theor. Biol.,* **84**, 93–101.

Sen, A.K., 1970, *Collective Choice and Social Welfare*, Holden Day, San Francisco.

Shapley, L.S., 1953, 'Stochastic games', *Proc. Nat. Ac. Sci.,* **39**, 1095–1100.

Shapley, 1959, 'The solutions of a symmetric market game', in: *Contributions to the Theory of Games, Vol. IV* (Ann. Math. Studies, No. 40), ed. A.W. Tucker and R.D. Luce, Princeton Univ. Press, Princeton, pp. 145–162.

Shapley, L.S., 1969, 'Utility comparison and the theory of games', in: *La Decision: Agregation et Dynamique des Ordres de Préférence,* Edition du CNRS, Paris, pp. 251–263.

Shapley, L.S. and M. Shubik, 1967, 'Concepts and theories of pure competition', in: *Essays in Mathematical Economics*, ed. M. Shubik, Princeton University Press, Princeton, pp. 63–79.

Shapley, L.S. and M. Shubik, 1969, 'On Market games', *J. Econ. Theory,* **1**, 9–25.

Shapley, L.S. and M. Shubik, 1972, 'The assignment game I: The core', *Int. J. Game Theory,* **2**, 111—130.

Shubik, M., 1959a, 'Edgeworth market games', in: *Contributions to the Theory of Games, Vol. IV* (Ann. Math. Studies, No. 40), ed. A.W. Tucker and R.D. Luce, Princeton University Press, Princeton, pp. 267–278.

Shubik, M., 1959b, *Strategy and Market Structure*, Wiley, New York.

Shubik, M., 1971, 'The dollar auction. A paradox in non-cooperative behaviour and escalation', *J. Conflict Resolution,* **15**, 109–111.

Shubik, M., 1972a, 'On the scope of gaming', *Mgmt. Sci.,* **18**, P20–P36.

Shubik, M., 1972b, 'On gaming and game theory', *Mgmt. Sci.,* **18**, P37–P53.

Shubik, M., 1975a, *Games for Society, Business and War*, Elsevier, Amsterdam.

Shubik, M., 1975b, *The Uses and Methods of Gaming*, Elsevier, New York.

Shubik, M., 1982, *Game Theory in the Social Sciences*, M.I.T. Press, Cambridge, Massachusetts.

Shubik, M., and R.E. Levitan, 1980, *Market Structure and Behaviour*,

Harvard University Press, Cambridge, Massachusetts.
Smith, B.T. and J.H. Case, 1975, 'Nash equilibria in sealed bid auction', *Mgmt. Sci.,* **22**, 487–497.
Sobel, M.J., 1971, 'Non-cooperative stochastic games', *Ann. Math .Stat.,* **42**, 1930–1935.
Sobel, M.J., 1982, 'Sequential marketing games', Working paper, Georgia Institute of Technology.
Stark, R.M. and M.A. Rothkopf, 1979, 'Competitive bidding: A comprehensive bibliography', *Oper. Res.,* **27**, 364–390.
Sweat, C.W., 1968, 'Adaptive competitive decisions in repeated play of a matrix game with uncertain entries', *Nav. Res. Log. Quart.,* **15**, 425–448.
Taylor, P.D., 1979, 'Evolutionary stable strategies with two types of players', *J. Appl. Prob.,* **16**, 76–83.
Taylor, P.D. and L.B. Jonker, 1978, 'Evolutionary stable strategies and game dynamics', *Math. Bios.,* **40**, 145–156.
Thomas, C., 1974, 'Design and control of metagame theoretical experiments', in: *Game Theory as a Theory of Conflict Resolution*, ed. A. Rapoport, Reidel, Dordrecht, Holland, pp. 75–102.
Thrall, R.M. and W.F. Lucas, 1973, '*n*-person games in partition function form', *Nav. Res. Log. Quart.,* **10**, 281–298.
Tucker, A.W. and R.D. Luce, 1959 (eds.), *Contributions to the Theory of Games, Vol. IV* (Ann. Math. Studies, No. 40), Princeton University Press, Princeton.
van der Wal, J., 1977, 'Discounted Markov games: Successive approximations and stopping times', *Int. J. Game Theory,* **6**, 11–22.
van der Wal, J., 1980, 'Successive approximation for average reward Markov games', *Int. J. Game Theory,* **9**, 1–12.
van der Wal, J. and J. Wessels, 1976, 'On Markov games', *Statistica Neerlandica,* **30**, 51–71.
Vickery, W., 1961a, 'Counterspeculation, auctions and competitive sealed tenders', *J. Finance,* **16**, 8–37.
Vickery, W., 1961b, 'Auctions and bidding games', In: *Recent Advances in Game Theory*, Mathematical Annals, 29.
von Neumann, J., 1928, 'Zur theorie der gesellschaftsspield', *Mathematische Annalen,* **100**, 295–320.
von Neumann, J., 1937, 'Über ein ökonomisches Gleichingssystem und eine Verallgemeinering des Brouwerschen Fixpunktsetzes, in: *Ergebnisse eines Math. Coll.*, ed. K. Menger, **8**,. 73–83.
von Neumann, J. and O. Morgenstern, 1944, *Theory of Games and Economic Behaviour* (2nd edn, 1947), Princeton University Press, Princeton.
Vorob'ev, N.N., 1977, *Game Theory, Lectures for Economists and Systems*

Scientists, Springer-Verlag, New York.
Williams, J.D., 1954, *The Compleat Strategyst,* McGraw-Hill, New York.
Wilson, R.B., 1967, Competitive bidding with asymmetric information', *Mgmt. Sci.,* **13**, 816–820.
Wilson, R.B., 1969, 'Competitive bidding with disparate information', *Mgmt. Sci.,* **15**, 446–448.
Winkels, H.M., 1979, 'An algorithm to determine all equilibrium points of a bimatrix game', in: *Game Theory and Related Topics*, ed. O. Moeschlin and D. Pallescke, North-Holland, Amsterdam.
Zeeman, E.C., 1979, 'Population dynamics from game theory', in: *Global Theory of Dynamic Systems*, Springer-Verlag, Berlin, pp. 471–497.
Zermelo, E., 1913, 'Über eine Anwendung der Mengenlehre ouf die Theorie des Schachspiels', Proceedings of the Fifth International Congress of Mathematicians, Cambridge, **2**, 501–510.

Index

Name index

Mathematics and its Applications

Continued from page 2

Butkovskiy, A.G.	**Green's Functions and Transfer Functions Handbook**
Butkovskiy, A.G.	**Structural Theory of Distributed Systems**
Cao, Z-Q., Kim, K.H. & Roush, F.W.	**Incline Algebra and Applications**
Chorlton, F.	**Textbook of Dynamics, 2nd Edition**
Chorlton, F.	**Vector and Tensor Methods**
Crapper, G.D.	**Introduction to Water Waves**
Cross, M. & Moscardini, A.O.	**Learning the Art of Mathematical Modelling**
Cullen, M.R.	**Linear Models in Biology**
Dunning-Davies, J.	**Mathematical Methods for Mathematicians, Physical Scientists and Engineers**
Eason, G., Coles, C.W. & Gettinby, G.	**Mathematics and Statistics for the Bio-sciences**
Exton, H.	**Handbook of Hypergeometric Integrals**
Exton, H.	**Multiple Hypergeometric Functions and Applications**
Exton, H.	***q*-Hypergeometric Functions and Applications**
Faux, I.D. & Pratt, M.J.	**Computational Geometry for Design and Manufacture**
Firby, P.A. & Gardiner, C.F.	**Surface Topology**
Gardiner, C.F.	**Modern Algebra**
Gardiner, C.F.	**Algebraic Structures: with Applications**
Gasson, P.C.	**Geometry of Spatial Forms**
Goodbody, A.M.	**Cartesian Tensors**
Goult, R.J.	**Applied Linear Algebra**
Graham, A.	**Kronecker Products and Matrix Calculus: with Applications**
Graham, A.	**Matrix Theory and Applications for Engineers and Mathematicians**
Griffel, D.H.	**Applied Functional Analysis**
Griffel, D.H.	**Linear Algebra***
Hanyga, A.	**Mathematical Theory of Non-linear Elasticity**
Harris, D.J.	**Mathematics for Business, Management and Economics**
Hoksins, R.F.	**Generalised Functions**
Hoskins, R.F.	**Standard and Non-standard Analysis***
Hunter, S.C.	**Mechanics of Continuous Media, 2nd (Revised) Edition**
Huntley, I. & Johnson, R.M.	**Linear and Nonlinear Differential Equations**
Jaswon, M.A. & Rose, M.A.	**Crystal Symmetry: The Theory of Colour Crystallography**
Johnson, R.M.	**Theory and Applications of Linear Differential and Difference Equations**
Kim, K.H. & Roush, F.W.	**Applied Abstract Algebra**
Kosinski, W.	**Field Singularities and Wave Analysis in Continuum Mechanics**
Krishnamurthy, V.	**Combinatorics: Theory and Applications**
Lindfield, G. & Penny, J.E.T.	**Microcomputers in Numerical Analysis**
Lord, E.A. & Wilson, C.B.	**The Mathematical Description of Shape and Form**
Marichev, O.I.	**Integral Transforms of Higher Transcendental Functions**
Massey, B.S.	**Measures in Science and Engineering**
Meek, B.L. & Fairthorne, S.	**Using Computers**
Mikolas, M.	**Real Function and Orchogonal Series**
Moore, R.	**Computational Functional Analysis**
Müller-Pfeiffer, E.	**Spectral Theory of Ordinary Differential Operators**
Murphy, J.A. & McShane, B.	**Computation in Numerical Analysis***
Nonweiller, T.R.F.	**Computational Mathematics: An Introduction to Numerical Approximation**
Ogden, R.W.	**Non-linear Elastic Deformations**
Oldknow, A. & Smith, D.	**Learning Mathematics with Micros**
O'Neill, M.E. & Chorlton, F.	**Ideal and Incompressible Fluid Dynamics**
O'Neill, M.E. & Chorlton, F.	**Viscous and Compressible Fluid Dynamics***
Page, S. G.	**Mathematics: A Second Start**
Rankin, R.A.	**Modular Forms**
Ratschek, H. & Rokne, J.	**Computer Methods for the Range of Functions**
Scorer, R.S.	**Environmental Aerodynamics**
Smith, D.K.	**Network Optimisation Practice: A Computational Guide**
Srivastava, H.M. & Karlsson, P.W.	**Multiple Gaussian Hypergeometric Series**
Srivastava, H.M. & Manocha, H.L.	**A Treatise on Generating Functions**
Shivamoggi, B.K.	**Stability of Parallel Gas Flows***
Stirling, D.S.G.	**Mathematical Analysis***
Sweet, M.V.	**Algebra, Geometry and Trigonometry in Science, Engineering and Mathematics**
Temperley, H.N.V. & Trevena, D.H.	**Liquids and Their Properties**
Temperley, H.N.V.	**Graph Theory and Applications**
Thom, R.	**Mathematical Models of Morphogenesis**
Toth, G.	**Harmonic and Minimal Maps**
Townend, M. S.	**Mathematics in Sport**
Twizell, E.H.	**Computational Methods for Partial Differential Equations**
Wheeler, R.F.	**Rethinking Mathematical Concepts**
Willmore, T.J.	**Total Curvature in Riemannian Geometry**
Willmore, T.J. & Hitchin, N.	**Global Riemannian Geometry**
Wojtynski, W.	**Lie Groups and Lie Algebras***